BROADCASTING FIDELITY

Broadcasting Fidelity

GERMAN RADIO AND THE RISE OF EARLY ELECTRONIC MUSIC

MYLES W. JACKSON

PRINCETON UNIVERSITY PRESS
PRINCETON & OXFORD

Published by Princeton University Press
41 William Street, Princeton, New Jersey 08540
99 Banbury Road, Oxford OX2 6JX

press.princeton.edu

ISBN: 9780691260723
ISBN (e-book): 9780691260846

British Library Cataloging-in-Publication Data is available

Editorial: Eric Crahan & Rebecca Binnie
Production Editorial: Jaden Young
Jacket / Cover Design: Benjamin Higgins
Production: Danielle Amatucci
Publicity: William Pagdatoon
Copyeditor: Leah Caldwell

This book has been composed in Arno

Printed in the United States of America

10 9 8 7 6 5 4 3 2 1

To Charlotte Myleen and Anna

For additional resources and a playlist of recordings
discussed in this book, see the "Resources" tab at
https://press.princeton.edu/books/broadcastingfidelity.

CONTENTS

ACKNOWLEDGMENTS

I BEGAN writing this book nearly a decade ago. I was fortunate enough to receive numerous research grants during that period. I was funded by a fellowship from the American Academy in Berlin and the Wissenschaftskolleg zu Berlin, the Alexander von Humboldt Research Award-Reimar Lüst Prize, a summer membership at the Max Planck Institute for the History of Science, and a summer research grant from the Institute for Advanced Study of the Technical University of Munich. I was able to receive a number of sabbaticals and leaves from New York University. In 2018 I accepted a professorship in the history of science at the Institute for Advanced Study in Princeton, New Jersey.

A number of scholars from various disciplines have kindly read earlier versions of this work and proffered wonderful and extremely helpful comments and constructive criticisms. I would like to thank the anonymous readers of the manuscript: Jonathan Sterne, Andrew Warwick, Jed Buchwald, Tiffany Nichols, Mark Goresky, Svitlana Mayboroda, Andrea F. Bohlmann, and Ken Alder, all of whom either read portions or all of the manuscript. I am also indebted to other members and colleagues at the Institute for Advanced Study and NYU's history department and Gallatin for their wisdom and collegiality. Trevor Pinch's pioneering work on the Moog synthesizer served as an inspiration of this work. Trevor tragically died before he could see a final version. I can only hope that this work will honor his memory. I would also like to thank Eric Crahan and Whitney Rauenhorst of Princeton University Press for their assistance.

The staff at the Institute for Advanced Study is stupendous. The librarians at the institute provided an invaluable service in procuring the books and articles that I needed. I would also like to thank my administrative assistant, Gabriella Hoskin, who has been both efficient and wonderfully pleasant to work with. The kitchen staff is legendary, and rightfully so. Whether lifting weights and playing football, being driven to and from Princeton University, or having things repaired in my home that I should have learned how to do from my father, I thank the staff members for their camaraderie. They make this place special: such an esprit de corps is rare as proverbial rubies these days. Let's hope this never changes. I would also like to thank the archivists of the Deutsches

Museum (Munich), Universität der Künste (Berlin), and the Deutsches Technikmuseum (Berlin) for their assistance over the years.

I am also indebted to those who posed numerous questions and commented on various lectures that I gave that formed the early versions of this book. They include audiences at the American Academy in Berlin, the Institute for Advanced Study in Princeton, Johns Hopkins University, the California Institute of Technology, Ca'Foscari University of Venice, the Deutsches Museum (Munich), International Centre for Theoretical Physics in Trieste, University of Pennsylvania, Northwestern University, Tel Aviv University, University of Toronto, the Max Planck Institute for the History of Science in Berlin and Hamilton College.

Life changes in mysterious and unpredictable ways. I got married during the writing of this book. My wife, Anna, has gone far beyond the call of duty in raising our baby daughter (now a preschooler) so that I could spend time thinking and writing. I thank her for that. Most of this book was conceived in the early hours of the morning with my then-infant daughter, Charlotte Myleen, sleeping on my shoulder. My regaling her with the narrative of the book often put her to sleep. While I was overjoyed that she finally fell asleep, I was always a tad disappointed to know that my work was the cause. I often joked that I would continue to rewrite the manuscript until she stopped soiling her diapers in protest of my thesis and/or narration. Alas, many diapers needed changing. It is to Anna and Charlotte Myleen that I dedicate this book.

ABBREVIATIONS

AEG Allgemeine Elektricitäts-Gesellschaft (The General Electricity Company)

HHI Heinrich Hertz Institut für Schwingungsforschung (Heinrich Hertz Institute for Oscillations Research)

RMI Reichsministerium des Innern (Reich's Ministry of the Interior)

RVS Rundfunkversuchsstelle (also known as the Funkversuchsstelle), the Radio Experimental Laboratory

THB Technische Hochschule Berlin, Technical University of Berlin

DMA Deutsches Museum (Munich) Archives

UDK-BERLIN Universität der Künste-Berlin Archives

Introduction

ON JUNE 20, those attending the Neue Musik (New Music) Berlin 1930 festival hosted by the Berlin Academy of Music listened to the debut of a new electric musical instrument: the trautonium.[1] The trautonium was named after its inventor, electrical engineer and physicist Friedrich Adolf Trautwein, who was assisted by Oskar Sala, a student of the renowned avant-garde composer Paul Hindemith at the academy. The performance featured a trio of trautoniums played by Sala, Hindemith, and Rudolf Schmidt. The reviews were mixed. Some felt the instrument was a laboratory joke, while others thought it was astonishing—the sensation of the festival—and that the concert marked the start of a new age of electroacoustic instruments. The numerous responses to the trautonium throughout its lifetime map nicely onto the myriad views concerning modernity. Those who embraced the instrument's so-called futuristic sounds saw it as reflecting a brave, new world, while others who did not share the optimism of such a future sharply criticized the instrument and insisted it neither was, nor ever would be, the music of the future. This book tells the story of that instrument.

The trautonium has enjoyed a lifespan of nearly a century, although its popularity has considerably waned.[2] It has gone through a number of instantiations, including the *Volkstrautonium* (people's trautonium), *Rundfunktrautonium* (radio trautonium), *Konzerttrautonium* (concert trautonium), and *Mixturtrautonium* (mixture trautonium). The instrument has been featured in a variety of entertainment genres, including classical music; *Gebrauchsmusik* (light music intended for amateurs); *Rundfunkmusik* (radio music, or music specifically composed for the radio); theater music; and sound effects for films and commercials. It has appeared in ballets, radio dramas, and operas, including several of Richard Wagner's—and at Bayreuth, no less. A number of leading composers—Paul Hindemith, Werner Egk, Paul Dessau, and Carl Orff—either wrote pieces for the instrument or used it as a substitute for other instruments. Renowned film director Fritz Lang used the trautonium in *The Indian Tomb*, which he directed in 1959. The trautonium's finest hour came in 1963, when

FIGURE 0.1. The first three musicians on the trautonium who performed at the Neue Musik Berlin 1930 festival hosted by the Berlin Academy of Music, June 20. Friedrich Trautwein is standing, Oskar Sala is seated on the left, Rudolf Schmidt in the middle, and Paul Hindemith on the right. *Source:* Kestenberg, *Kunst und Technik*, facing 112.

Alfred Hitchcock had Sala play the instrument to produce the sounds of birds screeching and flapping their wings in his classic horror film *The Birds*. I like to say that the trautonium is the most famous instrument you have never heard of but that you have most likely heard.

In the first part of the book, I trace the confluence of a number of scientific, technological, political, and musical communities that existed during the 1920s and '30s and that resulted in the invention of the instrument. One such community is centered around early German radio, whose origins date back to the first public broadcast on October 29, 1923, in Berlin. The same equipment, skills, and practices used in radio were also employed in building the trautonium; indeed, the origin of the trautonium is imbricated with the early history of radio. While I agree with scholar Paul Théberge that the digital age witnessed the destruction of the boundary between media and instruments, I argue that that dismantling was already occurring in early-twentieth-century Germany.[3]

The second community that was necessary for the trautonium's invention comprised physicists and electrical engineers working on electroacoustics. This newly formed community—as leading German acoustician and physicist Erwin Meyer insisted—was created to solve the problems then faced by fledgling radio. Early German radio broadcasts were plagued by acoustical distortions that hampered the broadcast, making it difficult for even the most discerning musical ears to differentiate between instruments or voices. Physicists and electrical engineers, with the assistance of physiologists and phoneticians, were charged with fixing these distortions, particularly the ones that affected frequencies at the upper and lower ends of the radio station's frequency range, or bandwidth. The engineers and physicists of the Weimar Republic transformed radio into an instrument for relaying music and the human voice. This would become important for propaganda purposes after the Nazis' rise to power in January 1933.

The third intellectual community that contributed to the invention of the trautonium was musicians, particularly those experimenting in the relatively new genre of electric music. Numerous composers of the 1920s and '30s—including Hindemith, Edgard Varèse, Olivier Messiaen, Joseph Schillinger, and Carlos Chávez, to name just a few—were fascinated by these new instruments. Some musicians, such as Max Butting, Kurt Weill, Ernst Toch, and Hindemith, composed works that were specifically suited for the new medium of radio. A number of musicians were also engineers. For example, electrical engineer and organist Jörg Mager, author of *Eine neue Epoche der Musik durch Radio* (*A New Epoch of Music by means of Radio*), invented the spherophone in 1926.[4] Most famously, around 1920, Russian-Soviet electrical engineer and cellist Lev Sergeyevich Termen—better known in the West as Leon Theremin—invented the theremin, of which he conceived while repairing his radio. The theremin was very popular in Germany.

The site where these various communities—electroacoustics, radio research, and electric music—came together was also the laboratory where the trautonium was invented: the Funkversuchsstelle, which opened on May 3, 1928, in the attic of the Berlin Academy of Music. Later, the laboratory would be known as the Rundfunkversuchsstelle (the Radio Experimental Laboratory, or the RVS). Funded predominantly by the Prussian Ministry of Culture and the Reich's Broadcasting Corporation (the Reichs-Rundfunk-Gesellschaft), the RVS hosted immensely fruitful collaborations between natural scientists, radio engineers, and musicians who would go on to improve radio broadcast fidelity, develop *Rundfunkmusik*, teach the use of radio equipment to music students, and eventually invent an electric musical instrument. The RVS was the crucible where the skills, theories, and practices of these scholars were forged.

Given my own passions and abilities, I shall approach this book as a historian of science and technology, albeit one with musical training and an interest in musicology. I have spent much time (perhaps too much) thinking about German history, specifically the various Germanies of the past century and their corresponding (and often antithetical) political views of modernity. Those three related areas of interest—history of science and technology; musicology; and the histories of the Weimar Republic, the Third Reich, and the Federal Republic of Germany—provide the foundation of this book.

What are this book's intended contributions? For natural scientists, engineers, and historians of science and technology, this study offers the history of the physics and electrical engineering upon which radio was predicated. It details the role of those disciplines in inventing a cadre of electric musical instruments after World War I. This is also a story about a scientific instrument, namely the harmonic analyzer, the history of which was hitherto largely unknown despite its importance to the history of electroacoustics and radio. By following the development of late-nineteenth- and early-twentieth-century disciplines, such as applied physics, physiology, phonetics, psychology, radio and electrical engineering, and electroacoustics, we begin to see the various ways in which scholars in those fields defined and understood important musical and scientific phenomena such as tone color (also known by its French term, timbre), fidelity, and the formation of speech sounds.

While much more famous as a musical instrument, the trautonium became a scientific instrument used to adjudicate between Hermann von Helmholtz's theory of resonance and Ludimar Hermann's subsequent theory of impulse (or shock) excitation to explain the creation of vowel sounds and the development of a musical instrument's tone color. Trautwein was convinced that the trautonium was the electrical counterpart to the human voice organs and certainly saw his instrument as settling the debate on the side of L. Hermann. Finally, the book details the status of natural scientists and engineers in the various Germanies. Many famously were blamed for Germany's defeat in World War I, felt alienated throughout the Weimar Republic, tried their best to ingratiate themselves with governmental officials by actively supporting the Nazis, and attempted once again to regain acceptance after the defeat of the Nazis, despite their active roles in World War II.

Since radio provided challenging intellectual and practical problems beyond the limited horizons of traditional physics and electrical engineering, it is important that I address the technical aspects of physics and electrical engineering. Omitting the technical knowledge would render the story woefully incomplete. I ask for the reader's patience as I guide them through the denser material. We need to appreciate and comprehend the labor and skill (both intellectual and manual) that these scholars brought to the problem. Two works in particular are relevant here: Wittje's important work on the history of electroacoustics and Yeang's outstanding tome detailing the technical history of the transformation of noise from an annoyance that electrical engineers attempted to ameliorate in electric sound reproduction to a subject relevant to understanding statistical detection, prediction, and the transmission of information.[5]

I hope the book will also appeal to musicologists. The trautonium was the quintessential modernist musical instrument, producing a new type of music that wished to distance itself both from classical compositions as well as those of Arnold Schoenberg, Alban Berg, and Anton Webern. This work also provides us with a better understanding of early radio and the music broadcast by important station managers such as Hans Flesch of Frankfurt, and later Berlin, by shedding light on the early years of Rundfunkmusik and its composers. It also contributes to the history of tone color during the 1920s and '30s, a period that witnessed an important metamorphosis in its meaning. By telling a hitherto unknown story of electric music and its relationship to electroacoustic and electronic music, musicologists can also begin to see the historically contingent processes of negotiation that defined those terms. The trautonium illuminates the relatively unknown debates between Trautwein and Sala on the one hand, and the pioneers of the Cologne Studio for Electronic Music—particularly Werner Meyer Eppler—on the other, as detailed in their correspondence. Their letters raise interesting questions about the definition of electronic music: Which aesthetic should be included and which should be excluded? What is the role of the composer in relation to the performer? Is there such a thing as an "authentic composition"?

Another relevant aspect of this book to musicologists interested in the twentieth century is the attempt to situate the trautonium within the longue durée of proto-synthesizers, as it shared a number of important features with them, including the synthesis of music and vowel sounds as well as the ability to imitate a large range of timbres and sounds. While many scholars realize that electric music predates World War II, we hear relatively little about instruments that shared some of the attributes of the synthesizers of the 1950s and '60s. RCA's Mark II (designed by Harry Olson and Herbert Belar), Max Mathews's work at Bell Labs on digital-computer music, and the Moog synthesizer were all postwar inventions.[6] When Mathews spoke on the history of synthesizers in 1985, he insisted that their origins were in the early 1950s with the work of Vladimir Ussachevsky and Otto Luening in New York City, Pierre Schaeffer in Paris, and Karlheinz Stockhausen in Cologne.[7] Oskar Sala wished to differentiate between his mixture trautonium and synthesizers for both entrepreneurial reasons and reasons of musical performance.[8] Unlike more modern synthesizers, the mixture trautonium did not possess sound envelopes or voltage-controlled filters or amplifiers. The functions of those devices were achieved by manually controlled circuits. As we shall see, Sala wished to produce many of the effects created by synthesizers himself.[9] In his view, the machine should never completely replace the human.

Finally, for German historians and those interested in German studies, this book discusses the interest of leading German intellectuals of the 1920s and '30s in German radio in general and the RVS in particular, including Bertolt Brecht, Kurt Weill, Walter Benjamin, Alfred Döblin, and Theodor Adorno. The trautonium's history during the Third Reich was an intriguing one. While one might think that the Nazis would consider electric music degenerate, the opposite was true. The trautonium could serve their purposes in creating an aesthetic of "steely Romanticism," contribute to *Hausmusik* for the *Volk*, and provide entertainment for mass gatherings. It can therefore shed light on the complicated relationships the Nazis had with music and technology. Claiming that they opposed everything "modern" is simply fallacious. This work also offers an account of the trautonium's contribution to industry and cultural films as well as television and cinema commercials—and music culture in general—in the Federal Republic of Germany.

The instrument is quite unique, even among the electric musical instruments of the period. Its sounds were produced by a glow-discharge (neon) tube and later thyratrons, rather than using frequency beats as was the case with the theremin and ondes Martenot, or tuned tube oscillators as was the case with the Coupleux-Givelet organ and the Hammond Novachord. The laboratory where the instrument was invented, the RVS, was also unique. Due to the lab being housed in a musical academy as opposed to an engineering company, the majority of its work was dedicated to music. The RVS was, in a sense, a mirror image of Bell Telephone Laboratories: engineers and scientists in this US laboratory focused their research on the transmission of speech, and music initially played a secondary role. The historian can use the RVS as a foil to the Bell Telephone Laboratories, which were created by AT&T and Western Electric. Finally, no other country possessed the caliber of such intellectuals in those numbers writing about early radio and its laboratories. All of these points begin to explain the distinctiveness of the trautonium and the context of its invention. While this book does not seek to support the German *Sonderweg*, it does describe a number of peculiar aspects of German music, science and engineering, and politics that go a long way to explain why the trautonium was an invention of the Weimar Republic.[10]

Musical aesthetics is one theme that runs through this work. Addressing the disciplines with which this book wishes to engage—history of science and technology, musicology, and German history—I investigate how scientists and engineers defined and measured musical aesthetics, how musicians defined and experienced those aesthetics, and how politicians shaped or quelled them. One example of musical aesthetics is tone color, defined as the quality of a sound that is unique to the instrument or voice that created it. That is to say, it explains the difference in tone between a cello playing a note at 220 Hz with a particular volume, and a piano playing that note with the identical pitch and volume. Timbre ties together the trautonium, electroacoustics, politics, and music. The instrument could imitate the tone colors of a number of traditional musical instruments as heard by the ears of skilled musicians and as seen by oscillograms it produced when compared with the oscillograms generated by those traditional instruments. Timbre also was relevant to the radio, since it was difficult to broadcast with sufficient fidelity: distortions in the broadcast's tone color were due in large part to the radio transmission and receiving equipment, namely microphones, loudspeakers, and amplifiers. A history of musical aesthetics of the 1920s and '30s is simultaneously a history of those radio parts; therefore, musical aesthetics were inextricably linked to electroacoustic theories, skills, and practices. Since radio created an important market for these devices, engineers and physicists busied themselves with rendering the requisite improvements, thereby improving broadcast, much to the appreciation of attentive audiences.

Emily Dolan has provided us with a wonderful account of timbre, from its origins with the works of Jean-Jacques Rousseau to its solidification in the nineteenth century.[11] She argues that Joseph Haydn's style of orchestration of the late eighteenth century must be understood through the emergence of the public's interest in various instruments' timbres. More recently, Dolan and Alex Rehding have coedited a collection of essays on timbre.[12] The volume illustrates how tone color has now become a key theme of research for historians (including historians of science and technology), musicologists, philosophers, science-and-technology-studies scholars, and sound-study scholars.

A history of tone color is also a history of fidelity, another example of musical aesthetics that ties the book together. A vast majority of physicists and engineers had initially defined fidelity of timbre as a static comparison of the oscillograms generated by harmonic analyzers depicting the relative amplitudes of the overtones of broadcast voices and instruments with those of the original sounds.[13] During this period, however, an ever-increasing number of physicists and engineers—including Trautwein—as well as physiologists and psychologists realized that timbre was not static and that it hung over the interval of playing a note: what we now call the sound envelope. The initial portion of playing a note (the attack) possesses different timbres than the decay, which in turn possesses different timbres from those of the release. Only by recapturing and consistently reproducing the entire process could one begin to speak of fidelity of tone color. In addition, engineers at Bell Telephone Laboratories began to show that tone color was also somewhat dependent on volume.

resigned, having only served for just over a hundred days. Political chaos spelt economic doom: on November 15, the value of a German paper mark plummeted to approximately 1 trillion Marks per US dollar. Pictures of Germans carrying wheelbarrows containing suitcases full of Reichsmarks to purchase food at a local market became iconic.

Weimar German radio was dedicated to being impartial (*Überparteilichkeit*, or literally being "above partisanship") and was also committed to underscoring the various dialects and cultures found throughout the Reich. No one language (such as *Hochdeutsch*), nor one culture among the German Volk was considered to stand above the rest. A "centralized decentralization," as it was referred to, was the necessary model for Weimar radio's success. Shortly before the rise of fascism, however, that had changed. German radio began to take political sides and infamously was used as an instrument of propaganda by Goebbels from 1933 until the end of the war. Under the Nazis, it was meant to unify the German Volk by stressing Hochdeutsch and various cultural characteristics shared by all Germans. The *Volksempfänger*, or radio receiver, was often referred to in the vernacular as "Goebbelsschnauze," or "Goebbels's snout," and was now cheap enough for the workers to afford.

Just because the pioneers of early Weimar radio wished it to be impartial, it would a mistake to see it as being free from politics. As a result of the explosion of radio's popularity throughout Europe and the United States in the late 1920s, various governments needed to impose restrictions on bandwidths to avoid unwanted interference due to the increase in the number of stations. These restrictions resulted in the cutoff of the overtones of high pitches and begin to explain why—in addition to the inefficiencies of the components of the transmitter and receiver—the soprano's voice did not come across as well in broadcasts as it did in live performances. Political and economic decisions that sacrificed the soprano's voice were made by men. Such a move was neither inevitable nor natural: there were alternatives.

The German nation-state also shaped engineering and scientific disciplines during the early twentieth century. The corresponding professions underwent rapid change from the late nineteenth century until well after World War II. By tracing the various discourses on modernity, particularly those espoused by leading engineers, one can begin to piece together their relationship with the German state and culture. By the second half of the nineteenth century, engineers were gaining in prestige, yet they still lagged behind those who were classically trained, the so-called *Bildungsbürgertum*, or the educated upper-middle class, whose education was based on the classical languages of Greek and Latin. The Bildungsbürgertum comprised the so-called free professions: high-ranking civil servants such as university professors, military officers, and church officials. A key to their education was German idealism as explicated in the works of Immanuel Kant, Johann Gottlieb Fichte, Johann von Goethe, and Friedrich Schiller. In contrast, engineers were trained in the so-called illiberal disciplines, which were taught at technical universities (*technische Hochschulen*) across the land. Members of the Bildungsbürgertum often viewed engineers, even those with advanced degrees, with disdain.

Given the importance of engineers to the rapid industrialization of the burgeoning nation, Kaiser Wilhelm II gave technical universities in 1899 the ability to grant the degrees of *Diplomingenieur* and *Doktoringenieur*. Not surprisingly, engineers with Doktoringenieur degrees wished to be treated as equals to those who possessed the equivalent of a PhD in the humanities. They, too, sought the coveted status of *Kulturträger*.[24] Trautwein, who received his Dr. Ing. from the Technical University of Karlsruhe in 1921, certainly felt he earned such a status. Technology was for him, as well as many other well-educated engineers of the period, an integral part of German culture. With the disastrous end of World War I, German scientists and engineers alike suffered from a crisis of identity throughout the Weimar Republic and were often viewed as being culpable for Germany's defeat. As Adelheid Voskuhl has pointed out, while engineers desperately sought to be members of bourgeois culture (*Bürgerlichkeit*), being influenced by right-wing, antimodern ideologies, they often simultaneously loathed its liberal attitudes and values. Industrialization, seen by many as one of the defining characteristics of liberal modernity, according to Voskuhl, was in reality the cause of its death.[25] Like a disproportionately high percentage of engineers, Trautwein joined the Nazi Party in 1933. Many of the educated elite were sympathetic to the Nazis' vision of a new and glorious future. A staunch supporter of Nazi ideology, Trautwein assisted the Nazi Party with his acoustical expertise on amplification for large, outdoor rallies. After World War II, Trautwein attempted to reestablish the role of educated engineers as Kulturträger by waxing poetic on both the philosophy of technology and the importance of technology to culture.

The Nazis' Reichsmusikkammer (Reich Chamber of Music), the musical state agency under the control of the Ministry of Public Enlightenment and Propaganda, subsidized a later version of the trautonium, the concert trautonium, built by Sala, who toured with the instrument throughout the Third Reich, in occupied territories such as Holland, France, and Hungary, and in allied countries, such as Italy, with support both of the Reichsmusikkammer and the *Kraft-durch-Freude* (Strength-Through-Joy) programs.

Various nation-states certainly played a role in the type of musical aesthetic that was acceptable after the defeat of the Nazis. Sala and, to a lesser extent, Trautwein did their best to cultivate the interest of the Allied powers in the trautonium after the end of the war. Sala actively sought out patronage from the Radio in the American Sector (RIAS) in Berlin. He also attempted to interest musicians in the Soviet sector of Berlin to build quartet trautoniums in the late 1940s. He subsequently found his niche in the capitalist industries of the Federal Republic of Germany.

This history of the trautonium is in part a story about the material culture of objects. As a material object of modernity, the trautonium challenges our preconceived notions about the isolation of various aspects of culture and science of the twentieth century. By offering a history of the trautonium, I hope to show how a musical instrument reconstituted the relationships between science, technology, politics, and musical aesthetics, thereby forcing us to rethink the notion of modernity.

The materiality of musical instruments, which owes much to Trevor Pinch and Frank Trocco's *Analog Days*—a pioneering work on the history of the Moog synthesizer—has become the subject of numerous investigations over the past two decades.[26] Emily Dolan and John Tresch's important work on organology traces the intersecting and divergent histories of music and science by looking at scientific and musical instruments. Musical instruments render the inner emotions and thoughts of a composer accessible to the outside world, while scientific instruments transport the external world into the thoughts of the inner world of a scientist's mind.[27] Similarly, the musicologist Rebecca Wolf has worked on the materiality of musical instruments, while the musicologist Alexander Rehding has written on the relationship between instruments and music theory.[28] Furthermore, Thomas Patteson's work has given us wonderfully contextualized accounts of several electric musical instruments, including the trautonium, during the 1920s, '30s, and '40s.[29]

The past twenty years have also witnessed the publication of many works linking the history of music with the history of science and technology. I seek to continue that trend by merging a history of music, specifically aesthetics, with the history of science and technology.[30] For example, Peter Pesic has written the longue durée history of the relationship between music and science, *Music and the Making of Modern Science*, and his more recent work, *Sounding Bodies*, elucidates the influence of both music and sound on the structure and content of biomedical sciences.[31] Historian of science Alexandra Hui demonstrates how leading physicists, physiologists, and psychologists dedicated themselves to understanding sound from a psychophysical perspective. She deftly argues how musical aesthetics were inextricably linked with the natural sciences.[32] Hui, Julia Kursell, and my coedited volume, *Music, Sound, and the Laboratory from 1750 to 1980*, proffer an account of how laboratory sciences changed the notion of sound over two hundred years. Newly invented laboratory techniques of sound detection and representation and the use of electricity and computers to generate sounds fundamentally altered acoustics as well as musical practice. The musicologist Kursell has written a number of important essays on nineteenth-century German science and music.[33] Theater and media studies scholar Viktoria Tkaczyk has recently shown how the elucidation of the function of the auditory cortex in the late nineteenth century influenced numerous academic disciplines in the natural sciences and humanities.[34]

While historians contributing to sound studies have argued for quite some time against separating aesthetics from science and technology, the theme is still unrepresented in the history of science and technology more broadly speaking. This work seeks to illustrate how one accounts for the mutual implications of aesthetics, science, and engineering. There have been a small number of important contributions on aesthetics and science provided by historians of science. Michael Dettelbach's influential essay is one of the first on this topic: he demonstrates how the Humboldtian aesthetic was based on precision measurement. I have offered examples along those lines with respect to physics and music. My *Harmonious Triads* explores how physicists such as Wilhelm Eduard Weber contributed to musical aesthetics in the nineteenth century. For example, Weber's work on compensated reed pipes, which were used to experimentally test the ratio of the increase in pressure and the increase in density of a sound wave, also led to the invention of organ pipes that could increase in volume without increasing in pitch. Organs now became expressive. Robert Brain's *The Pulse of Modernism* is an important study linking the origins of artistic modernism to physiological theories of perception forged in late-nineteenth-century French laboratories. "Physiological aesthetics" altered the way artists, poets, and musicians plied their craft and in so doing changed the notion of art itself. Deborah Coen interweaves a wonderful account of the relationship between science, politics, and aesthetic qualities in the visual arts and liberalism in nineteenth-century Vienna by tracing the history of the Erxleben family in her *Vienna in the Age of Uncertainty*.[35] John Tresch's *The Romantic Machine* is a magisterial tome addressing how the sciences and the arts, rather than being antithetical, were critical for uniting a deeply fractured nineteenth-century French society. Finally, Norton Wise argues for the importance of the aesthetic sensibilities involved in drawing and the visual arts to Hermann von Helmholtz's early work on physiology.

The trautonium itself has attracted the attention of scholars lately. Two recent works are doctoral dissertations in musicology and media studies: Benedikt Brilmayer, "Das Trautonium: Prozesse des Technologietransfers im Musikinstrumentenbau," and Christina Dörfling, *Der Schwingkreis: Schaltungsgeschichten an den Rändern von Musik und Medien*.[36] In a third, Peter Donhauser adroitly details the complex technological developments of the various instantiations of the trautonium from its origin to the mixture trautonium some twenty-two years later.[37] There has been, however, very little written on the instrument in English.[38]

Much like Aldous Huxley's *Brave New World*, where the protagonist Bernhard Marx does not appear until after the initial chapters, the protagonist of my story, the trautonium, does not immediately appear in my work either. Switching the roles just for the analogy, just as Marx disappears in favor of John in Huxley's work, the radio succumbs to the trautonium. I feel the trautonium's late appearance is justified, since I need to trace the various traditions that formed the instrument's context to appreciate how the instrument came about.

Chapter 1 details how early German radio was considered to be an experiment by one of its pioneers, Hans Bredow. Radio would teach and entertain the German people at a time of economic devastation and extreme political uncertainty in the aftermath of World War I. Brecht stressed that radio needed to be experimental in order to render transparent the arcane processes occurring daily in the Reichstag. He hoped the RVS would improve radio broadcasts, as he saw the apparatus as critical to the education of the German people. Radio created a new musical genre (Rundfunkmusik) and a spoken genre (*Hörspiele*, or radio dramas), which were invented because of the medium's popularity. Furthermore, new musical forms featuring electric musical instruments, such as the trautonium, were broadcast over the radio.

Chapter 2 traces the history of fidelity by investigating the challenges that physicists and engineers faced trying to improve the broadcasting of tone colors. Violins were often mistaken for flutes or clarinets. During opera broadcasts, the soprano's voice came across as dull. Such infidelity gravely threatened the young medium's future; therefore, it is not a surprise that there was a certain urgency in the research conducted on linear and nonlinear distortions that plagued broadcasts, as discussed in chapter 3. Throughout the 1920s and '30s, radio engineers and physicists used harmonic analyzers to study the effects of the components of the transmitter and receiver on the relative amplitudes of a sound's overtones. This provided scientists and engineers with a metric of fidelity that in turn could be used to determine the imperfections of the equipment that needed to be remedied. Scientists and engineers conducted their experiments in laboratories of major German electrical engineering companies such as Siemens & Halske, Allgemeine Elektricitäts-Gesellschaft (AEG), and Telefunken, collaborating with numerous governmental and university laboratories. Their major competitors were researchers at US companies, particularly Western Electric and American Telephone and Telegraph Company (AT&T), and Bell Telephone Laboratories (from 1925 onward), Radio Corporation of America (RCA), and General Electric (GE).

The RVS is the subject of chapter 4. Established in 1928, it was the site of experimentation of myriad issues for natural scientists, physiologists, phoneticians, engineers, and musicians. The RVS improved broadcast fidelity, invented a new instrument, and tested microphones, loudspeakers, and amplifiers. It contributed to the genre of Rundfunkmusik, taught students how to work with microphones, and it also hosted the experiments of Hindemith and his fellow Neue Sachlichkeit composer Ernst Toch on how changing the speed of a recorded sound alters the sound's timbre. Arnold Schoenberg stopped by the RVS and encouraged Trautwein and Sala to increase the range of the trautonium to the range of the grand piano. The musicologist, psychologist, and physiologist Carl Stumpf visited the RVS and was amazed by the number of overtones the trautonium could produce.

As discussed in chapter 5, the trautonium was invented during the height of Neue Sachlichkeit. Hindemith, one of the movement's leaders, was the first and by far the most important musician who composed pieces for the instrument. It was a period when electric musical instruments filled the ether waves of radio and concert halls. Debates erupted about this new genre of music. Was it an example of mechanical music? If not, how was it different? Could the new instruments be of assistance to what many felt was the stagnating genre of musical composition by unleashing new tone colors? Did musicians welcome or spurn the role played by natural scientists and radio engineers in establishing a new musical aesthetic?

Although the vocoder, invented in 1938 by Homer Dudley at Bell Telephone Laboratories, was much better at synthesizing speech, the trautonium predates it in synthesizing vowel sounds. The trautonium was taken up in debates among physiologists and phoneticians about vowel production. Drawing upon the research of Stumpf, the physicist Dayton Clarence Miller, and the radio engineer Karl Willy Wagner of the 1910s and '20s, Trautwein argued that the vibrations generated by impulse (or shock) excitation that produced musical sounds were the electrical equivalent of the vibrations created by air traveling through the speech organs giving rise to the vowel sounds. The damped and decaying vibrations formed the formants, or the dominating overtones of a musical note or speech sound that determined its timbre. Such investigations were critical to radio broadcasting fidelity.

After initially focusing on the origins of German radio and the trautonium much like a convex lens, the second portion of the book behaves like a concave lens, radiating out to trace the various trajectories of the instrument. Both Trautwein and Sala realized shortly after the National Socialists' rise to power that they needed to convince the Nazis to support their work on the trautonium, as detailed in chapter 6. Iverson has written on the Nazi past of physicists and engineers working on electronic music after World War II, particularly Werner Meyer-Eppler.[39] Trautwein was no exception. While Sala never joined the Nazi Party, he possessed business savvy and certainly benefited from the party's patronage. Goebbels thought that the trautonium could serve as a perfect instrument for Hausmusik and was delighted to hear about its potential use in mass rallies. He was deeply impressed by the instrument after being granted a private demonstration and performance by Sala in April 1935. The trautonium could play traditional works of Paganini, Beethoven, Mozart, and Bach for violin, flute, cello, and organ. Here imitation was key. But more modern pieces by Harald Genzmer and Ferrucio Busoni—the Italian composer who influenced the Italian Futurists, many of whom were sympathetic to fascism and supporters of Benito Mussolini, were also featured. A number of critics, amazed by the range of tone colors that the instrument generated, labeled it a "Wunderinstrument." Chapter 6 also tells the story about the role a particular engineer, namely Trautwein, played in supporting the Third Reich. In that respect, he was typical of many German engineers during the period.

After the war, Trautwein receded into the background, as discussed in chapter 7. He did, however, write a number of essays reflecting on the engineer's role in German culture. Reminiscent of the attempts by natural scientists and engineers to redeem themselves before a skeptical public after the humiliating defeat of World War I, Trautwein wished to carve out a space for electrical engineers contributing to music with a hope of restoring his and his discipline's reputation with the newly created state of the Federal Republic of Germany and its intellectuals. Sala, on the other hand, continued to experiment with the instrument. While he still used the trautonium to imitate more conventional instruments, he branched out into other genres, thereby enabling his instrument to unleash unique, futuristic, and uncanny timbres employed in radio dramas and theater pieces. The concert trautonium was often used by composers who wished to make a conscious break with the Nazis' legacy, for example Brecht and Paul Dessau's operas, *Die Verurteilung des Lukullus* (*The Condemnation of Lucullus*) and *Deutsches Miserere*. Sala's final invention, the mixture trautonium of 1952, produced sound effects for operas, such as the Grail bells in Richard Wagner's *Parsifal* and the hammering sounds of the goldsmith in *Das Rheingold*. While some critics praised the instrument's versatility, others were troubled by the monstrous cacophony it produced. By the time he retired in the 1990s, Sala had played his mixture trautonium for over one hundred radio and television commercials and three hundred films, a number of which were commissioned by various chemical and Big Pharma companies in West Germany.

The music the trautonium played was not the only type of new music filling the airwaves in postwar Europe. Pierre Schaeffer and his *musique concrète* cohorts in Paris were creating new types of sounds by manipulating tape. Meyer-Eppler and his colleagues, Robert Beyer and Herbert Eimert, created the Studio for Electronic Music in Cologne. As discussed in chapter 8, Trautwein and Sala competed with these groups, particularly the Cologne Studio, to help preserve their contributions and legacy to postwar music. Sala in particular thought hard about the ways that tape recording increased the types of tone colors and sounds that could be combined, as tape provided a new experimental system for the instrument. The competition was fierce, and the trautonium featured prominently in the crucial debates, such as the role of the performer as interpreter of a composition, the importance of experimentation and improvisation with the instrument while playing, and the use of tape recording while performing. Would the future of music consist of instruments made of vacuum tubes and electric circuits, or (later) transistors? Would electronic music be successful in eliminating the human in music altogether?

This is a story about twentieth-century science and the social spaces where it occurred, the metropolises. By the late nineteenth century, science had been reorganized and had become slightly more egalitarian due to less social stratification. While women and people of color were still massively underrepresented, many more people were participating in science compared to earlier in the century. Much of that was due to the rise of the engineering disciplines as well as heavy industry, and World War I certainly accelerated the transformation. Metropolises such as Berlin provided a venue where a vast multitude of skills and expertise from newly created industries and disciplines, or previously existing ones that had had no contact with each other, began to combine in extremely fecund collaborations. Physicists and electrical engineers were now working with physiologists and musicians. We all know about "Big Science"; however, this particular story is about the relationships between science, technology, and music—their complexities as well as their interrelatedness—that generated explicit collaborations that enabled the creation of new aesthetic concepts and technical possibilities.

The live performances were carried out in the recording room on the third floor of the Vox House, which happily hosted the radio station, seizing the opportunity to advertise their wares: Vox was a record label established in 1921. The broadcasting room was located in the attic, one floor above the recording room. It was split into two, with the larger portion being covered with violet crepe paper and the smaller used to house technical equipment.[4] A stool supported two address books, on top of which a microphone was precariously perched. Most of the live performers would go on to form the core of the Berlin Radio Orchestra. After the playing of "Das Deutschlandlied," the station wished its listeners a good night and reminded them to please ground their antennas for safety purposes.[5]

Radio now belonged to the German people, at least those who could afford a radio and pay the requisite radio-licensing fee.[6] That initial broadcast proffered a microcosm of the types of music that radio would go on to broadcast during the early years of the Weimar Republic: an eclectic combination of serious, highbrow music, including symphonic and operatic pieces, and popular folk songs and patriotic hymns. Radio, as discussed below, was meant to entertain as well as educate Germans.

1923 and Early Radio in the Weimar Republic

German journalist Sebastian Haffner, who was in exile in England, wrote in 1939 that "no people on earth has experienced what the Germans experienced in 1923."[7] Author Stefan Zweig, also in exile, was convinced that the world had "never produced similar lunatic asylums in such huge proportions."[8] In January 1923, French and Belgian troops marched into and occupied the Ruhr region of western Germany to force Germany to pay war reparations. In protest, German workers were encouraged by the German government to go on strike. Rioting broke out when France sent its own workers as replacements, and its military arrested German police and organizers of the strike. The Ruhr region had seen violence some three years earlier, when the Communist-backed workers, seeking to bring about a revolution, clashed with police and members of the *Freikorps*, which were right-wing, paramilitary groups.

The autumn was particularly ominous. In October the Rhenish Republic declared in the city of Aachen that it wished to secede from Germany and become a protectorate of France. In that same month, the so-called German October took place: it was the plan of the executive committee of the Communist International to foment a revolution in the Weimar Republic. In the states of Thuringia and Saxony and the city-state of Hamburg, the Communist Party of Germany entered into a coalition with the Social Democratic Party to create communist territories in central Europe. The German government quelled the resulting skirmishes with the assistance of the *Reichswehr* and the police. It subsequently suspended the legitimacy of left-wing state governments.[9]

While the Communists posed existential threats to the Republic, a fascist threat was gaining strength in Bavaria, where a national dictatorship was sought. Infamously, on the night of November 8–9, Adolf Hitler, General Erich Ludendorff, and members of the *Kampfbund* attempted a coup d'état, the so-called Hitler-Ludendorf Putsch, or the Beer Hall Putsch. The coup failed, and Hitler was sentenced to five years in prison, of which he only served eight months. In that same month, hyperinflation peaked in Germany with the devaluation of its currency: one dollar was the equivalent of roughly one trillion marks. On numerous occasions, Reich President Friedrich Ebert drew upon Article 28 of the Weimar Constitution, which gave him the power to restore order and combat inflation. It is fair to say that the Weimar Republic was perilously close to ending violently only after four years of existence.[10] Certainly hyperinflation and the liberal policies implemented to tackle it were deleterious to liberal parties. Many liberals were convinced that the only way the rights and freedoms of the individual could be protected was by implementing authoritarian decision-making policies.[11]

Amid the tumult, German public radio emerged. At the end of World War I, Hans Bredow, one of Germany's radio pioneers, became director of radio at the Reich's Ministry of the Post Office, which from late 1919 started collaborating with the three Berlin electrical companies on the construction of radio stations and equipment, Telefunken, Siemens & Halske, and Allgemeine Elektricitäts-Gesellschaft (AEG).[12] The Ministry of the Post Office was the primary governmental agency in charge of radio, and the Reich's Office of Telegraph Technology in Berlin was responsible for radio's scientific and technological development.[13]

On November 16, 1919, Bredow lectured to a packed general audience in Berlin's Urania on the future use of radio as a means of entertainment and instruction for all levels of German society.[14] While the audience was skeptical, radio's visibility and use were to increase dramatically. On December 22, 1920, the radio station at Königs Wusterhausen—an arc-transmitter station used by the military that broadcast in long wave, thereby reaching all parts of the German Reich and parts of Northern and Western Europe—transmitted a Christmas concert that included violin, cello, and clarinet solos. The concert could be picked up as far away as 2,000 kilometers.[15] On June 8, 1920, the radio station transmitted the opera *Madame Butterfly* from the Berlin State Opera. Via a 10 kW tube transmitter, music could now be heard by sailors on steamboats some 3,900 kilometers away.[16] Starting on May 13, 1923, Königs Wusterhausen began to broadcast weekly Sunday concerts.[17]

FIGURE 1.2. The network of German radio stations at the end of 1929: there were twenty-seven primary and secondary stations plus the station at Königs Wusterhausen outside Berlin. *Source:* "Nachtrag aus dem 'Rundfunk-Jahrbuch 1929,'" in *Rundfunk Jahrbuch 1930* edited by Reichs-Rundfunk-Gesellschaft (Berlin: Union Deutsche Velagsgesellschaft, 1930), 469.

under the RRG's purview. On February 26, 1926, the Post Office acquired the RRG, ensuring its success.[37] Chaired by Bredow, who was named radio commissioner, the RRG was now the parent company of radio in the Reich.

Purpose of Weimar Radio

So what exactly was the purpose of radio during the Weimar Republic? Prior to that historic broadcast on October 29, 1923, Bredow made his intensions for radio clear to his fellow citizens:

> In a time of serious economic emergency and political duress, radio will be made available to everyone. It will no longer exclusively serve economic purposes, but rather an experiment should be made to use this cultural

> advance, in order to bring some excitement and joy to the lives of the German people. . . .
>
> The German people are economically impoverished, and it cannot be denied that the intellectual impoverishment continues to advance. Who then can purchase newspapers and books today? Who can treat themselves to the joy of good music and entertainment as well as educated lectures? Relaxation, entertainment, and variety distract the mind from the heavy worries of everyday life, refresh and increase the joy of work; but a joyless people are reluctant to work. This is where the role of broadcasting comes in, and if it is possible in this way to bring artistic and intellectual high-quality performances of all kinds to all classes of the population, if a new field of activity is at the same time opened up for industry and thus job opportunities are created for workers, then the radio has a constructive effect, and the German people have a right to it.[38]

Bredow stressed time and time again: radio in the Weimar Republic had two primary goals, to entertain and educate. The German people were in desperate need of both.

While education and entertainment were the hallmarks of Weimar radio, being a propagandizing tool of a political party certainly was not. *Überparteilichkeit*—meaning impartiality, independent, or, literally, above partisanship—was a hallmark of Weimar radio. Because of radio's power to influence public opinion—a fact famously exploited by the Nazis—governmental officials of the Weimar Republic insisted that the nation-state and all the various state governments control radio. The first of the ten "Richtlinien für die Regelung des Rundfunks" ("Guidelines for the Regulation of Radio") of 1926 declared that "radio serves no party. Its entire news and lecture service is therefore to be organized strictly impartially (überparteilich)."[39] Such a stance resonated with the calls of other nations to embrace the spirit of internationalism.[40]

Early on, the Post Office tussled with the Ministry of the Interior over the ownership of radio. While Bredow, who represented the Post Office, insisted on impartiality, the RMI—represented by Ernst Heilmann, the leader of the Social Democrats in the Prussian Parliament, and his secretary Kurt Haentzschel—insisted that German radio be used to support the young and fragile Reich. Many saw such an argument as assisting the Social Democratic Party in gaining influence. Bredow won that particular battle, but the RMI still had a say in radio operations. The Post Office was responsible for the general legislative work, the collection of radio-licensing fees, establishment of the radio stations, the implementation of the technical development required for broadcasting, and the monitoring of economic management.[41] The RMI was in charge of *Drahtloser Dienst* (*Dradag* for short, or Wireless Service), a publishing service provider of political and economic news. In addition, the RMI, working in concert with the state governments, was responsible for drawing up and enforcing the basic laws of radio and the political and cultural questions arising from the programming. The state governments and the radio stations were responsible for the programming, fulfilling all requisite cultural duties, selecting the performing artists, and implementing and coordinating broadcast performances.[42]

Nearly all nations prohibited private citizens from sending and receiving radio signals during World War I. After the end of the war, this prohibition was lifted. Germans living in occupied territories such as the Rhineland and Ruhr regions, however, were not permitted to use radio receivers. Governmental officials of the Reich feared that Germans living in the unoccupied territories would exploit radio's potential to mobilize the masses with a view to plan revolts and attempt to secede. Officials realized that they needed to thwart any use of the radio for political purposes, particularly in instances where the influence was coming from another nation.[43] Bredow relentlessly insisted that radio must be "an apolitical instrument to spread culture . . . for the interests of everyone."[44] There was a so-called duty to strict objectivity out of respect for Germany's various "tribes" (*Stämmen*), religious affiliations, and social classes.[45] The news needed to be impartial as well. The watchful eye of the RMI ensured that Dradag did not deliver news that benefited one of the political parties.[46]

A sort of "centralized decentralization" (*zentralisierte Dezentralisation*) seemed to be the way forward to ensure impartiality.[47] Decentralized broadcasting made sense. After all, the states differed from one another quite markedly. Some were predominantly Lutheran, while Roman Catholicism was the major religion in others. Some were highly industrialized, while others were rural. They differed in political orientation as well. Most possessed their own German dialect and a unique cultural identity as reflected by local literature, art, and music. A type of radio federalism was the key to success.[48] After all, Germany had been united as a nation only since 1871.

The decentralized aspects of Weimar radio came under attack during the Republic's final years. While the right-wing parties had been calling for national and nationalist radio throughout the 1920s, by 1930 there was a concerted effort to create just that. They indefatigably sought to exploit radio as a tool for as nationalizing the German people and standardizing the German language, something that Weimar radio took pains to avoid.[49] German linguists and rhetoricians such as Theodor Siebs, Friedrich Karl (also written Friedrichkarl) Roedemeyer, Ewald Geißler, and Franz Thierfelder labored intensely to establish a unified *Hochdeutsch* ("high" or proper German), which stage and radio could promulgate.[50]

In 1930, Kurt Magnus, lawyer and one of Germany's early radio pioneers, penned an essay titled "Gegen Zentralisierung" ("Against Centralization") as a response to the growing nationalist sentiments.[51] He argued that radio programs popular with folks in Hamburg might not be as popular in Munich. And Munich-oriented radio programs would not find much following in Berlin or Königsberg, given the omnipresent tensions between the Bavarians and Prussians. "It is simply not in the interest of the German radio listeners to create a unified program for all German societies," he wrote.[52] He concluded by arguing that a fully centralized radio program, as was the case in Great Britain with the British Broadcasting Company (BBC) and in Austria, was simply inappropriate for Germany. He even remarked that centralized programming of the BBC was a "grave cultural mistake."[53] One central program would not satiate the various appetites of the entire German population; therefore, the regional broadcasting stations needed to provide their own programs to their listeners.[54] Decentralization of programming also enjoyed the advantage of generating a large number of programs without sacrificing quality. For example, in 1929, twenty-seven stations broadcast some 2,500 different programs, averaging seven different programs daily.[55]

Many of the broadcasting stations highlighted and fostered regional culture, including local artists and authors and regional dialectics. For example, Nordischer Rundfunk AG (Northern Radio Corporation, or NORAG for short) in Hamburg, which also served Bremen, Hanover, Kiel, and Flensburg, extolled the unique characteristics of the Germans living in that region, namely the *Niederdeutsche* (lower Germans), the *Niedersachsen* (Lower Saxons), and the *Hansadeutsche* (Hansa Germans). Programs could be heard in two languages, Hochdeutsch and *Plattdeutsch* (or Low German), which was one of the six major dialect families of the German language. And *Plattdeutsch* had its own dialects, including *Westfälisch* (Westphalian), *Ostfälisch* (Eastphalian, also known as *Calenberger Platt*, or Calenberg Low German), *Mecklenburgisch* (Mecklenburg), and *Nordniedersächsisch* (North Low Saxon), spoken between Bremen and Northern Schleswig.[56] The radio station catered to locals by broadcasting in other regional dialects such as *Niederfränkisch* (Low Franconian) and *Friesisch* (Friesian).[57] Another example of the importance of regional dialects and cultures was seen in the Stuttgart radio program from September 6, 1925, featuring the work of Karl Eichhorn, a local author whose works were largely unknown outside of Swabia. The Schwabian dialect was featured five days later with Michael Spätzle's comedy "Kutz gohscht na."[58]

A perusal of the programs of various German radio stations during the 1920s reveals a strong commitment to regional dialects and cultures.

> One finds in Stuttgart a strong emphasis on Allemannic and Bavarian in Munich. In Breslau the Silesian dialect is cultivated as is its native poetry. The Old Berlin dialect and Markish ethnicity (*Volkstum*) are kept alive on

> the Spree River. Saxon and Hessian have been well preserved in Leipzig and Frankfurt, as has the Rhineland dialect in Langenberg, Lower Prussian and Lower German in Königsberg and Hamburg.[59]

Every Friday the NORAG offered a Low-German evening, a weekly seminar titled "School of Low Germans (Niederdeutsche)," and once or twice a month a Low German youth hour.[60] Perhaps just as important, these types of programs ensured the survival of local-color works.[61] One finds numerous articles in the various annual volumes of the *Rundfunk Jahrbuch* by directors of the local broadcasting stations in Königsberg, Munich, and Cologne, for example, delving into the musical and literary cultures of their respective states and regions.

Radio pioneers such as Bredow and Carl Hagemann, the theater director and manager of Funk-Stunde Berlin from 1927 to 1929, saw radio in part as assisting the integration of the working (lower) classes into the *Kulturnation* of Germany. They felt that appealing to regionalism would help unite the social classes. Decentralized broadcasting also sought to unite rural and cosmopolitan Germany. That was a tall order: imagine appealing to the tastes of someone like Walter Benjamin in Berlin as well as a farmer from Upper Silesia.

In 1930 Bredow reiterated the importance of Überparteilichkeit of the German radio, without which "the radio can become an unmistakable danger to the inner freedom [of the Reich]."[62] Bredow linked decentralization with impartiality, precisely when cries for a national radio program was reaching a deafening crescendo: "None of the current radio directors desires to function as an exponent of a particular political orientation and is therefore convinced that impartiality is an appropriate form."[63] On September 22, 1930, Bredow repeated his commitment to the decentralization of German radio at a conference of German and Austrian radio station directors in Vienna. He felt strongly that decentralized radio helped lead to the development of a greater number of programs of higher quality. He, too, was contrasting German radio with the centralized radio of Austria and Great Britain.[64]

Who Were the Listeners?

So, who were the radio listeners during the Weimar Republic? And what types of radios did they use? Not surprisingly, while radio's popularity exploded during the 1920s, not everyone could access this new medium. Social class and proximity to major cities were key factors in determining the audience. Radios of the early 1920s were interesting-looking devices. Unless a family was wealthy, each person would sit around the radio with their individual headphones.[65] Initially, loudspeakers, which became much more common in the late 1920s, were too expensive and did not offer very good sound quality. While less expensive than loudspeakers, radios with headphones were still costly. As time went on and loudspeakers improved in quality and decreased in cost, radio listening became a much more communal cultural activity. A radio receiver with three tubes (*Röhrennetzempfänger*) represented the top of the line in 1925 and cost between 250 and 300 RM, or the equivalent of one month's salary for a skilled laborer or a clerk.[66] They were sensitive but difficult to operate.

Detektoren, or detectors—also known as crystal sets—were less difficult to operate and did not require electricity. They were also considerably cheaper than a radio set, with the average price running between 25 to 30 RM in 1924–25.[67] The problem, however, was that the detector's reception was not nearly as good as high-end radio receivers. They could only pick up the strongest ground waves, which traveled only 5 to 10 kilometers from broadcasting stations: they could not detect radio waves in the air.[68] Furthermore, they often suffered from interruptions caused by interference from electrical equipment, electric streetcars, electric motors, and healing devices (*Heilgeräte*).[69] Starting in late 1926, the Berlin electronics company Loewe developed a new radio receiver, the *Röhrenortsempfänger*, which cost around 40 RM and was meant to replace the detectors with its newly invented triple-tube design (*Dreifachröhre*).[70] It could receive more programs more clearly and with a greater volume than the detectors; however, it too could only detect ground waves. It was to become the most popular radio receiver in Germany from 1926 until well into the Third Reich.[71]

By the late 1920s, radios began to assume the look of furniture, as members of the upper classes sought to domesticate them. Some of these more sophisticated radios contained loudspeakers. In 1928 the high-end radios cost about 380 RM.[72] Families often used their Christmas bonuses to purchase them, or, as was true in 80 percent of the cases in 1931, they were rented monthly with the intent to buy.[73] As part of this domestication, directors of radio stations were now interested in attracting the attention of bourgeois female listeners.[74] In the years during and immediately following World War I, the types of jobs and roles women assumed in the workforce changed markedly, as men had either been killed or injured in battle. In January 1919, less than two months after they secured the right to vote, just under 9 percent of the Reichstag seats were occupied by women. During the early years of the Weimar Republic, women were emancipating themselves from a paternalistic society.[75] Radios were no longer perceived as toys for male tinkerers, as more women were listening, and broadcasting stations needed to include programs that catered to them as well. But such freedom proved to be ephemeral. By the end of the 1920s, leaders of the major political parties wanted women to return to their more traditional roles as guardians of family values, particularly during tumultuous times.[76] Women were seen as "carriers, or bearers of the household"

(*Trägerin der Hausgemeinschaft*), and radio reinforced this transition.[77] In addition, children were targeted by clever programmers, who included broadcasts of children's theater, craft lessons, children's songs, and fairy tales.[78] Mothers in particular welcomed the assistance in teaching and educating their children—or so it was claimed by the men. Conservatives wanted the radio to reacquaint women with their domestic roles.[79]

Since the average transmitting stations by 1926 only had a power of 0.85 kW, sound waves only traveled a small radius.[80] A 1927 report of radio listeners reckoned that 31 percent of the Reich's population lived in areas where ground waves could be detected with sufficient strength to hear the radio station: an overwhelming majority of radio listeners lived in large cities or in close proximity to them. Those living beyond the limited reach of the *Detektoren* needed to purchase the more expensive radios.[81] By the early 1930s, the difference in the number of radios in cities versus the number in the countryside diminished somewhat; however, it was still extreme.[82] In short, "over the first four years of broadcasting slightly under one-tenth of the population belonged to the radio audience: around two-fifths of those in the largest cities and less than one-thirtieth elsewhere."[83]

In addition, the monthly 2 RM licensing fee proved to be difficult for many to pay, particularly during the economic crisis of late 1929 and the early 1930s. In 1927–28 German working-class families only had an average of 5.68 RM per month to spend on social and cultural events. On the other hand, civil servants and salaried employees averaged 11.60 RM and 10.86 RM, respectively, for their leisurely activities.[84] Only 50 percent of German households were powered by electricity, with a much lower percentage in rural areas, where dwellers would need to purchase batteries, adding to the expense of radio listening.[85] Despite the meteoric rise in the number of radios and licenses purchased and the best efforts of the radio pioneers, not all of Germany experienced this new form of mass communication.

> The middle classes clearly dominated Weimar broadcasting audiences. Taken together, entrepreneurs, members of the free professions, civil servants, and salaried employees (*Angestellten*) constituted approximately two-thirds of all listeners in 1928 as well as in 1930. Almost one in two households of civil servants and employees owed radio sets in the late 1920s, while the same was true of only one in seven working-class families.[86]

Even though in 1930 radio stations vastly increased their transmission power with a view to reach a larger audience, only the middle classes seemed to benefit from the improvements.[87] Although radios were becoming increasingly cheaper and easier to use, the working classes still did not flock to them by the end of the Weimar Republic.

What Did They Listen To?

How were the goals of German radio, namely entertainment (*Unterhaltung*) and instruction (*Belehrung*), defined and executed? What counted as entertainment? Which aesthetic should radio promulgate? And in which subjects should instruction occur? While the first public broadcast on October 29, 1923, lasted only one hour, by the end of the year Radio-Stunde Berlin (later Funk-Stunde) was being broadcast daily from 4:30 p.m. to 10 p.m. A year later the broadcast averaged nearly eight hours a day. After the regional radio companies and Deutsche Welle had been well established, the total number of German broadcasting hours in 1927 was 37,670. That number increased to 54,975 in 1931.[88] In 1929 Funk-Stunde Berlin broadcast over five thousand hours, with the daily broadcast ranging between fourteen and eighteen hours.[89]

On average, German radio stations broadcast music just under two-thirds of that time. All sorts of genres were featured: operas, operettas, symphonies, solo instrumental and vocal works, choral pieces, *Schlager* (hit songs), dance music, and musical variety shows. Musical aesthetics that appealed to a wide range of tastes were featured. Two of the more progressive stations, Südwestdeutscher Rundfunk (Frankfurt am Main) and the Funk-Stunde Berlin, featured works by avant-garde composers such as Igor Stravinsky, Kurt Weill, and Paul Hindemith.

Weimar radio played a critical role in acquainting Germans with the new genre of electric music catering to the aesthetics of avant-garde modernists. For example, in 1929, Hans Flesch, at the time director of Funk-Stunde Berlin, was the first to set up an electroacoustic and electric music studio in a radio station.

> Today we still cannot conceive of what this still unborn creation can look like. Maybe the term "music" isn't right for it. Perhaps one day the peculiarity of the electrical vibrations will create something new with their conversion into acoustic waves, which has to do with sounds, but nothing to do with music.[90]

He admitted, "But it helps us if we encourage the creative forces to deal with our instruments and try to reconcile artistically their productivity with the strange possibilities of electrical wave conversion."[91] Because of his interest in electric music, Flesch was on the board of directors of the Rundfunkversuchsstelle, the Radio Experimental Laboratory (RVS).[92] Hindemith was fascinated by electric music ever since he visited the Donaueschingen Music Festival in the Black Forest in 1926.[93]

Breaking down the averages among the radio stations, of the 64 percent of the broadcasts dedicated to music, about one-quarter comprised symphonies and chamber concerts, vocal music, operas, and operettas. The remaining three-quarters was dedicated to *Unterhaltungsmusik*, or light music, such as popular live concerts, dance music, recorded concerts, *Schlager*, and musical variety shows. The 36 percent not devoted to music included the news, the new genre of *Hörspiele* (radio dramas), theater performances, children's festivals and fairy tales, and sporting events. Also broadcast were instructional lectures on myriad subjects in the natural sciences, social sciences, and humanities, such as language training, gardening, trade unions, chess playing, legal issues, stamp collecting, economics, and the medical field.[94]

In 1926 Funk-Stunde Berlin broadcast a total of 1,350 lectures: technology and the economy comprised 30 percent, literature and the arts 17 percent, land economy 12 percent, the natural sciences 11 percent, art history 11 percent, foreign languages 6 percent, sports 4 percent, and 9 percent on miscellaneous topics.[95] Two years later, that station broadcast music slightly less than average, or 63 percent. Südwestdeutscher Rundfunk had the lowest percentage of music broadcast with 58 percent, while Süddeutscher Rundfunk in Stuttgart came out on top with 68 percent. Those numbers represented a notable increase over the percentages just two years earlier, when the average amount of music transmitted was only 44 percent of the total broadcast. As will be discussed in chapter 3, the technological improvements resulting in the greater fidelity of radio broadcasting played a key role in the dramatic increase. By 1930, Stuttgart still led all German stations, with 69.3 percent of its programming dedicated to music. Berlin was at 64.5 percent, while the average of all nine German stations was slightly less, 64.4 percent.[96]

From October 1927 to October 1928, Berlin radio offered its audience twenty-one opera broadcasts from its three opera houses. Nine oratorios were also successfully transmitted as well as broadcasts of nine *Oper-Sendespiele* and twenty-five *Operetten-Sendespiele*, or shortened versions of operas and operettas modified for radio. Seventy-four orchestral concerts were featured, the overwhelming majority of which were "symphonies of serious music" and "overtures of folk music."[97] In addition, there were twelve so-called nations' evenings, which were dedicated to the music of various countries. Chamber music was growing in popularity and was featured during fifty-eight evenings. Quartets were also broadcast, and those performing included the best of the 1920s: the Amar Quartet, the Deman Quartet, the Guarneri Quartet, the Havemann Quartet, and the Rosé Quartet. Soloists on piano, harmonium, organ, violin, and many other instruments were commonplace. The soloist would often briefly discuss the piece played. Vocal soloists were also featured, as were numerous choirs. Several stations, including Funk-Stunde Berlin and Südwestdeutscher Rundfunk, had their own orchestras and choirs.[98] From October 1927 to October 1928 all German stations witnessed an increase in opera broadcasts from the radio station from just under 600 hours to over 1,200 hours, opera broadcasts from the opera houses from about 550 to just under 700 hours, and broadcasts from concert halls from over 400 hours to over 600 hours.[99] From January to March 1930, the number of hours the Berlin station dedicated to playing concerts increased from 125 hours to 175 hours compared to the same time period a year earlier. The number of hours dedicated to broadcasting operas, however, had fallen from thirty-seven hours in that time frame in 1929 to only ten hours in 1930. Similarly, the number of hours of broadcast operettas had fallen from just over fifteen hours in the first quarter of 1929 to just shy of seven hours during the first three months of 1930.[100]

Funk-Stunde Berlin and the Radio Experiment

Berlin was seen as the most progressive German broadcasting station.[101] This was particularly true starting in 1929, when Hans Flesch became the radio director.[102] Flesch was trained as a physician, and in his early career he carried out X-ray research. His claim to fame, however, was his contributions to Weimar radio. Born on December 18, 1896, in Frankfurt am Main to a wealthy, liberal, and respected Jewish family, Hans was well connected to Frankfurt's artistic and musical communities. On May 15, 1924, Hindemith married the younger sister of Hans's wife; hence, they were brothers-in-law and remained close friends for over two decades. On April 1, 1924 at 8 p.m., Frankfurt's radio station broadcast for the first time, and Flesch was its director and took charge of the programming.

On June 1, 1929 Flesch left Frankfurt to become the director of Funk-Stunde Berlin, which enjoyed the largest number of listeners of any German station. Berlin was the envy of German radio, with much of its programming being taken up by the other regional stations. His Frankfurt programming was seen as quite radical, and in Berlin he was accused of being a Bolshevik dictator whose programming decisions were becoming evermore extreme. He was dedicated to educating the Berlin public to new forms of music, such as the works of Stravinsky, Schoenberg, and Hindemith, as well as the new genre of electric music.[103]

Flesch insisted that radio's charm was its technological achievement, or what he referred to as the miracle of mediation. It was not a substitute for live performances, but a mediator (*Vermittler*) of them.

> It can never replace a visit to a concert or opera. That would be as ridiculous as saying that a good record of Caruso could replace the unforgettable and that the expert (yes, even the expert) could be happy to hear musical nuances from a good gramophone, from which he could claim: "Yes, that was Caruso." Radio is applicable to the intellect (*Geist*), not the emotions.[104]

Krenek, Anton Webern, Franz Schreker, Charles Ives, Olivier Messiaen, and Edgard Varèse. Bekker defined "new music" as "a rejection of the current state of our musical culture, hoping for its renewal from the spirit of a modern view of the meaning and essence of music in general."[124] This new musical genre challenged what Bekker saw as the antiquated octave of Western music based on twelve semitones by featuring fixed-tone instruments that were based on quartertones, such as a modified harmonium. He saw microtonality as the way forward. It also included compositions based on whole tone scales, which Claude Debussy and Schoenberg used to great effect. But Bekker wanted more than just the abolition of the twelve-semitone scale: he wanted new melodies that no longer harkened back to the classical or romantic traditions. He encouraged composers

> to find a new melodic style, which is equivalent in formative power to that of the old polyphonic art, without imitating it, and which is therefore only related to it in kind, and now, stimulated internally by the formal wealth of polyphonic as well as harmonic art, a new type of form-generating force emerges. That is the task of what I call new music.[125]

A portion of the so-called new or modern music belonged to the genre of Neue Sachlichkeit, which saw itself as the antidote to Expressionism. Two of its most prominent composers were Hindemith and Toch. While the movement is much better known in the visual arts, the musical movement extolled the precision of the machine.[126] As German composer and member of the Neue Sachlichkeit movement Hans Heinz Stuckenschmidt said, "The increasing desire of our times for precision and clarity illuminates more and more the inability of humans to be regarded as interpreters of works of art."[127] He sought to eliminate the human subject from music, rendering the art form truly objective. For him, humans should completely entrust to machines the task of interpreting music.[128] Being resolutely opposed to Romanticism, Neue Sachlichkeit incorporated different types of machines in its music, particularly the Welte-Mignon reproducing piano and electric music instruments.[129] Machines were particularly important to Hindemith's music and aesthetics, as he composed pieces for player organ and player piano. He used a chronometer (*Synchronisationapparat*) built by the German engineer Carl Robert Blum for the music he composed for the cartoon, *Felix der Kater im Zirkus* (*Felix the Cat at the Circus*).[130] He also composed pieces for the trautonium, as discussed in chapter 5. Neue Sachlichkeit included various forms of *Gebrauchsmusik* and *Lehrstücke*, or "lesson pieces," a dramatic form that was didactic in nature. The master of the *Lehrstück* of the period was, of course, Brecht.[131] Ironically, while wanting to appeal to the general public, their works were enjoyed by a small percentage of the population, generally the well-educated.

Neue Sachlichkeit embraced wild musical experimentation, including compositions with sound machines, propellers, turbines, and electric generators. Stuckenschmidt conducted record experiments with László Moholy-Nagy in the Bauhaus in Dessau. They punched a second hole in records so that they rotated unevenly, thereby generating a "vou-vou-vou" sound that produced a new type of music. Stuckenschmidt also experimented with the paper rolls of electric pianos.[132] As we shall see in the next chapter, a new genre, Rundfunkmusik (radio music), was created and quickly became an important genre in the Neue Sachlichkeit movement.[133] Kurt Weill, Franz Schreker, and Ernst Krenek belonged to this movement, which also counted among its members the musicians associated with Die Novembergruppe, or the November Group, which took its name from the German Revolution of November 1918, namely Max Butting, Hanns Eisler, Philipp Jarnach, Heinz Tiessen, Vladimir Vogel, Stefan Wolpe, and Gustav Havemann. Politics and aesthetics were inextricably linked.

Conceived primarily by Hans Bredow during the 1910s and early 1920s, German public radio was born on the eve of October 29, 1923. In many respects, radio was an experiment of the grandest sort. It was intended to possess and extol the virtues of the young Republic: democratic, liberal, and decentralized, and it was meant to be a *Kulturträger*, or bearer of culture. Alas, despite its amazing popularity and growth, not everyone could enjoy the new medium: that privilege was reserved for the middle and upper classes as well as those who lived in major cities. Not only was the audience restricted, so too were the types of music that could be broadcast. Before all of the problems of broadcasting could be ameliorated, composers and radio station managers needed to experiment to determine which instruments worked well together, where the microphones should be placed, what new compositions worked best for the radio, and how compositions for more traditional genres, such as symphonies and operas, should be tailored to a radio audience.

2

High Infidelity

WHEN WE TURN on a radio today to listen to a live performance—or even a recorded one—of a cellist, we can recognize the instrument even with minimal musical knowledge. With a bit more musical training and a good ear, we might be able to identify the performer and piece, such as Yo-Yo Ma or Jacqueline du Pré playing Edward Elgar's Cello Concerto in E minor, Op. 85. Or, we might be listening to a soprano singing Antonín Dvořák's "Měsíčku na nebi hlubokém" ("Song to the Moon") from act 1 of *Rusalka*, and we are able to tell if it is Anna Netrebko or Renée Fleming singing. Or, for those who prefer other musical genres, one can hear that Jimi Hendrix is playing a guitar in "All Along the Watchtower," or one can differentiate between Cass Elliot and Michelle Phillips' voices in "Creeque Alley" sung by the Mamas and the Papas. The ability to discern such information from attentive listening, which today might seem natural or normal, belies the massive amount of work by physicists, radio engineers, physiologists, and musicians during the 1920s and '30s.

Let's transport ourselves back to the 1920s. As Edward L. Nelson, a radio engineer at Bell Telephone Laboratories, commented:

> As long as music and entertainment continue to hold a prominent place on broadcasting programs, fidelity of transmission will probably remain the most sought-for characteristic, not only for the radio transmitter itself, but for all of the apparatus units in the system.[1]

Radio broadcasts, for example, had difficulties transmitting the soprano's voice with a sufficient degree of fidelity. A well-worn joke of the period went something like this:

> Not so long ago, an elderly gentleman whose aristocratic appearance was impaired by a distinct frown, entered a radio store and addressed the clerk as follows—"You sold me a receiver some time ago. I wish you would send someone to my home to fix my set. I hear nothing but screeching sopranos. Here is my address. Thank you."[2]

As a 1928 *New York Times* article explained: "True reproduction of a musical concert or the voice by radio depends on so many factors that it is difficult and expensive to provide the fidelity that many prospective customers are taught to expect."[3] The phenomenon was certainly well known to German engineers and physicists. The renowned acoustician Erwin Meyer noted:

> Speech and language should be broadcast with fidelity [*naturgetreu*, literally "faithful to nature"] . . . A whole series of scientific investigations had to be carried out in order to recognize which conditions had to be met to guarantee a fidelitous reproduction (*naturgetreue Wiedergabe*, or literally "a reproduction faithful to nature") of sound processes of any kind.[4]

Obviously, music aficionados were also well aware of the problem. German musicologist Georg Schünemann, the director of the Staatliche akademische Hochschule für Musik (Berlin Academy of Music), complained that "orchestral pieces were distorted by the radio such that in one piece the basses seemed to be absent, in another the timpani [and] the violins sounded like clarinets."[5] This clearly was not a restricted phenomenon, but one that afflicted early radio globally. So, what in the world was going on?

The transmission of musical instruments and the human voice provided scientists and engineers with intriguing and challenging research problems. As German telegraph engineer Hans Carl Steidle explained in 1925:

> The problem of the electric broadcast of lectures and musical performances leads to a frontier (*Grenzgebiet*) of technology and art, which gives the telegraph engineer a number of new tasks to solve that fall outside the framework of general rail transport technology.[6]

As a result of the formidable challenges posed by radio, the disciplines of acoustics and electrical engineering were fundamentally changed. According to electrical engineer Günther Lubsznynski:

> It is certainly not said enough that acoustics, which had previously been the stepchild of physics, has surely not only come to experience many new stimuli (*Anregungen*) through radio, but has also become quite "modern."[7]

Meyer went further: "A new branch of technical acoustics, electroacoustics, came into being as a result of the impulse from the radio."[8] A new profession was born, the "musical engineer," now known as the sound engineer.

> Two counterparts (*Gegengebiete*), which previously had no relation with each other, now engage with one another, namely the electrical—particularly the radio—engineer and the musician. Out of this marriage

> the "musical engineer" is born, which is seen as a paradox today, but tomorrow might be taken for granted (*Selbstverständlichkeit*).[9]

This is, then, a story about the invention of a cultural medium, the radio, as well as the simultaneous creation of a new branch of a scientific and engineering discipline, electroacoustics.[10]

The history of broadcasting—and sound reproduction in general—is simultaneously a history of fidelity. As Jonathan Sterne has so persuasively argued, sound fidelity has just as much to do with the social functions and interactions of humans with machines as it does with the relationship of a recorded or broadcast sound with the original source.[11] That source is just as embedded in sociocultural relationships—including techniques, skills, practices, sounds, and machines—as its reproduction. Raymond Williams reminded us that we need "to break from the common procedure of isolating an object and then discovering its components. On the contrary, we have to discover the nature of a practice and then its conditions."[12] Questions of sound reproduction were being raised precisely at a time when physicists, physiologists, and radio engineers in various laboratories around the world were beginning to define and measure sound fidelity, thereby giving it not only a physical meaning, but an aesthetic one of perfect reproduction as well.

The global success of radio during the 1920s has been well documented. The importance of the new medium to so many human endeavors was clear at the time to major corporations and governments alike. One major imperfection, however, needed to be remedied. As seen from the quotations above, broadcasting fidelity of that decade suffered from distortions that arose from the radio station transmitter, the receivers in the homes, and cable lines. Listeners, even those with trained musical ears, had problems differentiating between a violin, flute, and clarinet. Even distinguishing between various spoken voices of a radio drama proved a formidable challenge. Various regional dialects, whether in the United States or Germany, were clearer than others. This chapter focuses on these problems of distortion that plagued early radio in Germany.

Over fifty years ago, the renowned historian of technology Lewis Mumford argued that two different types of technologies—authoritarian and democratic—have coexisted throughout history. Democratic techniques were small-scale, and the skills and practices were controlled by artisans and farmers. Authoritarian techniques were much larger in scale, more recent, and involved coercion and slavery to construct human machines, such as the military and bureaucracy.[13] Some thirty-five years ago, Langdon Winner famously insisted that artifacts themselves possess politics.[14] More recent studies discuss the biases of voiced-based personal assistants, such as Google Assistant and Siri, against those who speak with accents;[15] racial disparities that exist in automated speech recognition;[16] and the inequalities present in voice-biometric systems based on racial and gender differences.[17]

This chapter raises important sociopolitical issues that still resonate today. It tracks the history of objects and analyzes how their construction was shaped by the scientific, technological, economic, aesthetic, and political decisions made by human actors. By uncovering those choices, we can and indeed must do our best to ensure access and make amends to those who have been victims of discrimination. Fidelity of human voices and musical instruments, it turns out, had (and still has) an important political component.

The Status of Engineers during the Weimar Republic

Physicists and engineers jumped at the opportunity to offer assistance to arts and culture from the outset. They also wished to ingratiate themselves with the nation's politicians and general public since they had been blamed, in part, for Germany's defeat in World War I.[18] Engineers in particular were attempting, with limited success, to gain acceptance and autonomy.[19] Ever since the late nineteenth century, engineers were fragmented. Some were educated at technical universities, which, by the end of the nineteenth century, were permitted to grant *Diplomingenieur* and *Doktoringenieur* degrees. Clearly, not all engineers or technicians were good enough to complete those degrees.[20] The ones that did, particularly the *Doktoringenieur* degree, sought the same status as doctors of philosophy in the humanities and natural sciences. Alas, traditional members of the *Bildungsbürgertum* did not respect degrees from technical universities, which were seen as training the hands, not cultivating the mind.[21] While Karl Marx wrote at length on social stratification based on capital, from the proletariat to the bourgeois capitalist, Max Weber famously wrote on the status of professions based on the symbolic capital earned by education.[22] One thinks of Weber's adage of the specialist in the age of capitalism: "Professionals (*Fachmenschen*) without spirit, pleasure-seekers without a heart, this nothingness imagines that it has climbed to the pinnacle of humanity that has hitherto never been reached."[23] Similarly, the engineer Hans Castorp in Thomas Mann's *Magic Mountain* (1924) could not quite understand the arguments of the Italian man of letters, Lodovico Settembrini, who associated technology with morality. It was the humanist, not the engineer, who could link cultural with technological progress.[24] Similarly, Mann's *Buddenbrooks* (1901) is a masterful story, based on his family over several generations, about the splitting of the *Besitzbürgertum*, or propertied middle class, from the traditional Bildungsbürgertum.[25] Mann, more so than any other author, had his fingers on the pulse of social relations and the role of technology in society during the early twentieth century.

In 1909 the Verband Deutscher Diplom-Ingenieure (VDDI) was created to organize academic engineers, who sought, among other things, the protection of their titles (*Standesbezeichnung*).[26] They felt belittled both by the leaders of industry, who wanted cheap labor and not overpaid intellectuals, and the members of traditional free professions. In addition, engineers suffered, like most other disciplines, from the devaluation of their intellectual labor after the war.[27]

Historian of technology Mikael Hård insists that liberal modernity had been under threat since the end of the nineteenth century. After the horrors that technology had wrought on civilization during World War I, liberal modernity continued its downward spiral.[28] The patriotism of members of the Bildungsbürgertum—doctors, lawyers, professors, military officers, senior civil servants, and ministers—was extreme as World War I broke out. But as time went on, tensions rose between members of the Besitzbürgertum, many of whom profited from the war, and the Bildungsbürger and *Kleinbürger*, both of whom began to feel material deprivation. Many feared that they would descend into the existence of the proletariat.[29] As numerous scholars have argued, the Weimar Republic witnessed a group of highly educated engineers shaken by the public's disdain for technology. The engineers responded by insisting on the cultural value of their disciplines.[30] Carl Weihe, editor of the VDDI's renamed journal, *Technik & Kultur*, insisted that "technology cannot be separated from culture," which "constituted its basis, support, handmaiden, distributor, and participant."[31] The journal included articles on technology and culture, particularly the philosophy of technology, authored by such engineering luminaries as Eberhard Zschimmer and Friedrich Dessauer. By 1932 it was espousing a radical right ideology that resonated with the Nazis.[32] As Adelheid Voskuhl has shown, German engineers began to philosophize during the Weimar Republic about the role of technology in culture.[33] While engineers desperately sought the status of Kulturträger (Wolfgang Mommsen referred to them as *das ältere Honoratiorenbürgertum*, or the traditional bourgeoisie of dignitaries and Fritz Ringer referred to them as the mandarins) by commenting on cultural issues, they simultaneously loathed and were envious of the Bildungsbürgertum.[34]

Infidelities

During the 1920s, physicists and engineers began inventing and improving devices that could compare and contrast the original live sound source with its reproduction. Comparing live sounds with their various forms of reproduction was, of course, critical to numerous enterprises. For example, the Edison Company created Tone Tests, or recitals in which an Edison Diamond Disc Phonograph generated the music. Audiences in concert halls were invited to compare the live musician with the recorded sound.[35] As Emily Thompson has argued, the practice of listening to a phonograph needed to be accepted as the cultural equivalent of listening to a live performance. While such a comparison might seem natural to us today, that was far from the case at the time. The Edison Company invested much time, money, and effort in teaching the American popular audience that a careful listener could intelligently compare a live performance with its reproduction.[36]

One often reads about the problems plaguing the fidelity of radio broadcasts. For example, back in late May 1923, during the weekly Sunday concerts, engineers and musicians at Königs Wusterhausen noted that while broadcasts of soloists were generally successful, duets or small ensembles did not fare nearly as well.[37] The radio listeners heard chatter, noises, and cluttered tones. The musicians, physicists, and electrical engineers needed to experiment with the placement of microphones when featuring two or more players. In one case involving a duet of a violin and cello, the engineers placed the (apparently and surprisingly docile) cellist and his microphone in the lavatory. Not surprisingly, that did not work. In the end, they decided to use a telephone with a funnel. Unfortunately, the musicians needed to play extremely closely to the telephone.[38]

Georg Schünemann summed up the problem in 1929. Recall his complaint that radio distorted orchestral instruments and pieces. One type of fidelity was the inability to differentiate between violins, flutes, and clarinets. Indeed, in 1929, Hermann Backhaus, a Siemens & Halske engineer who researched the timbre of violins, reiterated that the common observation that violins sounded like clarinets and flutes had been known since the early days of radio broadcasting.[39] Another type of infidelity involved pitches at the lower range of the cello and particularly the double bass: they often seemed to disappear during a piece. The low frequencies needed much more energy to be heard given the human ear's difficulty in hearing those pitches. If they were heard at all, they were often confused with horns.[40] In short, pitches at both ends of the radio bandwidth were difficult to transmit with a sufficient degree of fidelity.[41] Schünemann's solution to the problem of infidelity was to broadcast exclusively Rundfunkmusik, or music composed and played specifically and exclusively for radio. Because of its technical limitations, he initially thought that radio could never replace going to the opera or a live concert, but it could broadcast music uniquely composed for it.[42] Rundfunkmusik should only feature musical instruments whose broadcast sound was the same as the original. In 1929 Berlin composer and music critic Max Butting noted that radio failed to reproduce the sounds as one heard them in a concert hall. This was particularly true of orchestral music.

There were therefore numerous types of infidelities during the 1920s. The one that caught the attention of musicians, natural scientists, and radio modified telephone cables composed of rubber encasing copper wire. Cables could transmit a larger frequency range with greater uniformity by using additional circuits and newly invented equalizers, which corrected for distortions. Since the lower frequencies could be sent with greater fidelity than the higher ones, the equalizers ensured that the broadcast of the lower frequencies were not too strong. Longer cable lines contained loading coils, which are induction coils inserted at certain intervals with conductors of a transmission line. This enables a broadcast of higher frequencies with minimal distortion.[96] In December 1927, installation had begun of new long-distance cables that possessed a frequency range of 50 to 7000 Hz.[97] By the end of 1928, 5,300 kilometers of the new cable had been laid. In 1930, there was an improved line for music transmission in every German long-distance cable, a total of 9,000 kilometers.[98] The line was equipped with amplifiers for the aforementioned frequency bandwidth at intervals of 75 kilometers. One such long-distance cable connected Berlin, Leipzig, Plauen, and Dresden with a branch to Lobositz (now called Lovosice in the Czech Republic), and Prague, and another connected Plauen, Nuremberg, Stuttgart, and Karlsruhe with branches to Freiburg and Frankfurt. The music lines of the old cables were discontinued, as the music lines in the newer ones could broadcast a larger frequency range and were less likely to suffer from cable interference. Prussian Ministerial Councilor Karl Höpfner was confident "that German broadcasting will have an efficient broadcasting network at its disposal in the foreseeable future," guaranteeing a superb transmission of music.[99]

Despite these technological advances in cable technology, governmental regulations restricted bandwidth size. The bandwidth comprises two sidebands, the upper and lower, and the carrier frequency. Sidebands are the sum and difference of the carrier frequency and modulation frequency. For example, radio stations with a maximum sideband of 5000 Hz would have a transmission bandwidth (or channel) of 10,000 Hz. Back in the early 1920s scientists and engineers generally believed that it was critical that the sideband between 30 and 5000 Hz be transmitted with equal efficiency in order to broadcast tones of "very high-quality music."[100] To transmit 30 Hz as the carrier frequency, one needs to have 30 Hz less than and 30 Hz greater than that carrier frequency, and the same is true for 5000 Hz. The reason is modulation, or the variation of the amplitude of transmitted radio waves due to variations of air pressure resulting from a sound, such as a voice or music. The modulated wave no longer possesses a single frequency, but rather is the sum of three different wave trains of constant amplitude: the frequency of the original unmodulated radio wave, also known as the carrier frequency, or $P/2\pi$, a frequency $Q/2\pi$ greater than the carrier frequency, and a frequency $Q/2\pi$ less than the carrier frequency, where $Q/2\pi$ is the modulation frequency. In order to transmit all the intermediate tones, two side bands of frequencies are required, the so-called upper and lower sidebands.[101]

Europe was experiencing a growth in the number of radio stations. In 1923 there were eighty-seven radio stations with an additional thirty-seven new stations ready to start broadcasting.[102] Some form of regulation was needed to space out the stations along the medium wavelengths of radio in order to avoid interference.[103] In early April 1925 ten European nations met in Geneva to deal with the problem of interference generated by radio stations broadcasting at frequencies that were too close to one another given their various levels of power. The delegates decided that European engineers needed to collaborate to ameliorate the problem. In July 1925, at a meeting chaired by BBC chief engineer P. P. Eckersley, the distance set between stations was 10,000 Hz, which corresponded to a sideband of 5000 Hz, on medium-wave radio. It came into effect on November 14, 1926.[104] However, in 1928 the Brussels Plan, which took effect on January 13, 1929, called for a reduction in bandwidth to 9000 Hz, which corresponded to a sideband of 4500 Hz on medium-wave radio and 10,000 Hz on all stations broadcasting above 1000 kHz.[105] The Brussels Plan was reinforced by the European Radio-Electric Conference in Prague of 1929; however, now only stations above 1400 kHz were permitted the 10,000 Hz bandwidth.[106] That reduction in bandwidth to 9000 Hz was the bare minimum required for a musical broadcast.

German engineers and physicists lobbied to increase bandwidth to capitalize on the fruits of their new technologies.[107] In 1931, for example, German electrical engineer and radio pioneer Eugen Nesper called for an increase in bandwidth to at least 13,000 Hz, which would contain two sidebands, 6500 Hz each.[108] Political decisions, in his view, should not stand in the way of the technological advances made to improve the radio-listening experience. Alas, their pleas were not heard.

Other Infidelities and Sopranos

The phenomenon of sopranos' voices disappointing radio audiences was well known in the 1920s. An article appeared in the October 1928 issue of *Scientific American* titled "Why Is a Radio Soprano Unpopular?" written by the journal's associate editor of radio engineering, John F. Rider.[109] Trained as a radio engineer, Rider was the author and publisher of over 125 books on radio and television. As has been argued previously, various forms of sound technology discriminate against women, particularly women of color.[110] Early radio was no exception.

During the 1920s, the soprano voice was extremely popular on stage. There were numerous famous stage sopranos of the period, including Grace Moore, overtones no longer entirely resembled their original sound. The receiver sent a sound to the loudspeaker, which possessed little fidelity to the original.[120] Sopranos were disproportionately affected. In short, US radio stations of the 1920s were either not permitted or were incapable of transmitting and faithfully responding to the overtones of the soprano's voice.[121]

The broadcast station was not the only technical problem for the fidelity of the soprano's voice. The third reason for the unpopularity of the radio soprano was the radio receiver, often itself a source of distortion. First, the portions of the receiver responsible for the selection of the desired broadcasting station could produce an effect called sideband suppression, which limits the overtones and their relative intensities being passed through the receiver to the loudspeaker. The second potential problem in the receiver was the audio amplifier. If the design of the audio amplifier was mediocre or subpar, then the amplification of the relative intensities of the overtones was distorted, resulting in an altered tone color. A vast majority of receivers did not employ equipment of the requisite quality for use in audio amplifying. The third problem was the loudspeaker, which—because it possessed the lowest quality of all the other components—was considered the greatest contributor to infidelity. Most speakers did not respond well to the overtones of the higher pitches, and the response curve of most loudspeakers was not at all uniform over the audio frequency band.[122] The reproduction of the radio's loudspeaker for sopranos often resulted in "a shrill unmusical shriek."[123] As the chief engineer of the Stromberg-Carlson Telephone Manufacturing Company explained, switching from battery-powered radios to fully electric receivers resulted in an alternating current. This created distortions that, when heard on cheap loudspeakers, resemble audible hums. Rectifying and filtering systems could ameliorate the problems; however, these were expensive in the 1920s. He continued, "Many radio listeners condemn a radio soprano as a poor radio performer because of the harshness and unnaturalness with which this voice comes through."[124]

Finally, human taste, the consumerism of the 1920s, and the desire for immediate gratification were also critical. The public's demand for the rapid production of radios that matched their interests had resulted in equipment that was not true to the soprano's voice.[125] The public seemed to prefer a greater fidelity for the lower pitches of the bass section of the piano and tones produced, for example, by the cello, viola, bassoon, and trombone as well as the upper-bass, baritone, tenor, and alto ranges of the human voice. This came with a sacrifice of the higher pitches of the soprano, flute, piccolo, and violin.[126] This was a classic example of the struggle between selectivity and quality of a receiver. While one clearly wanted both superb selectivity and quality, both required state-of-the-art equipment back in the 1920s. Quality in this instance meant that all frequencies in the audible range needed to be reproduced with equal loudness, if they were equal in the original sound. Selection meant choosing a specific range of frequencies while suppressing others. The more selective a circuit was, the more efficient it was at suppressing unwanted frequencies. A compromise was necessary: a lack of selectivity could lead to the presence of background signals from other stations. Too much selectivity could lead to a hollowness of sound.[127] These choices were made by men, not women, as women, though often hired as radio operators, were rarely hired as broadcasters since their voices were perceived as "flat" or "shrill." Fewer still were employed as radio engineers.[128]

It is interesting to note that women's voices have noticeably deepened since the end of World War II. Cecilia Pemberton has studied the voices of Australian women from 1945 until the early 1990s and found that their fundamental frequency had flattened by 23 Hz, from 229 Hz, or approximately the A# below middle C (A3 in modern musical notation) to 203 Hz, or approximately G# (G3). After ruling out a number of issues, the team speculated that perhaps women purposely lowered their voices to "project authority."[129]

Preferences for men's voices over women's became a technical issue. This section of the chapter illustrates how human decisions about technology affect social relations. In this case, radio technology reproduced the gender hierarchy that plagued much of the US and Europe.

Rundfunkmusik

Radio's broadcasting shortcomings affected music both at the micro-level (namely, the fidelity of the reproduction, such as with the tone colors of instruments and voices) and at the macro-level (by determining the types of compositions that were best suited for this new medium). Radio broadcast, with varying degrees of success, music that was originally composed for concert halls or salon rooms. During the 1920s, a genre of performing art, called *Rundfunkkunst* (radio art) was created. It featured radio dramas (Hörspiele) and music (Rundfunkmusik) composed specifically and exclusively for radio performance.[130] A leading supporter of Rundfunkmusik, Max Butting stressed: "The possibilities of radio are not limited to presenting a series of music reproductions which are substitutes for the original, like the reproductions of paintings in a portfolio of collected works."[131] Rundfunkmusik appealed to the musicians affiliated with Neue Sachlichkeit.[132]

Writers and composers of radio dramas and radio music needed to now think carefully about how to create works without an audience present and therefore without immediate reactions to their oeuvre. They also needed to appreciate that the environment of a radio audience differed considerably from that of more established venues. Often radio listeners sat alone with their headsets. Or they were with immediate family members sitting around their radio sets in familiar settings. Not being influenced by the pageantry of a live performance, they could concentrate on the music itself. In addition, Butting maintained that the person sitting alone with their radio was

> more subjective and more prone to fantasy than the concert-goer who is closely bound to the actuality of his surroundings. The concert-goer experiences music plus reality, the radio listener music plus fantasy.[133]

Authors of radio dramas and composers of radio music were also required to consider the acoustics of radio stations' broadcasting studios, which were far smaller than theater stages. Furthermore, the acoustics of those studios were different from those of seventeenth-century salons or nineteenth-century concert halls. Composers were forced to reconsider the size of their orchestra or choir as well as the types of instruments used and their relative placement in the studio. Often, instruments that complemented each other in live performances did not work well for a radio broadcast.[134] As Butting asked, "Should not music written for the radio be based on a technic of composition (instrumental, polyphonic, etc.) which will take into account the technical factors and be calculated automatically to ensure good reproduction?"[135] He proffered an example. The editing of a musical march in Berlin's Radio Experimental Laboratory (RVS)—discussed in further detail in chapter 4—required the reduction of the soft-sounding brass instruments from six (in the original nineteenth-century score meant for the outdoors) to two for the radio, while the harder sounding brass instruments with less resonance needed to be reduced from six to four.[136]

The trouble with Rundfunkmusik, it turns out, was that it was meant to be

> music for all, but a large portion of modern music is music for the few. A powerful movement has taken place in music that has produced works, which in their character are essentially different from that which lives in the musical consciousness of the masses. All musicians are influenced by these works, particularly the most interesting and creative ones. Their creations are not "popular"; they only find an echo within a small circle. Who, however, writes for radio must adapt to the level of understanding, which is common to all classes of society. With all art, with all the subtlety of its quality, this music must also have the characteristic of becoming "popular."[137]

Rundfunkmusik, much like Neue Sachlichkeit, needed to appeal to the masses much more so than previous musical genres. Butting was keenly aware of radio's importance to the culture of the Republic.

> The artists realize that they have a great responsibility. With broadcasting a government monopoly in Germany, the great power has been given to them; they can influence the taste of the people; they nearly all feel that they have a cultural mission . . . [138]

Butting reminded his audience that "radio is able to reproduce a relatively large quantity of subtleties, but in many cases they are not the ones that the composer has thought about. Most of the works that are broadcast over radio are the ones that are not composed for this purpose, and technology must try to represent them as well as possible."[139]

The broadcast of the orchestra in Wagner's *Ring der Nibelungen*, for example, was not what it sounded like in an opera house.[140] The massive, overpowering, and pure orchestral sound so characteristic of Wagner's operas was lost over the radio. "Thick" instrumentation, a result of various groups of instruments reinforcing each other, came across as "thin." Instruments were no longer perceived as blended parts of an overall organic whole, but rather as individualized pieces.[141] Wagner's operas were not the only works that suffered from radio broadcasting. The doubled thirds of the woodwinds in Johannes Brahms's symphonies, for example, lost their unique characteristics as a result of radio's insufficient fidelity. Similarly, Richard Strauss commented in his enlarged and revised edition of Hector Berlioz's *Grand traité d'instrumentation et d'orchestration modernes* that the second violins, which were meant to be less brilliant-sounding and intense than the first violins, sounded as if they were a different type of instrument. That distinction could not be detected over radio.[142] Butting added that "no matter how careful and technically correct a broadcast may be, the tones emanating from the loudspeaker are different from the original."[143]

Wiesbaden theater director Carl Hagemann felt that radio music was an ephemeral, albeit necessary, genre until acousticians and engineers perfected the broadcasting of live musicians.

> We would like (and have the appropriate trust in our acousticians and technicians) to be given the means and possibilities to transmit any and all types of music with the same effect, i.e., the original tone, to the receiver as much as is possible. As long as that is not the case, or at least not to a satisfactory degree—as long as we are not able to capture effectively the music as a whole, that is, sound flawlessly on the radio and broadcast all over the earth, we will be content with a specific type of radio music, namely that which is adapted to the current state of radio technology.[144]

The journalist Frank Warschauer agreed. Rundfunkmusik would only exist as long as the microphone technology could not pick up all the nuances of sound.[145] Georg Schünemann explained at the opening of the Berlin Academy of Music's RVS in May 1928 that radio must have its own music, composed or edited for its own purposes, because it was not technically able—and indeed never would be able—to send pieces of any musical structure via the transmitter.[146] As discussed in chapter 4, the RVS researched the types of music best suited for radio broadcast.

Most of the major composers dedicated to Rundfunkmusik were active members of the Neue Sachlichkeit, including Butting, Hindemith, Weill, Franz Schreker, and Hermann Scherchen. As a station manager, Flesch was attuned to the technological shortcomings of radio and therefore underscored the importance of Rundfunkkunst, having written the first radio drama to be broadcast in Germany, titled "Zauberei auf dem Sender" ("Magic on the Broadcasting Station"), which debuted on the Frankfurt station on October 24, 1924. In 1927 Hindemith received a contract from the Frankfurt station to compose a three-movement work for organ and chamber orchestra, which debuted in January 1928 and was subsequently broadcast by other regional stations. These Rundfunkmusik compositions, though first dubbed as "experiments," could be heard with increasing frequency on German radio starting in January 1929. Often commissioned by the radio stations themselves with the financial backing of the Reichs-Rundfunk-Gesellschaft (Reich's Broadcasting Corporation), new works by leading composers were featured monthly, including Toch and Weill. Perhaps most famously, in 1929 Brecht, Weill, and Hindemith teamed up for a thirty-five-minute lesson for the radio titled "Lindbergh-Flug" ("Lindbergh Flight")—an homage to Charles Lindbergh's historic 1927 transatlantic flight.[147] It captured listeners' imaginations and created new pictures in their mind.

> The fight with the storm, the emotions of the aviator are presented more vividly than is possible on the stage, on the screen, or in a narrative. The fantastic, the conceptual, in short everything which stretches to the horizons of human imagination and intellect, are never so effective as over the radio, which adds something unreal to anything this apparatus brings us.[148]

Much like Weill, Brecht took an interest in radio from its advent, not least because of his interest in radio dramas. In an open letter published on Christmas Day 1927, Brecht addressed the recommendations of the director of the Funk-Stunde Berlin. Brecht wrote that he hoped radio would be used to democratize Germany, thereby enabling all Germans to closely observe Reichstag meetings. In so doing, radio would frustrate attempts of lawmakers to obfuscate their procedures and decisions. Much like his literary works, Brecht wanted radio to force the audience to think. He therefore recommended that disputes and debates between prominent leaders be broadcast and advocated for an end to the "gray monotony of the daily bill of fare of house music and literary courses."[149]

The author Alfred Döblin, who was employed by Funk-Stunde Berlin and famous for his piece, *Berlin Alexanderplatz,* envisaged a somewhat different role for German radio. In a lecture titled "Literatur und Runkfunk" ("Literature and Radio"), delivered at a workshop on poetry and radio in Kassel-Wilhelmshöhe on September 30, 1929, he complained that German radio was "something vulgar, for entertainment and instruction of a crude kind."[150] He called on writers to engage more with this new medium to increase the dissemination of important literary works. Radio was, in his eyes, "a transforming medium" that could bring high-brow literature to the masses.[151]

The young Walter Benjamin, famous for his works on of the influence of technology on mass culture in the 1930s, wrote and directed over eighty pieces for radio for the Frankfurt and Berlin radio stations from 1925 to 1933, including children's tales, lectures, travel pieces, stories, radio plays, and school radio.[152]

Hamburg composer and music critic Robert Müller-Hartmann, who worked with the Nordischer Rundfunk AG in Hamburg, agreed that not all music was suitable for radio. He felt that classical, and classically oriented Romantic music, should be prioritized. Joseph Haydn, Wolfgang Amadeus Mozart, and Franz Schubert were much more enjoyable to listen to over the radio than Wagner and the young composers of the 1920s. His reasoning: less extravagant and lavish composers, that is to say, those who practiced restraint, limited the scores of the instruments to enhance their natural functions and pure tone colors. This was not just a matter of aesthetic taste but of acoustics as well: large orchestras with their distinctive sound combinations could create disturbing overtones that enveloped and distorted the music. All of the so-called clever and overly sophisticated special effects of the orchestra, for which Wagner was so (in)famous, were often lost in the broadcast.[153]

Theodor W. Adorno

One philosopher in particular was concerned with radio's problem with broadcasting live music. A decade before Theodor W. Adorno—Walter Benjamin's foil and friend—started critiquing jazz, he was a contributor to the Viennese music journal *Musikblätter des Anbruch* (Musical Sheets of the Dawn), becoming a member of the editorial board in 1929 and successfully lobbying to rename the journal *Anbruch.* He was particularly fascinated with "light music" and kitsch, seeing them as a product of specific class interests—a reactionary music form rather than a modern one.[154] Adorno wished to neither praise nor bury kitsch, but to understand it. He also proposed that the journal feature a column dedicated to critiquing the work produced for new media, including the radio: "It must also address problems concerning radio and possibly even provide regular views of the most important broadcasts of modern music (here too, critique!)."[155]

chapter, the term possessed different definitions during the 1920s. The various definitions of sound fidelity were inextricably linked to the failure of early radio. For radio listeners, fidelity meant that one could differentiate between various instruments. Listeners could also distinguish between the voices and sound effects of radio dramas as well as understand all the words, and they would be able to hear the lower notes of a double bass and the overtones of the soprano. Composers began to think that a new form of music needed to be written since radio broadcasts distorted sounds in different ways than concert halls did. Radio engineers and physicists of the 1920s—who we shall see in the next chapter had their own definitions of fidelity, Naturtreue—were presented with the daunting task of enabling each instrument and each voice to be transmitted without distortion. In addition, the history of fidelity is one in which not all actors, human and nonhuman alike, were treated equally. Radio technology of the period discriminated not just against certain musical instruments, but women's voices as well. Certain economic, political, technological, and aesthetic choices were made mostly (if not exclusively) by men, and those choices favored deeper voices at the cost of the soprano. While a number of scholars have questioned the now-antiquated assertion that radio was a wonderful democratizer of culture, we now can also appreciate that its transmission of all voices and instruments was also rather inequitable.

3

Analyzing Distortions and Creating Fidelity

DURING THE 1920S, most electrical engineers and physicists sought to create what they referred to as "objective" measurements of timbre, rather than the "subjective" judgments of skilled musicians. For them, removing the human was paramount to establishing an international standard for the science of tone color. Sound analysis was rarely based on the use of tuning forks, resonators, and interference tubes, but it was now powered by electric devices such as harmonic analyzers and oscillographs.[1] But with more research during that decade, a number of engineers and natural scientists, such as Erwin Meyer in Berlin and Harvey Fletcher at Bell Telephone Laboratories, and psychologists and physiologists, such as Carl Stumpf and Wolfgang Köhler in Berlin, began to realize that timbre was a tad more complex than had been previously thought. While the oscillograms of the relative strengths of the overtones revealed much, it did not reveal the complexity of tone color

A number of countries mobilized their natural scientists and engineers to solve the infidelities of early radio; however, the United States and Germany led the charge. This chapter explores how natural scientists and engineers markedly improved the fidelity of reproduction. Whereas the previous chapter focused on the infidelity of broadcast tone color, this chapter discusses how its fidelity was greatly improved. Laboratories in universities, companies, and governmental institutions labored on the topic, as radio was rapidly becoming the most important cultural medium of the early twentieth century. Whereas experimentation in the previous chapter referred to tinkering with musical compositions and radio programs, this chapter focuses on experiments conducted in laboratories that contributed to the aesthetic of tone-color fidelity.

The first step in tracing the creation of timbre fidelity was defining timbre as a function of the frequency components of a sound's overtones. The second step was developing extremely responsive circuitry to allow for the requisite precision. The third was improving the equipment of the broadcasting stations and the receiver components, such as the microphone, amplifier, and loudspeaker based on those measurements of sound. Those three sections structure this chapter. For natural scientists and engineers, this was the trajectory of quantifying fidelity of tone-color reproduction in particular, and musical aesthetics in general.

Helmholtz, Hermann, and Stumpf

In order to continue the story from the previous chapter, a brief foray into the history of sound synthesis and analysis during the mid- to late-nineteenth century is necessary precisely because the aforementioned German physicists, physiologists, and engineers saw themselves as a culmination of a particular historical lineage. While the history of speech synthesis and analysis is impressive, my actors of the 1920s—including Friedrich Trautwein and Oskar Sala—turned to the nineteenth-century works of Hermann von Helmholtz and Ludimar Hermann for the historical basis of their research. Helmholtz famously showed most thoroughly and convincingly that each vowel is characterized by harmonics (harmonic overtones)—or the integer multiples of the fundamental pitch. Helmholtz, Germany's leading nineteenth-century physiologist and physicist, was the quintessential Kulturträger, or bearer of culture, and member of the Bildungsbürgertum. Born in Potsdam to a headmaster of a humanist gymnasium, Helmholtz was well read in German idealism, including the works of Immanuel Kant and Johann Gottlieb Fichte. He also possessed a thorough knowledge of the works of Johann Wolfgang von Goethe, and—if we are to believe Johannes Brahms and the renowned violinist Joseph Joachim—he was a talented amateur pianist.

Helmholtz wanted to see if he could artificially generate vowel sounds (and, to a lesser extent, musical tones) by using tuning forks of the harmonics.[2] He detailed this in his 1859 essay, "Ueber die Klangfarbe der Vocale" ("On the Tone Color of Vowels") and in his 1863 tome, *Die Lehre von den Tonempfindungen als physiologische Grundlage für die Theorie der Musik* (*On the Sensations of Tone as a Physiological Basis for the Theory of Music*). He developed an ingenious instrument, the synthesizer, to generate compound sounds from simple tones of eight tuning forks. One tuning fork sounded at the fundamental frequency, which provided the tone's pitch, and seven others sounded at pitches corresponding to the harmonics. By changing the relative intensities (amplitudes) of the forks corresponding to the various overtones, he synthetically produced all of the German vowel sounds.

Helmholtz analogized the vocal cords to the reeds of reed pipes. Both acted as "tongues" producing a series of discontinuous and sharply separated pulses of air. In the case of the membranous, organic tongue, the resulting vibrations are rich in harmonics. By using his resonators, he identified sixteen harmonics of a front (bright) vowel sung by a bass voice at a low pitch. These vibrations traveled to the oral cavity, which behaved like a resonating chamber.[3] Some of the harmonics were strengthened by resonance and formed what would later be called formants; these enhanced harmonics gave vowels their distinctive sounds.[4]

He then decided to see if his tuning forks could reproduce the sounds of an organ using that same method.[5] By employing tuning forks corresponding to the harmonics of a fundamental and its harmonic partials, he could artificially generate a tone color, which previously had been thought unique to a particular sounding device. He concluded, as a result of these tuning-fork experiments, that the differences in musical quality of tone (the tone color, or timbre) was due to the strength of the partials.[6] According to Helmholtz, these partials were always harmonic with respect to the fundamental tone.[7]

Nearly thirty years after the publication of *Tonlehre*, the physiologist Ludimar Hermann conducted his research on formants, a term he coined in the 1890s to refer to the portions of an acoustical spectrum where the overtones have the greatest intensities (thereby determining the timbre of the speech sound). The son of a printshop owner, Hermann attended a humanist gymnasium in Berlin, where he subsequently studied the natural sciences and medicine, receiving a medical degree in 1859. He became a professor of physiology in Zurich and then in Königsberg, East Prussia (now Kaliningrad, Russia). A student of the renowned Berlin physiologist Emil du Bois-Reymond, with whom he had fallen out, Hermann became the director of the Physiological Institute in Königsberg.

His contribution to human physiology in general, and speech sounds in particular, garnered him fame among physiologists. Hermann introduced an optical method for depicting speech sounds.[8] He initially photographed the vibrations of a paper membrane against which a particular vowel was spoken and then photographed the surface of a phonograph's grooves after a recording of various vowels. The images in both cases were projected onto a small mirror, which then magnified the images and projected them onto a photographic plate. He then used Fourier analysis to create a Fourier series and calculate the resulting curves. Fourier series applies to signals that are repetitive, such as a sawtooth waves. These nonalgebraic curves can be represented by a sum of sine and cosine waves, each term given an amplitude weighting. One can start with a fundamental harmonic term, then add as many more as one needs to represent the measured signal as accurately as required, or as makes any difference to what is actually heard.[9] By means of this experimental technique, Hermann identified many more overtones than Helmholtz had previously. He was also able to determine those overtones' relative strengths. In so doing, he provided a much more detailed composition of vowel sounds.

According to Hermann, the characteristic tone of each vowel sound did not need to be harmonic with respect to the fundamental tone, as Helmholtz had insisted: inharmonic overtones also played a role in forming the vowel sounds. Hermann's physiological explanation of formant formation also differed from Helmholtz's view of resonance. For Helmholtz, the fundamental tone of vowel sounds was formed in the larynx, and the certain characteristic harmonics were enhanced by resonance by the shape of vocal cords and the organs in the oral cavity. Hermann, on the other hand, was convinced that intermittent puffs (or pulses) of air produced by the glottis, or the portion of the larynx consisting of the vocal cords and the opening between them, initiated the generation of eigentones (tones produced by a vibrating system, which are also acoustical transients, or short-duration sounds occurring at the beginning of a wave train). They were subsequently damped and decayed until they were restarted by a second puff (or pulse). These puffs of air did not necessarily follow each other with the same periodicity; therefore, they could be inharmonic with respect to the fundamental frequency.[10] These damped oscillations were what Hermann called formants: they characterized the vowel sounds' timbres.[11] During the first decade of the twentieth century, a number of natural scientists drawing upon Hermann's work insisted that the formants responsible for determining the timbre of musical instruments were also damped oscillations of the eigentones.[12] Hermann did, however, agree with Helmholtz that the formant for each vowel possessed a precise frequency regardless of the pitch of the fundamental tone that is spoken or sung.

Helmholtz and Hermann were seen as championing two competing theories of vowel production. In 1909, the Berlin physician who founded phoniatrics (or phoniatry—the study of speech organs), Hermann Gutzmann Sr., contrasted the "theory of overtones or resonance," of which he was a staunch supporter, with "the theory of formants."[13] In short, Ludimar Hermann's theory of formants was based on the impulse (or shock) excitation theory (*Stoßerregungstheorie*) in which new frequencies (the transients), both harmonic and inharmonic, were generated, damped and decayed in the oral cavity. Helmholtz's theory, meanwhile, was predicated on the oral cavity strengthening certain harmonics that were already present in the complex tone, by means of resonance.

According to renowned American acoustician Harvey Fletcher, the main difference between the theories was their methods of describing and representing the motions in physical terms. Hermann's impulse theory emphasized what was occurring at the anatomical level and therefore was more informative to phoneticians and physiologists interested in the mechanism of vowel production. Helmholtz's harmonic theory was more useful to engineers who wished to design telephone systems that could accurately transmit speech.[14]

By the late 1920s, most natural scientists felt that Helmholtz's theory was generally accurate; however, Hermann's theory correctly accounted for a special case, namely where the chest register (or chest voice) is used.[15]

Starting in 1913, Carl Stumpf took up research on formants. A student of influential philosopher and psychologist Franz Brentano, Stumpf's research was dedicated to elucidating the sensation and perception of tones. In 1894 he became director of the Institute of Psychology of the University of Berlin, and four years later he founded the influential journal *Beiträge zur Akustik und Musikwissenschaft* (*Contributions to Acoustics and Musicology*). Basing his work on resonance and interference, he generally supported Helmholtz's main thesis regarding the structure of vowels and the general theory of hearing.[16] He did, however, argue that both Hermann and Helmholtz put too much emphasis on single characteristic tones for the formation of the vowel sound. Formants were "not a single tone, but in general a section of the tonal region, which contributes in a significant way to the characteristic of the vowel."[17] They did not possess precise frequencies, as Helmholtz had argued with his tuning forks and resonators, but rather were regions of resonance, including the main formant (*Hauptformant*) and several adjacent bands (*Nebenformanten*), all of which contributed to the unique vowel sound.[18] The section of the formant, which included the most decisive tone of the sound, formed a strong maximum that was surrounded on both sides by a gradually decreasing volume of the overtones. For Stumpf, the oral cavity, as a result of its ability to damp vibrations, did not merely resonate with a single tone with a precise pitch, but resonated with numerous tones of neighboring pitches.[19] Finally, contra L. Hermann's claim, Stumpf found no experimental evidence of inharmonic overtones.[20] He was able to determine the fixed formants characteristic of each vowel and their structure based on harmonics existing within the audible range.

In 1926 Stumpf published the results of his acoustical research. Of particular relevance to this chapter was his work on the analysis of consonants and the analysis and synthesis of vowels and pitches of musical instruments using tuning forks, flue pipes (also called labial pipes, *Lippenpfeifen*), and particularly his interference apparatus, which comprised numerous tubes that generated the necessary partials with the correct amplitudes. All of the devices were provided by his Institute of Psychology and the Berlin Academy of Music.

It is fair to say that Stumpf not only supplemented Helmholtz's view of timbre, he also modified it. Stumpf was not convinced that tone color could be reduced to the depictions of the relative amplitudes of a sound's overtones. Rather, he felt that timbre was a "Komplexeindruck" (complex impression) that necessitated an understanding of the role played by the psychology of human perception.[21]

Synthesen von Instrumentalklängen.

Grundton c = 128 Schw.

	Klarinette	Tenorposaune	Horn	Viola
g^4	3			
$\overline{fis}^4$	1			
e^4	5			
d^4	6			
c^4	6			
h^3	4			
$\bar{b}^3$	5			
$\bar{a}^3$	6			6
g^3	8			4
$\overline{fis}^3$	9			4
e^3	8	6	2	8
d^3	6	6	4	6
c^3	5	12	6	10
$\bar{b}^2$	4	16	0	16
g^2	5	20	10	20
e^2	6	16	12	18
c^2	1	13	12	14
g^1	10	10	10	10
c^1	1	8	12	10
c	7	0	10	2

Grundton c^1 = 256 Schw.

	Trompete	Klarinette	Tenorposaune	Horn	Viola	Neue Flöte	Alte Flöte
g^5	2						
e^5	4						
c^5	4						
$\bar{b}^4$	4	4					
$\bar{a}^4$	4	6					
g^4	5	8					
$\overline{fis}^4$	4	6					
e^4	6	6					
d^4	6	12	4				
c^4	6	8	5			2	
$\bar{b}^3$	6	12	6			2	
g^3	8	11	8		4	4	
e^3	8	10	6		6	4	
c^3	10	12	12	6	10	6	5
g^2	12	20	20	10	12	10	4
c^2	10	6	12	16	12	8	6
c^1	8	14	10	12	10	6	5

Grundton c^2 = 512 Schw.

	Trompete	Klarinette	Horn	Viola	Neue Flöte	Alte Flöte
g^5	4					
e^5	6					
c^5	8					
$\bar{b}^4$	8	5				
g^4	10	6		6	6	
e^4	10	6		10	6	
c^4	12	10		8	8	
g^3	12	16	5	8	12	8
c^3	20	8	10	12	16	12
c^2	18	16	20	8	16	16

Grundton c^3 = 1024 Schw.

	Trompete	Klarinette	Viola	Neue Flöte	Alte Flöte
g^5		5			
e^5	4	4			
c^5	8	6	8		
g^4	8	9	8	8	
c^4	16	14	15	10	10
c^3	14	12	12	16	16

FIGURE 3.1. Carl Stumpf's synthesis of musical instrument tones at various frequencies (128, 256, 512, and 1024 vibrations per second, Hz) with the requisite amplitudes using his interference apparatus, which contained numerous tubes in which frequencies were selectively eliminated by means of the interference of wave trains. The resulting pure tones were brought together to generate the pitch of a musical instrument. The instruments analyzed included the clarinet, tenor trombone, horn, viola, trumpet, a new flute, and an old flute. Stumpf also synthesized vowels using this technique. *Source:* Stumpf, *Die Sprachlaute*, 388.

"Objective" Sound Analysis

Whereas the sound analyses of Helmholtz, Hermann, and Stumpf were considered "subjective," subsequent work by acousticians (and, slightly later, radio engineers) were deemed by my historical actors to be "objective," since their analyses were based solely (or so they claimed) on the output of newly invented scientific instruments that generated waveforms.[22] One no longer relied on throat tickles, as Helmholtz had. Initially, these waveforms were analyzed mechanically by registering the visual representations of the sound-pressure curve with the assistance of a membrane and a mirror. The visual representation seemed to trump the musical ear. The most famous and effective early example of this method was the one employed by American physicist Dayton Clarence Miller.

During the 1910s and early '20s, Miller, of the Case School of Applied Science (now Case Western Reserve) in Cleveland, Ohio, was researching the tonal qualities of various musical instruments and vowel sounds using oscillograms.[23] Following Helmholtz's cue, he specifically explored the ratios of the amplitudes of the various harmonic partials generated by numerous instruments and vowel sounds. According to Western Electric and Bell Telephone Laboratories' physicist, Irving Bardshar Crandall, Miller's work on sound analysis was "the beginning of modern physical research on speech sounds."[24]

Miller drew upon Fourier analysis to determine the relative ratios of the harmonics. To analyze a curve, one needed to determine the particular coefficients of the Fourier equation. It was a tedious process costing much time and effort. Miller, however, decided to use an area-integrating machine, or planimeter. A certain type of these contraptions, known as harmonic analyzers, determined the areas under curves in such a way that the numerical values on the dials generated the Fourier coefficients.[25] Harmonic analyzers were invented by Lord Kelvin in 1878 to measure the graphical representations of daily changes in temperature and pressure. Some of the most accurate ones were designed by German-born London mathematician Olaus Henrici, who headed a laboratory of mechanics at Central Technical College (later to become Imperial College, London), where he also designed other types of planimeters and calculating machines. Henrici's harmonic analyzers were often used in the early twentieth century to measure the fundamental tone and the harmonics of complex sound waves. The actual harmonic analyzer that Miller employed was constructed by Gottlieb Coradi, a Zurich precision instrument maker. The harmonic analyzer was an improved version of Henrici's devices. Miller only needed to place a tracing of the sound wave he wished to analyze underneath the harmonic analyzer. He then moved a mechanical stylus to trace the waveform. The readings on the instrument's dials yielded the phase

FIGURE 3.2. The harmonic analyzer, devised by German-born London mathematician Olaus Henrici, built and improved upon by the Zurich instrument maker Gottlieb Coradi. Dayton C. Miller used this instrument to determine the overtones and the ratios of their relative amplitudes of various musical instruments. *Source:* Miller, *The Henri Harmonic Analyzer*, 289, reproduced in Miller, *The Science of Musical Sounds*, 96.

and amplitude of ten Fourier harmonic components.[26] With his harmonic analyzer, Miller measured the distribution of energy among the several harmonics of the various vowel sounds, such as the "a" in "father," produced by eight different voices. In so doing, he was convinced that there was neither a fixed harmonic nor a fixed pitch that characterized a vowel, but, pace Stumpf, the vowel was characterized by a fixed region or regions of certain overtones that are reinforced by resonance.[27] He also measured the amplitudes of the harmonics of the middle and lower range of one of his own flutes made by the renowned early-nineteenth-century Bavarian flute maker, Theobald Boehm, at various volumes and the relative amplitudes of the harmonics of the four open strings on a violin.[28] In addition, he charted the harmonics and their relative amplitudes for the clarinet, oboe, tuning forks, flute, and French horn.[29]

None of the aforementioned sound analyses, however, dealt with the conversion of sound waves into electrical form and their reconversion to sound with minimal distortion. Such a technique, of course, was critical to the technologies of telephony, telegraphy, recording, and radio.

Bell Telephone Laboratories

After Miller's research, the mechanical methods of analysis in physics and engineering laboratories succumbed to electrical methods involving the use of microphones, loudspeakers, amplifiers, oscillographs, and electrical harmonic analyzers. With the advent of electrical analysis and reproduction of sound, according to the engineers and physicists of the period, the notion of fidelity began to shed some of its subjective definitions.[30]

During the 1920s, harmonic analysis became a powerful technique for studying sound waves that had been converted to electrical waves. As German physicist Erwin Meyer pointed out, this was the origin of electroacoustics.[31] In the United States, much of the relevant research in the late 1910s and '20s was conducted in the laboratories of American Telephone & Telegraph (AT&T) and Western Electric. In 1913 Western Electric hired physicist Irving Bardshar Crandall to analyze and measure the component frequencies of electric currents. His approach was straightforward: understand how systems set into motion vibrate and apply that knowledge to speech analysis and hearing.[32] His investigations included the voice and its transmission through telephone (and later radio) equipment to the ear.[33] Other young physicists and engineers joined the Western Electric Company at this time, including R. C. Wegel, C. F. Sacia, and, most famously, Harvey Fletcher. After a brief period of studying underwater sound transmission and detection during World War I, they returned to their work on speech and hearing, publishing the fruits of their labor in the 1920s.[34] Their research primarily focused on the frequency sensitivity of the human ear, including the minimum amount of energy needed to hear tones at various frequencies.[35] They also investigated the auditory masking of one tone by another by studying the dynamics of the inner ear.[36]

Part of this research was predicated on extremely precise instruments to measure both the energy magnitudes and the frequency components of the sound waves that were converted into electrical waves.[37] High-quality oscillographs and electric harmonic analyzers were necessary to obtain depictions of the compound sound waves, whether those waves were generated by voices or musical instruments.[38] Crandall oversaw the design of Western Electric's original oscillograph, while Sacia conducted the experiments. Fletcher wrote that "an instrument for recording speech sounds [the oscillograph] was finally obtained which was nearly free from distortion than any which had yet been used."[39]

With this device in hand, scientists and engineers could now investigate how vowel sounds were formed. Indeed, vowel research represented the next phase of sound studies at Western Electric. Crandall and Sacia analyzed spoken vowel sounds as recorded by oscillographs. Sound spectra for four women and four men of thirteen vowel sounds were generated illustrating the relative amplitudes of the different component frequencies. Speech sound waves, which were picked up by a telephone's microphone, were converted into electrical waves. These electrical waves were amplified by a vacuum tube and sent to a recording apparatus, namely a modified oscillograph, which recorded the waveforms of various speech sounds free from distortion from 100 to 5000 Hz. The sound waves were recorded on a rotating film drum upon which the oscillograph's vibrator was placed: the result was an oscillogram.[40] Sacia also investigated the relative magnitude of power of speech. Such data were crucial to engineers seeking to avoid load distortion.[41]

In 1925 AT&T and Western Electric formed the Bell Telephone Laboratories, located at 463 West St., Manhattan. It was undoubtedly the leading US research site to study the fidelity of sound transmission.[42] Bell Telephone Laboratories' first director, electrical engineer Harold DeForest Arnold, summarized in 1928 the "philosophy of the investigation of hearing" conducted in the laboratories in those early years before the United States entered to the war, until the late 1920s. This philosophy included obtaining accurate physical descriptions and measurements of the mechanical operation of human ears; the testing of the ability to discriminate between sounds in order to discern the smallest perceptible distortion; and designing both instruments and experimental systems.[43] A particular emphasis was placed on researching speech to determine to what extent small variations or imperfections could affect intelligibility.[44] While scientists and engineers at Bell Telephone Laboratories did work on the fidelity of music broadcasting, most of their emphasis was initially on the spoken word.[45]

Physicist R. L. Wegel and engineer C. R. Moore of Western Electric (and later Bell Telephone Laboratories) invented a harmonic analyzer employed for the study of sound. The analyzer converted sound into electrical waves and could measure and photograph all the frequency components of an electrical wave at a particular moment in time.[46] In 1924 they published the details of their apparatus and their techniques for depicting all the components of an electrical wave.[47] A condenser transmitter transformed the sound waves into an electrical wave. This copy was sent to a network possessing a sharply tuned circuit whose tuning frequency was controlled by changing the circuit's capacitance. A maximum response of the circuit occurring at each tuning frequency corresponded to a component of the wave. An automatic photographic recorder registered the frequency and amplitude of each of the wave's components. The entire record was generated once the analyzer had swept through all the frequencies of its two frequency ranges, namely 20 to 1250 Hz, and 80 to 5000 Hz. It took approximately five minutes to generate this record.

In 1927, Moore copublished an article with A. J. Curtis, an electrical engineer, describing the use of the voice harmonic analyzer. The analyzer employed a resonating element with a fixed frequency that could uncover the components of a sound wave by heterodyning them with oscillations generated by variable frequency oscillators.[48] It was quite different from the electrical harmonic analyzer invented by Moore and Wegel some three years earlier. Their voice harmonic analyzer was initially invented to improve telephone transmitters that required precise measurements of frequency components.

German Physics and Radio Engineering

German physicists and engineers at universities, state-supported radio research laboratories, and electrical engineering companies (including Siemens & Halske, Telefunken, and AEG) were conducting similar research on how to improve fidelity via broadcasting human speech sounds and musical sounds. As one of Germany's leading acousticians of the period, Erwin Meyer, remarked, radio transmission and recording of speech and music necessitated a knowledge of the total intensity of a complex sound, or the so-called sound spectrum, which included its highest and lowest frequencies and the distribution and amplitudes of its overtones.[49]

One of the early pioneers of the application of physics and electrical engineering to the improvement of radio broadcast was Karl Willy Wagner. The origin of his fascination with radio and telegraphy dates back to when he lived for several years in the house of Johann Philipp Reis, one of the inventors of the telephone. Wagner was born in 1893 in Friedrichsdorf, just north of Frankfurt am Main, the son of Wilhelm Wagner, a merchant.[50] Wagner attended a

Volkschule and then a *Realschule* in Homburg, finishing in 1899. Realschulen differed from humanistic gymnasiums, where the curriculum was based on the classics: Realschulen stressed a practical, hands-on education. Compared to his colleagues, Wagner received quite a different education. His uncle, Gustav Gauterin, taught Wagner the mechanical trade in his workshop in Friedrichsdorf. Wagner went on to study electrical engineering at the Technikum Bingen, passing his engineering exam in 1902.

From 1904 to 1908, he worked as a research engineer in a Siemens-Schuckert's high-voltage laboratory in Charlottenburg, just outside Berlin. In 1909 he was employed as a telegraph engineer at the Telegraphes-Versuchsamt (Telegraph Experimental Office, TV) of the Reich's Post Office in Berlin, where he worked on improving long-distance telephone connections. Investigating electric current and dielectric fatigue in long-distance cables, he completed his doctoral degree in physics at the University of Göttingen in 1910 with a dissertation titled "Der Lichtbogen als Wechselstromerzeuger" ("The Arc as an Alternating-Current Generator") and habilitated—or qualified as a university senior lecturer—at the Technical University of Berlin, or TH Berlin, two years later. In the academic year 1912–13, he became a professor at the Physikalisch-Technische Reichsanstalt (National Metrology Laboratory), where he directed the electrical engineering laboratory.

During World War I he improved radio stations for signaling planes and ships. In the early years of the war, he investigated the ways in which speech sounds suffered when one tried artificially to restrict their frequency bands while transmitting the human voice. This project was part of his research on establishing a procedure for scrambled telephony for the Kaiser's Navy. Toward the end of the war, he worked at Telefunken. After the war, he worked in the Post Office and became director of the TV, and from 1923 to 1927 he was president of the office that replaced the TV, the Reich's Office of Telegraph Technology (Telegraphentechnisches Reichsamt) where he researched acoustical formants. In August 1927 he became professor of oscillations research at the TH Berlin in Charlottenburg and became founding director of the Heinrich Hertz Institute for Oscillations Research (HHI), which opened three years later. In 1936, he was dismissed as director of the HHI, as he did not fire his Jewish colleagues and allegedly embezzled research funds. From 1943 to 1945 he served as a scientific adviser to the Office of Group Research, Inventions, and Patents (Amtsgruppe Forschungen, Erfindungen und Patente) in the Supreme Command of the Navy of the Third Reich.

In 1924 Wagner published an important paper on the problems associated with improving the fidelity of transmitting speech and music.[51] While it owed much to D. C. Miller's and Carl Stumpf's research[52] on vowels, Wagner drew upon electrical techniques such as resonant (or LC) circuits, rather than older acoustical tools such as tuning forks, interference tubes, and flue pipes, to analyze and synthesize vowels and sounds of musical instruments.[53]

When a frequency band is artificially constricted, speech undergoes changes. To study such changes, Wagner used a resonant circuit possessing low-pass filters located in a telephone set comprising a recording device located in an adjacent room, an amplifier, and a loudspeaker.[54] He spoke all of the (German) vowel sounds into the recording device and then gradually lowered the transmission's frequency by adjusting the circuit. He then compared the original spoken sound to the transmitted sound. He noted that if one spoke the German vowel E, its fidelity was greatest at 3100 Hz. It transformed into the vowel sound ÖE at circa 2300 Hz, and then to ÖO at 2070 Hz, and finally to a pure O at circa 1300 Hz. At a frequency of 130 Hz, all vowels sounded like U. Altering the vowel sounds by eliminating the overtones—thereby eliminating certain formants—via the filters of the circuit simulated what a broadcast did to voices and musical instruments back in the 1920s. He determined that in order to achieve perfect fidelity, one needed to transmit over a range of frequencies between 100 to 10,000 Hz, particularly for music broadcasts. Such a range was impracticable at the time. It turns out the range from 4000 to 10,000 Hz could be cut off without human speech being rendered unintelligible; hence, engineers working on telephones focused on the range of 300 to 3000 Hz.[55] As Wagner pointed out,

> the experiments show, and experience has taught us, that long-distance telephones and telephone lines must evenly pick up, forward, and reproduce vibrations up to around 3000 Hz if really good communication is to be achieved.[56]

He then turned his attention to transmitting music. He claimed this posed a greater challenge than transmitting the human voice since the range of frequencies was greater for musical instruments than for voices, and a greater degree of fidelity was required. The key was to ensure that the relative amplitudes of the higher order overtones be preserved. Drawing upon the work of Miller (and often using the same diagrams taken from Miller's *The Science of Musical Sounds*), Wagner illustrated the ratios of amplitudes of the overtones of the various musical instruments and a tuning fork.[57] He discussed how instruments, which are rich in overtones, suffer from broadcasting by using the aforementioned LC circuit with low-pass filters that eliminated a number of those overtones. That is why the violin occasionally sounded like a flute: eliminating higher overtones destroyed the formants generated by bowing.[58]

In August 1927, to continue research that aimed to improve, among other things, radio broadcasting, the Prussian Ministry for Science, Art, and Popular Education (henceforth the Prussian Ministry of Culture) addressed the issue of increasing the collaborations between musicians and engineers by creating the Heinrich Hertz Institute for Oscillations Research (HHI), whose mission also included the invention of electric musical instruments.[59] In early 1925, the Heinrich-Hertz-Gesellschaft (Heinrich Hertz Society) came up with a plan to improve radio in Germany. They were responsible for investigating household sources of radio disruptions such as vacuum cleaners.[60] The society worked closely with the Elektrotechnischer Verein (Electro-Technical Association) and colleagues at the TH Berlin who, for several years, had sought ways to improve the future training of radio engineers.[61] In March 1927, the Post Office, the Prussian Ministry of Culture, the TH Berlin, the Association of German Electrical Engineers, and the major firms of the German electrical industry formed the Studiengesellschaft für Schwingungsforschung (Research Association for Oscillations Research) with a view to create a research institute, the HHI, which was officially founded in February 1928. Its founding director, K. W. Wagner,[62] was assisted in its creation by Gustav Engelbert Leithäuser, director of the Reich's Office of Telegraph Technology[63] in Berlin and director of the Deutscher Amateur-Sende- und Empfangsdienst (German Amateur Sender and Receiver Service).[64] A new building was erected near the TH Berlin to house the HHI. It was to carry out research in the theory and experimentation of oscillation phenomena relevant to heavy current engineering, telegraphy, telephony, radio engineering and communications, acoustics, and mechanics.[65] It focused its efforts on researching radio and television technologies, architectural acoustics, and electric music thereby providing us a glimpse of how radio was thought of at the time.

At the same time Wagner was working on fundamental frequencies and overtones for radio broadcast, physicist Ferdinand Trendelenburg's laboratory at Siemens & Halske was developing harmonic analyzers that converted speech sounds into electrical waves, similar to the techniques used at Bell Telephone Laboratories. Born in 1896, Trendelenburg, the son of renowned Leipzig surgeon Friedrich Trendelenburg—the personal physician to King Friedrich August III of Saxony—was a pioneer in electroacoustic research.

Like most children of the Bildungsbürgertum, he attended a humanist gymnasium, the Thomasschule in Leipzig. He began studying physics and mathematics at the University of Edinburgh; however, his study was interrupted by World War I. He served in an artillery regiment and ended up as an officer on the Western Front, where it wasn't so quiet. In 1919 he resumed his studies at the Friedrich Wilhelm University (now the Humboldt University) in Berlin and at the University of Tübingen. He finished his doctoral dissertation under the tutelage of physicist Max Reich in Göttingen on how thermophones work. He qualified for a senior lecturing position in 1929 in Berlin, where he became a professor in 1935 and an honorary professor in 1940. During World War II, he worked on controlling torpedoes for the Third Reich Navy. After the war, he built up the Siemens-Schuckertwerke, which had relocated from Berlin to Erlangen.

As part of his *Habilitionsschrift*, or a postdoctoral thesis required for a senior university lectureship, in Berlin, he published a two-part essay in 1924 and 1925 that detailed his apparatus and the procedure for analyzing the components of sound waves associated with speech.[66] Given the interest that research on the human voice and its reproduction had sparked among physicists and physiologists, Trendelenburg stressed that an accurate instrument and technique for measuring sound waves were now necessary. His research yielded the "objective judgment" (*objektives Urteil*) needed to quantify "the fidelity of the reproduction" (*naturgetreue Wiedergabe*).[67] The key was to compare the original sound—such as vowels and consonants—with the sound exiting an apparatus that was reproducing or remotely transmitting the sound, such as a microphone or loudspeaker. The aim of the paper was to create a new method for researching sound by determining the composite structure of the vowels.[68] He compared the composition of vowel sounds as determined by Stumpf's subjective method—as he called it—using tuning forks and resonators, with his own objective method of analyzing the vowel sounds exiting a condenser microphone created by Hans Riegger in Siemens & Halske's research laboratory. The condenser microphone converted the acoustical vibrations of the vowel sounds and sibilants into frequency modulations of a high frequency wave. The modulations were then amplified by means of a high-frequency amplifier circuit and subsequently sent to an oscillograph, the recordings of which were photographed. His research supported Stumpf's work on the formant frequencies that were characteristics of the vowels. He concluded that a frequency range from 50 to 5000 Hz was necessary for "the reproduction of speech that was fidelitous [literally, faithful to nature]."[69] In short,

> [t]he arrangement allows an objective judgment to be made about the quality of the voice recording or reproduction by comparing the sound (*Klangbild*) provided by the natural sound with the sound (*Klangbild*) of the artificial one.[70]

The second part of his essay was dedicated to offering a similar study of the consonants L, M, N, and R, comparing and contrasting them with vowels.[71] He was convinced that vowels were the products of purely periodic processes. The vowels consisted of the fundamental tone, which originated in the vocal cord, and the harmonics that were amplified by the resonance produced in the oral cavity. The consonants, on the other hand, were mixtures of harmonic and inharmonic tones.[72]

During the mid-1920s to the early 1930s several new methods of objectively analyzing a sound had been published in German journals.[73] They were based on the rather straightforward and highly accurate method of applying the so-called search tone (*Suchton*) or search frequency (*Suchfrequenz*).[74] Physicists and engineers at Western Electric and Bell Telephone Laboratories used electric harmonic analyzers that added a constant sinusoidal frequency to the frequency mix being analyzed. This mixture (and the combination of the frequencies) generated so-called beats (or heterodynes) at a frequency that was the difference between the two initial frequencies. This was also called a difference tone.[75] The amplitude of those difference tones could be calculated: this would yield the intensities of the overtones of the frequency mix. Rather than using a constant frequency, the Germans preferred to use a slightly changeable singular frequency with a constant intensity, the so-called search tone, discussed in greater detail below. From that search tone and the overtones of the sound wave being investigated, the heterodynes (or difference tones) were created. The intensity of those heterodynes was proportional to the strength of the overtone. One could thereby obtain the spectral distribution and amplitude of each overtone.[76] This technique, which was seen as the most advanced of the period, was apparently also much easier to use than the one employed at Bell Telephone Laboratories. While Wegel and Moore's electrical harmonic analyzer significantly cut down the time to obtain the visual recording of a sound wave with all of its components to five minutes, the methods using a search tone could photographically depict all overtones of a sound within a minute.[77]

One of K. W. Wagner's esteemed colleagues at the HHI was Erwin Meyer, who also made important contributions to improving the fidelity of radio broadcasting.[78] Born in Königshütte in Upper Silesia (now Chorzów, Poland) in 1899, his father, who was the postmaster of Königshütte, kept his son up to date on the importance of the new broadcasting technology to the post office. Growing up in the household of a civil servant, Meyer had an education trajectory that was rather typical of most scientists. He received a classical education by attending a humanist gymnasium. Immediately upon graduation in December 1917, he was drafted into military service in World War I. As a result of the horrors he experienced on the front, he wished to pursue an education in medicine. Alas, his parents lacked the requisite finances, so he studied physics, receiving his doctoral degree in the subject—specializing in acoustics—in December 1922 from the University of Breslau. He studied with the highly respected physicist Erich Waetzmann, who was well known for his contributions to physiological acoustics and combination tones working on the forces that sound waves exert on membranes.

Like so many scientists of the period, Meyer was required to do relevant research for the Third Reich after the outbreak of war. He conducted research at the HHI on the development of sonar ranging and sound propagation and absorption of sound in shallow ocean water. Clearly, this work was relevant to German U-boats. After the war, he settled in the British-occupied sector of Germany. In 1947 he became a professor of physics at the University of Göttingen, turning down two professorships in the United States out of a hope that German science would recover after the war. He became the founding director of Göttingen's Drittes Physikalisches Institut (Third Physics Institute), which merged the former Institute for Applied Mechanics and Institute for Applied Electricity.[79]

Shortly after finishing his dissertation, Meyer attended a lecture in Breslau on the importance of the radio, given by Hans Salinger, an official with the Central Post Office in Berlin. Salinger's radio, however, was damaged en route, and Meyer donated his receiver for the talk. Salinger was so impressed by the radio set Meyer had made that he recommended to K. W. Wagner that the young *Privatdozent*, or a scholar who has not yet achieved the title of professor but is permitted to teach at a university, be hired at the Reich's Office of Telegraph Technology. Wagner concurred and hired the young acoustician in November 1924 as a scientific research assistant to investigate technologies relevant to the telephone and the fidelity of broadcasting music over the radio.

Shifting his focus from the physics of hearing and the ear, Meyer immediately set out to improve metrological techniques relevant to sound given the development of, and continued improvements to, vacuum tubes. Of importance to this particular book was his work on loudspeaker distortion, the calibration of capacitor microphones, and sound analysis. His *Habilitationsschrift* submitted to the TH Berlin in December 1928 dealt with methods of sound analysis. In April 1929 Wagner appointed him head of the acoustics department of the HHI.

While at the HHI, Meyer collaborated with Gerhard Buchmann to produce a type of spectral atlas of all the various types of musical instruments by employing the search-tone method.[80] The sound to be investigated was passed through a microphone that was attached to an amplifier. A low-pass filter, which only permitted the lowest frequencies of the sound down to 20 Hz to pass through to a galvanometer, was placed behind the amplifier. A sinusoidal voltage generated by a heterodyne oscillator acted on the amplifier. The frequency of this oscillator could be altered continually over the frequency range of human hearing. If the search tone of the oscillator was allowed to run through the frequency range while the sound to be examined simultaneously passed through the microphone (changing it into an electric frequency mixture), the galvanometer indicated when the search tone uncovered a component, or overtone, of the sound. The difference tones (or heterodynes) generated by the rectifier between the relevant sound components and the search tone could then pass through the low-pass filter. The deflection of the galvanometer was proportional to the strength of the sound component over which the search tone passes; therefore, the amplitude of the frequency could be measured quite precisely.[81]

With this technique in hand, Meyer produced the sound spectra of a number of musical instruments between the frequency range of 30 to 15,000 Hz, including grand pianos, a hammer clavier (an old piano forte), a harpsichord, and a clavichord. Three grand pianos were provided by their manufacturers, Bechstein, Grotian-Steinweg, and Blüthner. Older keyboard instruments were provided by the Berlin Academy of Music. These spectra illustrated the ratios of amplitude of up to as many as fifty overtones in some cases.[82] Older keyboards possessed more overtones than the modern grand pianos. In addition, Meyer demonstrated unequivocally that the intensity with which the hammer hit the string (and thereby increasing the volume) actually altered the instrument's tone color.[83] A forcefully struck key producing a loud sound possessed more overtones than a softly struck key. Changing the volume while playing the piano, then, slightly altered its tone color.

Plucked instruments, such as the harp, lute, banjo, and zither, had far fewer overtones than keyboard instruments, thereby accounting for their softer timbre. The point on the string that was plucked also influenced the number and intensities of the overtones. Meyer also analyzed the lowest notes on the bowed instruments, namely the bass, cello, viola, and violin.[84] The fundamental tone was at times absent when lower pitches were played: the overtones determined the pitch heard by the ear, even though that pitch did not physically exist.[85] Like all other instruments, the overtones of the higher notes of bowed instruments produced fewer overtones than the lower notes.

Meyer also studied the sound spectra of a number of woodwinds. The contrabassoon, also known as a double bassoon, was shown to have some fifty overtones, most of which possessed very high frequencies. For example, 20 percent of the total amplitude of the overtones of clarinets was near 10,000 Hz. Clarinets and bass clarinets were built in such a way that the even-numbered overtones were weak. Flutes with cylindrical borings had a much sharper timbre than those possessing a conical boring.[86]

The overtones of brass instruments, particularly trumpets, generally had the same amplitude throughout their sound spectra. Percussion instruments, such as the various sets of drums, castanets, bells, xylophones, and triangles, were seen to be unique because their overtone distributions resembled those of various types of noise. Many of their spectra contained curves, rather than vertical lines, indicating inharmonic overtones. And since their frequencies ranged from below 100 and well above 5000 to 6000 Hz, the broadcasting fidelity of these instruments was usually very poor.[87]

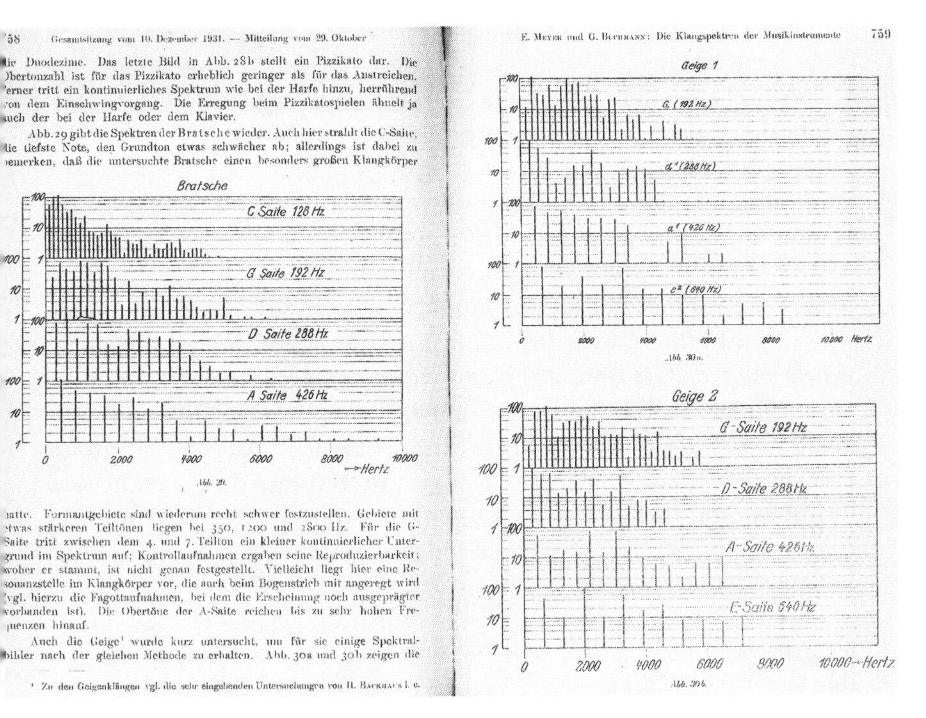

758 Gesamtsitzung vom 10. Dezember 1931. — Mitteilung vom 29. Oktober

die Duodezime. Das letzte Bild in Abb. 28b stellt ein Pizzikato dar. Die Obertonzahl ist für das Pizzikato erheblich geringer als für das Anstreichen. Ferner tritt ein kontinuierliches Spektrum wie bei der Harfe hinzu, herrührend von dem Einschwingvorgang. Die Erregung beim Pizzikatospielen ähnelt ja auch der bei der Harfe oder dem Klavier.

Abb. 29 gibt die Spektren der Bratsche wieder. Auch hier strahlt die C-Saite, die tiefste Note, den Grundton etwas schwächer ab; allerdings ist dabei zu bemerken, daß die untersuchte Bratsche einen besonders großen Klangkörper

Abb. 29.

hatte. Formantgebiete sind wiederum recht schwer festzustellen. Gebiete mit etwas stärkeren Teiltönen liegen bei 350, 1200 und 2800 Hz. Für die G-Saite tritt zwischen dem 4. und 7. Teilton ein kleiner kontinuierlicher Untergrund im Spektrum auf; Kontrollaufnahmen ergaben seine Reproduzierbarkeit; woher er stammt, ist nicht genau festgestellt. Vielleicht liegt hier eine Resonanzstelle im Klangkörper vor, die auch beim Bogenstrich mit angeregt wird (vgl. hierzu die Fagottaufnahmen, bei dem die Erscheinung noch ausgeprägter vorhanden ist). Die Obertöne der A-Saite reichen bis zu sehr hohen Frequenzen hinauf.

Auch die Geige[1] wurde kurz untersucht, um für sie einige Spektralbilder nach der gleichen Methode zu erhalten. Abb. 30a und 30b zeigen die

[1] Zu den Geigenklängen vgl. die sehr eingehenden Untersuchungen von H. Backhaus l. c.

E. Meyer und G. Buchmann: Die Klangspektren der Musikinstrumente 759

Abb. 30a.

Abb. 30b.

FIGURE 3.3. Erwin Meyer and Georg Buchmann's analysis of the amplitudes of a musical instrument's partials at particular frequencies using a harmonic analyzer. The pitches are given in Hz, and the instruments whose values are depicted here are the viola (left) and two violins (right). They also determined the partials of other orchestral instruments. *Source:* Meyer and Buchmann, "Die KlangspRektren," 758–59.

Meyer proffered a critical remark toward the end of his article:

It is certainly not the case that the tone color, the spectrum alone, allows an instrument to be fully recognized. The dynamic progression, the intensity progression while playing the note, is also significant. For example, for the piano a strong attack and gradual decay are characteristic. Reverse the process—that is, allow the sound of a piano's timbre to resonate gradually and then suddenly discontinue—one gets the impression that the sound was generated by an accordion. One can be convinced of that rather easily by playing a record of a piano piece backward. Another example has become well known in a number of circles this year, namely the electric piano of Vierling and Nernst, in which the vibration of the strings is picked up electrically and rendered audible by the amplification of a loudspeaker. If one suppresses the process of attack by suitable measures and allows the

> tone to swell during the beginning, then one absolutely gets the impression that one is hearing a harmonium—although the string's vibration is the cause of the sound. With a sufficient increase in volume, one gets the impression that one is listening to an organ and not a piano.[88]

He was arguing that timbre could not be reduced to the relative strengths of the static overtones: volume plays a key role, and the overtones (and therefore the timbres) change over the duration of the note. In 1927 he wrote in his contribution to the *Handbuch der Physik* on hearing: "Tone color is not only determined by the relative strength of the overtones. The absolute strength of the entire sound (*Gesamtklang*) plays a considerable role because of the nonlinearity of the ear."[89] By the 1920s, researchers were becoming aware—for the first time—of what would later be called the sound envelope, or acoustical envelope. The attack, equilibrium, and decay were characteristic of the musical instrument and played a critical role in its tone color. As will be discussed in chapter 5, Friedrich Trautwein discussed the concept in conjunction with his invention of the trautonium.

Given all of the research during the 1920s and '30s on the formation of formants and tone color, the term "formant" had taken on a new meaning by the 1930s. Whereas Ludimar Hermann had coined the term some four decades earlier to refer to the damped frequencies of transients (the eigentones) that puffs of air generated by the glottis to explain vowel sounds, it now also referred to a region of overtones in musical instruments, which gave them their unique tone colors.[90] As British physicist James Jeans observed, "Clearly the formant has much to do with the characteristic timbre of the instrument: some writers even claim that the timbre of the instrument is completely determined by it."[91]

While working on formants at the HHI and the TH Berlin, engineer and electric musical instrument maker Oskar Vierling noted this definitional metamorphosis.[92] Vierling was born in Straubing in 1904. After receiving a certificate of upper secondary school (*Obersekundareife*), he attended the Ohm Polytechnic in Nuremberg. Upon finishing his courses in 1925, he was immediately hired at the Reich's Office of Telegraph Technology. He went to Berlin in 1929 to study physics, receiving his doctorate in 1935 and habilitation two years later, both at the TH Berlin. He worked as K. W. Wagner's assistant at the HHI. In 1938 he received a professorship in radio technology and electroacoustics at the Technical University of Hanover. During World War II he worked at a laboratory built in the Feuerstein Castle in Ebermannstadt, Bavaria, for armament research for the Wehrmacht. He developed and tested acoustically controlled torpedoes, and he worked on encryption and the acoustical detonation of mines. He was a member of the Nazi Party.[93]

During the late 1920s, Vierling was the assistant of Jörg Mayer, an inventor of electric musical instruments. He later worked alongside the Nobel laureate in chemistry Walther Nernst on the electric Neo-Bechstein grand piano.[94] In 1928 he obtained a German patent for the imitation of sounds using an electric current, whereby a vibration rich in overtones acted as a resonant circuit, thus enhancing those partial overtones present in the resonance range.[95] Drawing upon this research, in 1936 he published a paper on the electric simulation of vowel and musical instrument sounds. Echoing the works of Helmholtz, Wagner, and Stumpf, he argued that formants arose from the effect of resonance on the overtones, which fall within the resonance region, and are therefore strengthened.[96]

In order to artificially create sounds via electricity, Vierling recommended using the spectrum of the harmonic partials as the starting point. He was able to generate a sound rich in overtones by using the LC circuit described in his earlier patent. The voltage on the LC circuit was amplified and produced a pattern on the oscillograph. A spectrum of harmonic partials was also generated. He then compared the diagrams of the original sound to the one produced electrically to see how closely they matched.

Microphones, Loudspeakers, and Amplifiers

The early 1920s through the early 1930s was a critical period for the production of radios. The new market required more powerful instruments with less distortion. The techniques of harmonic analysis discussed above provided a powerful tool for testing the efficacy of radio equipment and ameliorating various types of distortions. Broadcast sound could be distorted by three main types of radio equipment: microphones, loudspeakers, and amplifiers. The key was to analyze the original sound produced by a spoken or sung voice or musical instrument and compare it to the sound after it exited these electrical devices to see how, if at all, the sound was being distorted. This was accomplished by using oscillograms generated by harmonic analyzers. Once that was ascertained, the electrical and radio engineers and physicists could then improve the apparatus's components and design to ensure that all of the overtones were present in their correct ratios.

Microphones underwent extensive improvements throughout the 1920s. It is interesting to note that the first article in the first edition of the popular German radio journal *Funk* discussed the early experiments with different types of microphones for the transmission of operas, concerts, and theater pieces.[97] Engineers and physicists labored tirelessly to mitigate the linear and nonlinear distortions that plagued earlier devices. Linear distortions involve a change in signal amplitude or signal phase for each frequency component without new frequencies being added to the original source by the equipment. They do not involve changing the shape of the individual sine components of a signal or sound. In contrast, nonlinear distortions add new frequency components to the original signal, such as overtones resulting from their interaction with the original frequencies. They occur when the energy limits of a broadcast are exceeded, for example, by supplying too much volume to increase its range.[98]

Recall from the previous chapter, the bandwidth of German radio stations in the 1920s and '30s was 9000 Hz, which meant a sideband of 4500 Hz.[99] This range included C8 in modern notation, or the highest key on the piano (Figure 3.4 shows a piano keyboard that only goes to A7) and the upper frequencies of the violin and piccolo at around 3480 Hz[100] as well as the C2, which is the lowest note on a cello—the open C string—at around 65 Hz and the bottom range of a bass voice. The upper pitches of a tenor were around 500 Hz, while a soprano could reach around 1100 Hz.[101] In addition to the frequency band, volume was also important. As will be explained in chapter 5, the frequency band of human hearing is not linear. A tone sounding at 100 Hz must be amplified considerably for the ear to hear it with the same volume as a tone sounding at 3000 Hz.[102]

Microphones needed to transmit all the overtones of sounds in the aforementioned frequency ranges with the same magnitude as the original sound.[103] Or, in terms used by physicists and radio engineers, the response curve of the apparatus must be flat: there must be no linear distortion.[104] When graphing the amplitude of sound in frequency (usually measured in decibels or volts per dynes cm^2) on the ordinate (y axis) and frequency (measured in Hz) on the abscissa (x axis), the graph should be flat throughout the range of human hearing. During this period, a deviation of 5 dB from 100 to 8000 Hz was generally deemed to be "flat." Very few microphones possessed such a small deviation. To put it more bluntly: what goes into the device needs to be the same as what comes out of it. This is a classic engineering definition of radio broadcasting fidelity. Plotting frequency against amplitude was often used as the method for determining the quality of the radio components. The other method employed to measure the degree of distortion involved harmonic analysis of various input tones throughout the audible frequency range to measure the distortion of the overtones.[105] Harmonic analyzers were employed to test complicated cases of nonlinearity of carbon microphones. In 1928 Meyer detailed a procedure whereby a harmonic analyzer was used to measure the nonlinearity of carbon microphones based on difference tones with a view to improve the device. Gerlach and Grützmacher also invented methods using heterodyning and harmonic analysis for the measurement of distortion caused by microphones, loudspeakers, and amplifiers.[106]

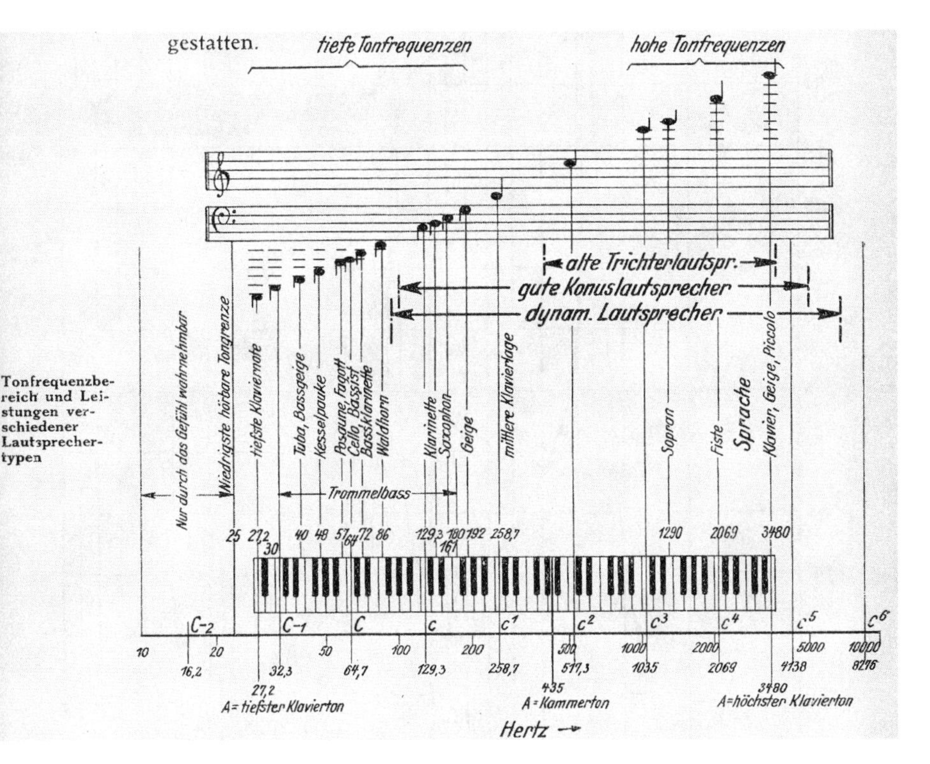

FIGURE 3.4. The frequency ranges (in Hz) of orchestral instruments and voice types. The keyboard of the piano is shown as a reference guide. Note that the highest pitch of this keyboard sketch is A7, while the highest pitch on modern pianos is C8. The figure also shows the ranges of three types of loudspeakers: an old funnel loudspeaker, whose range included 400 Hz to 3500 Hz; a good cone loudspeaker, from 100 Hz to just below 5000 Hz; and a dynamic loudspeaker from 90 to 8,000 Hz. Research on these ranges was critical for radio fidelity. *Source:* Nesper, *Kompendium*, 196.

Since the carbon microphone possessed a high sensitivity and was simple to build, it was used globally in the transmitter in telephone systems at the start of the twentieth century.[107] One of its major flaws, however, was nonlinear distortion, namely the production of unwanted combination tones, which were first detected in 1913 by German physicist Erich Waetzmann, Meyer's dissertation adviser.[108] Early carbon microphones also suffered from linear distortion. There was much variation in the pressure amplitude over a range of frequencies.[109] As a result, high overtones—such as those of the highest pitches of a violin or flute or the sounds of small bells, were not reproduced with a high degree of fidelity. Sound waves produced by high frequencies

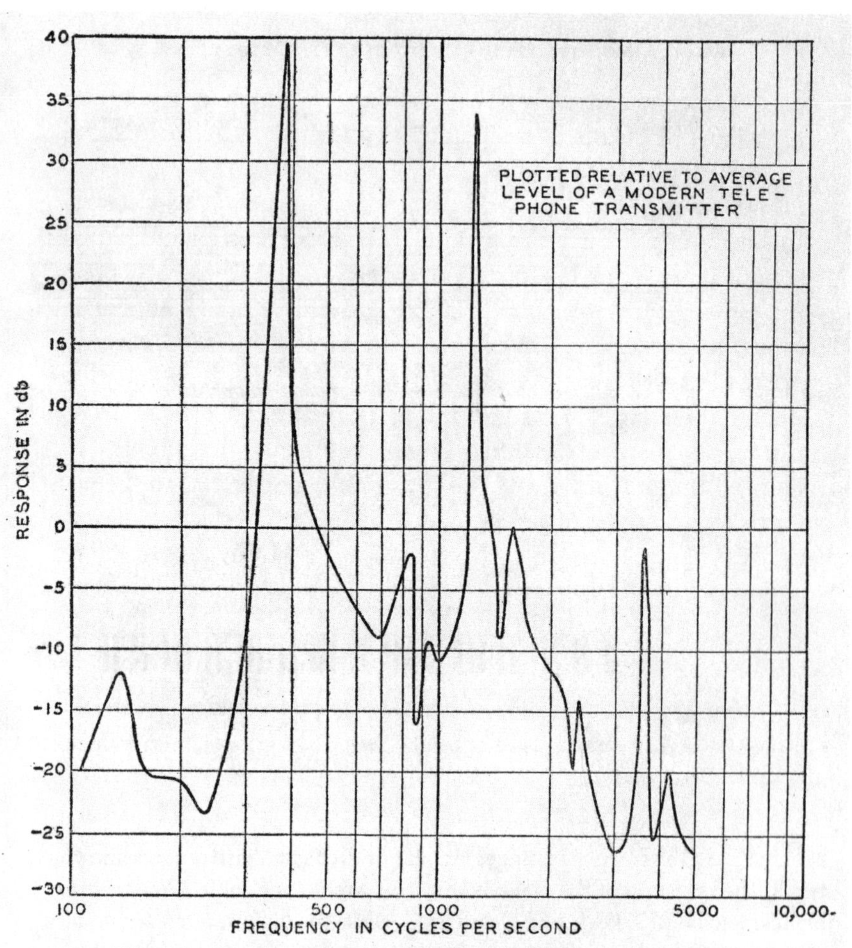

Fig. 26. Response frequency characteristic—Blake single contact transmitter of 1878.

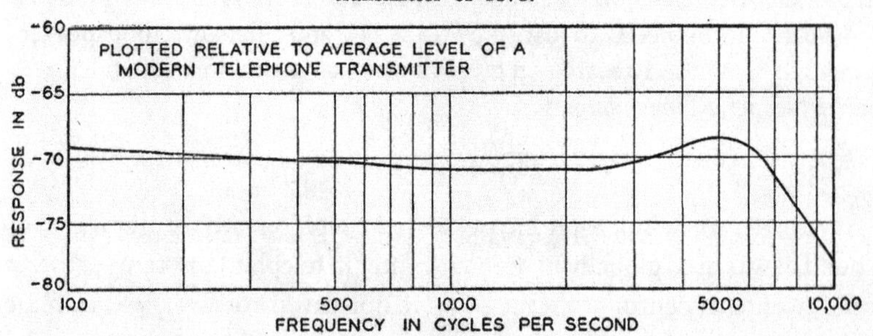

Fig. 27. Response frequency characteristic—condenser microphone.

187

FIGURE 3.5. A comparison of the response frequency characteristic of a Blake carbon microphone of 1878 (top) used in telephones with a Western Electric condenser microphone of 1931 (bottom). Notice the difference in the plots: one wants a horizontal line throughout the frequency range of 100 to 8000 Hz. Clearly, the Western Electric condenser microphone was vastly superior, with a deviation of circa 5 dB within that range and only 3 dB from 100 to 5000 Hz. Indeed, the Western Electric condenser microphone set the standard at the time. *Source:* Frederick, "The Development of the Microphone," 187.

could not penetrate far enough into the microphone's carbon granules; therefore, they could only produce a small amount of pressure and resistance variation. The microphone was therefore not very sensitive; basically, its singular use was as a telephone transmitter that could not pick up musical sound well.[110]

In addition, the earliest carbon microphones also suffered from electrical and mechanical packing. In electrical packing, the carbon particles are subjected to a voltage that causes them to cohere to each other, thereby drastically decreasing the microphone's sensitivity. Mechanical packing results from the settling and compressing of the carbon particles.[111]

By the early 1920s, in response to the needs of radio broadcasting, carbon microphones were being vastly improved. Eugen Reisz, initially of the Allgemeine Elektricitäts-Gesellschaft (AEG), made a number of enhancements to carbon microphones. Perhaps the most important improvement took place in 1923, when he and Georg Neumann solved the packing problem by firmly pressing a thin layer of carbon granules, approximately two millimeters thick, into a wooden chamber enclosed in a marble block. This arrangement improved the fidelity of the amplitudes of the higher frequencies.[112] As a result, higher pitches suffered only minor distortions, and the frequency range of the broadcast was extended. It was popular during the early days of radio.[113] In fact, Neumann's experiments in late October 1923 at the Vox-House Berlin—famous for hosting the original German public radio broadcast—were so successful that Reisz provided many of the microphones for German broadcasting stations.[114]

While advances were being made on the carbon microphone, other types of microphones were being invented or improved, such as the condenser microphone, ribbon microphone, moving-coil microphone, and piezoelectric microphone. Germany and the United States were among the market's major suppliers. Condenser microphones did not suffer from distortions as much as carbon microphones did. They therefore could operate over a much broader range of frequencies. Invented in 1916 by Western Electric physicist Edward Christopher Wente as the "telephone transmitter," condenser microphones possessed a capacitor, which converted acoustical energy into an electrical signal.[115] It was composed of two metal plates, a so-called back plate and a diaphragm, that detected changes in air pressure caused by a sound, such as a voice or musical instrument. As the diaphragm oscillated, the distance between it and the back plate varied, resulting in a change in voltage of the capacitor. That voltage, of course, was the electrical signal. Vastly improved vacuum tubes made by Harold D. Arnold of AT&T provided the requisite amplification. In 1922 Western Electric made their condenser microphones commercially available once Irving Crandall made numerous improvements to Wente's initial design.[116] The microphone was now a hundred times more sensitive than the original model. Condenser microphones debuted in recording studios throughout Germany and the United States in 1925.[117]

The company Georg Neumann GmbH began to market condenser microphones in Berlin in 1928.[118] Neumann had left AEG to start his own company with Erich Rickmann.[119] In 1924 Hans Riegger of Siemens & Halske invented a condenser microphone with a high frequency circuit. It was also used in a number of German radio broadcasting stations along with condenser microphones manufactured by Reisz and Vogt.[120]

In 1924 Walther H. Schottky and Erwin Gerlach of Siemens & Halske invented the ribbon microphone.[121] They, too, were used in German radio stations.[122] It possessed a light metal ribbon, usually made of aluminum, which was suspended between the poles of a permanent magnet, causing the ribbon to oscillate and generate an electric current. Unlike other types of microphones, they did not use diaphragms. Their advantage was in their flatter responses at high frequencies, meaning that, unlike carbon microphones, they reproduced higher pitches with a greater degree of fidelity. Unlike condenser microphones, they could be used without a preamplifier mounted on them.[123] While their quality was superior to condenser microphones for the upper frequencies, they tended to be large and cumbersome.

In 1924 an article in a German radio magazine underscored the superiority of the Siemens's ribbon microphone over a carbon microphone. Whereas the fidelity of string instruments picked up by a carbon microphone was shoddy—indeed, they resembled woodwinds—the Siemens's ribbon microphone produced a perfect broadcast whether it involved solo instruments or a large orchestra. These microphones were particularly well suited for the piano, the broadcasts of which did not always enjoy the highest level of fidelity. The article even claimed that such a reproduction "rivaled the direct listening in the concert hall."[124] It seemed that radio was now on its way to reproducing sound with a high degree of fidelity.

We know that while the Funk-Stunde Berlin radio station initially used a microphone from Telegraphon-A. G. of Berlin, in February 1924 it began to use a C. Lorenz-manufactured cathodophone microphone, which operated by electric current moving through ionized air thereby eliminating the need for diaphragms or other mechanical devices, and a ribbon microphone from Siemens & Halske. In October 1924, Funke-Stunde Berlin switched over to the carbon microphone of Reisz, which was still in use in 1930.[125]

Americans were also investigating ribbon microphones. Harry F. Olson of RCA was impressed by the uniformity of sound intensity (or sound pressure) of ribbon microphones over a wide frequency range.[126] The so-called velocity microphone was a new version of the ribbon type invented by RCA Victor Company in Camden, New Jersey, in 1932. Its ribbon was composed of duralumin,

FIGURE 3.6. Microphones used by the Funk-Stunde Berlin radio station in 1924. *Source:* Reichs-Rundfunk-Gesellschaft, ed., *Rundfunk Jahrbuch 1930*, 51.

which vibrated with the minute variations of the air particles that were set in motion by sound waves. It was particularly popular since it did not produce either artificial whistles or hissing sibilants and speech sounds due to resonance.[127] By the early 1930s, the ribbon microphone was far less bulky and enjoyed one distinct advantage over the condenser microphone: the amplifier could be placed at a greater distance from the transmitter without any loss in efficiency.[128]

Invented in the early 1920s, moving-coil microphones were—like ribbon microphones—another example of dynamic microphones. In 1931 Wente and Thuras of Bell Telephone Laboratories developed a moving-coil microphone that possessed an impressive uniform sensitivity from 45 to 10,000 Hz.[129] It was marketed as the Western Electric 618A Electrodynamic Transmitter.[130]

Intense research was also underway to improve loudspeakers, and once again radio provided the impetus. Up until the early 1920s, the difficulty to exceed the amplification of a telephone receiver was a serious obstacle. Other than a horn channeling the waves, resonance was the only way to enhance the sound.[131]

In 1924 Schottky and Gerlach invented the ribbon loudspeaker by reversing the effect of their newly invented ribbon microphone. In their new loudspeakers, which were hornless, electromagnets were used to increase amplification. At the same time, Hans Riegger of Siemens & Halske constructed another type of electrodynamic loudspeaker, namely the Blatthaller, which used duralumin, rather than pertinax, as the membrane.[132] A year later, Chester W. Rice and Edward W. Kellogg of General Electric filed a patent for a new type of loudspeaker, which they had been working on for several years, that would go on to dominate markets in the United States and Europe. It was a moving-coil loudspeaker that contained a small electrodynamic coil-driven diaphragm in a baffle that created the sound and which possessed a broad midfrequency range of uniform response.[133] They stressed that "voices and music do not sound natural unless reproduced at approximately the original level of intensity, even though the reproduction may be free from all wave-form distortion."[134] This was another definition of fidelity and necessitated an improved amplifier in the loudspeaker. Their improved amplifier design increased the power transmitted to the loudspeakers. In 1926 RCA started using that design in their famed Radiola radios. As a result of the labors of Schottky, Gerlach, Rice, and Kellogg, the key question for loudspeaker designers was no longer how loud, but how faithful, was the reproduction.[135] Much like Rice and Kellogg, Riegger and Wente both independently discovered that a small coil-driven mass-controlled diaphragm in a baffle resulted in a superior loudspeaker.[136]

In 1927 Meyer published an important improvement on testing loudspeakers.[137] Frequency curves of loudspeakers had previously been created with the assumption that the sinusoidal current produced only a pure tone. At low frequencies, however, this was not the case: powerful overtones often occurred with the fundamental frequency. Meyer invented a sound compensation method whereby the strengths of the overtones were measured in relation to the strength of the fundamental tone by means of harmonic analysis. The quotient, known as the distortion factor, yielded the magnitude of the nonlinear distortion.[138]

Bell Telephone Laboratories were also improving loudspeakers during this time. E. Wente and A. L. Thuras created a telephone receiver of the moving-coil variety that could be adapted to the horn of a loudspeaker.[139] It possessed a frequency range of 60 to 7000 Hz, a large portion of which enjoyed a conversion efficiency from electrical to sound energy of 50 percent. Such an efficiency was far greater than contemporary cone or horn loudspeakers, which possessed a conversion efficiency of only about 1 percent. Since it could handle a constant input of 30 watts in contrast to 5 watts, which had been the largest power capacity hitherto, this loudspeaker unit could generate up to three hundred times the sound output of any previous loudspeaker. It was commercialized in 1926–27 as the Western Electric 555W speaker and was used as a part of the Vitaphone and Movietone sound systems for talking motion pictures.[140] In 1930 Thuras filed a patent (granted nearly two years later) for the so-called bass-reflex principle, which prevented sounds with very low frequencies from being lost.[141]

By 1933 Western Electric and the Jensen Company marketed tweeter units (also known as treble speakers), which extended the frequency range to 12,000 Hz, thereby ushering in the age of two-way (or dual-range) systems, or ones that divided up the range of frequencies into two portions.[142] RCA's dual systems were implemented in Radio City Music Hall starting in 1932. A year later Western Electric developed the three-way loudspeaker incorporating two tweeter units and four Jensen direct-radiator loudspeakers.[143]

In 1934 E. C. Wente and A. L. Thuras described an interesting experiment they had conducted in April 1933. Using a triple-range speaker, they researched a radio broadcast of the Philadelphia Orchestra in the city's Academy of Music that was transmitted to newly built Constitutional Hall in Washington, D.C. Drawing upon Harvey Fletcher's work on stereophonic recording and reproduction, the performance was based on three-channel stereophonic sound.[144] They noted that "lacking the visual diversion of watching the orchestra play, such an audience centers its interest more acutely in the music itself." As a result, the broadcast "requir[ed] a high degree of perfection in the reproducing apparatus both as to quality and as to the illusion of localization of the various instruments."[145]

The two engineers wanted to test the quality of the microphones set up in Philadelphia's Academy of Music and the loudspeakers in Constitutional Hall by determining the power a loudspeaker needed to deliver the requisite frequency range without any perception of distortion. Their work focused on the microphones and loudspeakers themselves. They chose a horn-type loudspeaker over a direct-radiating one as it generally possessed higher values of efficiency throughout a broader frequency range.[146] Apparently the loudspeakers possessed a uniform response throughout the entire frequency range of the orchestra, a greater output capacity with no linear distortion, and a uniform distribution of sound at all frequencies throughout a wide solid angle. They were pleased with the results of their moving-coil microphones, as the sound emanating from the loudspeakers was generally the same as the sound picked up directly in front of the microphone. Finally, they felt that the sensitivity of their microphone was superior to other types, except perhaps carbon microphones.

As was the case with microphones, loudspeakers also needed to reproduce the overtones with the same relative amplitudes as the original sound. They were required to possess a broad, flat frequency curve over the radio frequency range measured in Hz when plotted against the sound pressure amplitude measured in dynes per cm^2 or in decibels, which is to say that there should be minimal change in the sound pressure amplitude as the frequency changes.[147] In addition, the sounds of all the instruments needed to be transmitted with their correct amplitudes: if the trumpets were blaring loudly while performing a piece, and the violins playing softly, that difference in amplitudes needed to be conveyed by the loudspeaker. The pressure of the sound waves emanating out of the loudspeaker was often measured using the Rayleigh disk technique in which a small disk was suspended from a thread at a particular angle to the approaching sound wave emanating from the loudspeaker. The twist in the thread was then measured to determine the degree of torsion needed to maintain the angular position against the sound wave.[148] This method was labor intensive, as one needed to measure the amplitude for each frequency; therefore, Meyer and Grützmacher invented a method that photographically recorded the loudspeaker curve.[149] In addition to testing the pressure of the loudspeakers throughout the range of radio broadcasting frequencies, they used harmonic analyzers to compare the overtones and their relative strengths of the original sound with the sound generated by the loudspeaker.[150]

As Meyer informed us in 1939, older funnel loudspeakers of the late 1920s had a range of approximately 425 to 3500 Hz. Large, high-quality conical loudspeakers with magnetic membranes of the early 1930s had a range of around 100 to 5000 Hz, while dynamic loudspeakers of the period enjoyed the largest frequency range, from about 80 to 8000 Hz. The lowest keys on the piano, the lowest open string on the cello, and the lowest open string on the double bass fell below that range (see Figure 3.4). Certain types of loudspeakers reproduced the frequencies of the original sound, but they also broadcast the overtones with amplitudes that were much greater than the original. It was difficult to reproduce a satisfactory tone below 150 Hz.[151] To radiate over eight-and-a-half octaves, from 30 to 12,000 Hz, at least two loudspeaker units were required, as physicists and engineers in the United States had also realized. For the lower range, up to 6000 Hz, electrodynamic speakers were optimal. A small-horn, high-frequency cone speaker with a solid suspension used as a centering device, a piezoelectric speaker, or an electrostatic speaker were preferable for the upper range from 6000 to 12,000 Hz.[152] Major suppliers of loudspeakers during this period included Siemens & Halske, AEG, Telefunken, and the smaller Berlin firm C. Lorenz, and the American companies Western Electric (AT&T), General Electric (GE), and the Radio Corporation of America (RCA).[153]

Finally, amplifiers were critical to radio broadcasting, and they were also being improved upon throughout the 1920s. The ability of triodes—a type of vacuum tube—to amplify sound had been known since the work of Lee de Forest back in 1906. More advanced vacuum tubes were being developed, including various forms of tetrodes and pentodes by numerous companies throughout Europe and the United States. Once again, radio provided the market impetus. It turns out that if a highly efficient microphone was used for broadcasting, the amplifier volume range could be reduced significantly. High-efficiency loudspeakers also reduced the volume required from amplifiers. The performance of any amplifier needed to be measured by its voltage amplification, its effect upon the circuit to which it is connected, and its distortion of the waveforms.[154] Linear and nonlinear distortions were again the issues. Various types of vacuum tubes were necessary, depending on whether battery-operated or line-powered receivers were used, whether high- or low-frequency amplification was desired, whether or not the vacuum tubes were being used for transformer coupling, and the size of the room or auditorium where sound would be amplified.[155] Amplifiers needed to supply the loudspeakers with the necessary levels of energy that would result in the volume being at least as loud as the original performance.[156] They also needed to be free from internal disturbances, such as feedback from their constituent parts, including transformers, retardation coils, and capacitors, that could cause severe frequency distortion. While measuring a performance of the Philadelphia Symphony Orchestra, Harvey Fletcher determined that the amplifier must be equally sensitive to at least a 75-dB volume range to cover the loudest and softest moments of a performance. Finally, the total harmonic component, or the percentage of the output that contained harmonic overtones, needed to be 1 percent or less of the fundamental tone. The percentage could be quickly and accurately ascertained by using a highly sensitive harmonic analyzer. Another method to measure distortion in amplifiers included using an oscillograph, which could test for uniformity of amplification across the frequency spectrum.[157]

To conclude, during the mid-1920s and early 1930s harmonic analysis had become a powerful tool used by numerous physicists and electrical engineers interested in, among other things, improving radio broadcast fidelity. Harmonic analyzers yielded the composition of complex sound waves. Radio engineers and physicists compared the analysis of the sound generated directly by a human voice or musical instrument with the analysis of the sound reproduced over the radio, specifically, the amplitudes of an original tone's overtones with those of its reproduction. With those analyses in hand, they could now improve the components of microphones, loudspeakers, and amplifiers. Tuning forks, flue pipes, and interference tubes yielded to harmonic analyzers. One might be tempted to argue that the skilled ear of Helmholtz or Stumpf was no longer required, as machines appeared to level the proverbial playing field by rendering musical training superfluous. Mechanical objectivity seemed to trump the fallible human senses. Scientists and engineers began to refer to these experiments as being "objective," since the results were deemed to be less dependent on the "subjective" human body.[158] It turns out that the electro-mechanical, however, did not always supersede the human. Since the human ear is not a machine, mechanical objectivity was not always a coveted attribute of physiological research. The role of the human ear in defining timbre was becoming an increasingly important topic of study for physiologists, physicists, and engineers.

4

The RVS: Radio Experiments

THE 1920S were a fascinating decade for Berliners. In the immediate aftermath of World War I, it seemed unfathomable that the city would soon become the world's third largest municipality. Despite the political and economic turmoil in the wake of the war, there was some cause for optimism. The capital was receiving an architectural facelift thanks to the influence of Walter Gropius, Ludwig Mies van der Rohe, and Erich Mendelsohn. German cinema was flourishing, featuring what would later become classics, such as *Dr. Mabuse der Spieler* (*Dr. Mabuse the Gambler*) and *Metropolis,* both directed by Fritz Lang. Bertolt Brecht and Kurt Weill were entertaining cinema and theater goers with powerful political and moral tales, while similar messages from the pen of journalist and cultural critic Walter Benjamin could be read in the city's newspapers. The capital could boast the residences of some of the world's leading scientists: Albert Einstein, Max Planck, Lise Meitner, Erwin Schrödinger, Otto Hahn, Leo Szilard, Max von Laue, Gustav Ludwig Hertz, Otto Heinrich Warburg, and Fritz Haber. In this intellectual climate, Funk-Stunde Berlin was beginning to fill the airways with news, radio dramas, educational lectures, and music.

In a report issued in December 1928, the Reich's Imperial Ministry of the Post Office in Berlin, which was responsible for the development of German radio, bemoaned the fact that:

> the delineation of the responsibility for the technical and artistic operations of the Reich's Broadcasting Corporation [the Reichs-Rundfunk-Gesellschaft, or RRG] was previously regulated in such a way that the necessary intimate collaboration between artists (*Künstler*) and engineers (*Techniker*) was not sufficiently supported. Various areas of electrical engineering research—such as microphones, amplifying and regulatory devices, connecting cables, and transmitters—all fell under the domain of the German Post Office and were managed by its civil servants. The artistic offerings, on the other hand, as well as the acoustical design of the recording rooms, the lineup of the musicians, the layout of the microphones, the presentation of the transmitting

> rooms as well as the broadcasting from the concert halls and opera houses, were, on the other hand a matter for the broadcasting societies.[1]

Something needed to be done to bridge the chasm.

The Prussian Ministry of Culture[2] and the Reich's Office of Telegraph Technology addressed the issue of increasing the collaborations between musicians, engineers, and physicists. Recall from the previous chapter that Karl Willy Wagner was appointed founding director of the Heinrich Hertz Institute for Oscillations Research (HHI), a creation of the Prussian Ministry of Culture and the Reich's Central Post Office. Its mission included the production of electric musical instruments. Later that year, the ministry, in consultation with the RRG, also founded the Funkversuchsstelle (later named the Rundfunkversuchsstelle, or Radio Experimental Laboratory, RVS) of the Berlin Academy of Music: it would become a focal point for the intense collaborations.[3]

> The Prussian Ministry of Culture, which is closely following the musical tasks of broadcasting, made contributions available from the funds received from the Reich's Broadcasting Corporation so that practical work could begin.[4]

As we shall see, the German nation-state led the charge with some assistance from the leading German electrical engineering companies.

As mentioned in chapter 1, it was also a period of a new aesthetic, Neue Sachlichkeit, that sought to reach a larger percentage of the public than earlier musical genres by linking modern music with society.[5] It was committed to the genres of radio music, *Zeitoper* and *Gebrauchsmusik*. Composers such as Paul Hindemith, Igor Stravinsky, and Kurt Weill were pushing the envelope of what constituted music. They saw themselves as following the calling of Ferruccio Busoni to create a new form of music based on, among other things, atonality.[6] Neue Sachlichkeit was an aesthetic of experimentation, mechanical-machine music, and modernity.[7]

The last two chapters discussed how radio broadcasting caused a distortion of tone color and how the disciplines of physics and engineering analyzed and ameliorated those problems. This chapter looks at the RVS, the space where physicists, electrical and radio engineers, phoneticians, physiologists, and musicians collectively plied their crafts in Berlin, all dedicated to improving the fidelity of radio broadcasts, teaching music students how to work with technological devices such as microphones, and inventing musical instruments, specifically, the trautonium. In this laboratory, the contours of science, technology, musical aesthetics, and politics overlapped in historically informative ways.

This chapter sheds light on a fascinating period in Germany history. Much, of course, was up for grabs. The outcomes were neither natural nor inevitable.

Through the implementation and fusion of various forms of cultural, technical, and scientific knowledge and practices, a new musical instrument was created as were new musical aesthetics and genres.

The Origins of the RVS

The RVS of the Berlin Academy of Music was a truly unique space. As the American radio engineer and musician R. Raven-Hart pointed out:

> Many such laboratories exist, but they are almost without exception organized and run from the engineering point of view, either by Universities and the like or by commercial interests. Here, on the other hand, a musical organization has established a radio department, with the result that the orientation is entirely different, and, at any rate from the musician's point of view, more interesting.[8]

Raven-Hart was particularly interested in the collaboration between Friedrich Trautwein and Hindemith. The critical difference between the research conducted at Bell Telephone Laboratories (as discussed in the previous chapter) and the research at the RVS was the former initially stressed the spoken human voice in the 1920s, while the latter was actively engaged in researching musical sounds from its inception.[9]

By 1926 the Weimar Republic was collectively taking a deep breath and looking forward to erasing the memories of hyperinflation and the numerous political crises during the politically unstable years from 1920 to 1923. The first inkling of a future musical institute dedicated to improving radio broadcasting can be traced back to June 1926, when Aloys Lammers, state secretary of the Prussian Ministry of Culture and member of the Deutsche Zentrumspartei (German Center Party, which was a Catholic political party), wrote to musicologist Georg Schünemann, then the director of the Berlin Academy of Music, underscoring how painfully clear it was that German radio lacked the requisite artistic, technological, and scientific foundations.[10] In that same year, Ludwig Kapeller, the chief editor for *Funk*, a leading periodical for radio listeners, complained that the RRG was simply not interested in art and culture. "Where are the 'Reich's Office for Radio Culture,' the acoustic-artistic 'laboratories', and the schools for speech and radio drama, experimental stages and academies of taste for broadcasting?"[11] M. Felix Mendelssohn, a writer of radio dramas and composer of radio operas, expressed his concern that postal officials had too much say in the future of radio at the expense of musicians. While he was overjoyed to hear the news of the planning of the RVS, he was not optimistic that the theoretical work there would bear any practical fruit.[12]

Lammers felt that a thorough study was needed to ascertain the proper conditions for broadcasting. An "experimental laboratory [should] be built as a part of the Berlin Academy of Music, which would be dedicated exclusively to answering these questions" relevant to the physics and engineering of radio broadcasting.[13] Schünemann responded to Lammers's letter by stating that he had already discussed the matter with Gustav Engelbert Leithäuser, president of the German Amateur Radio Society and member of the Reich's Office of Telegraph Technology, and conceded that radio suffered from many technological problems that could only be solved in collaboration with musicians. Those problems included the broadcasting of instruments, the change in tone color generated by broadcast reproductions, and the various effects of instrumental groups on each other.[14] Students at the academy playing in the orchestra and singing in choirs could be put at the disposal of the testing laboratory. While he hoped to include the building and playing of electric musical instruments, he was not quite sure what the costs of building those instruments would be. The other projects were given priority. Finally, such a laboratory would enable the broadcasts of the various concerts and choral and opera productions of the academy.[15]

On November 20, 1926, those who attended a meeting at the Prussian Ministry of Culture agreed that Berlin desperately needed a radio experimental laboratory.[16] The goal was establishing a close collaboration in the areas of electrical and radio engineering, physics and acoustics with the HHI, specifically with K. W. Wagner, who sat on the RVS's board of directors from the beginning.[17] Orchestral broadcasting in Germany lagged behind London radio, as evinced by the broadcasts from the Royal Albert Hall. In addition, engineers in Prague and Warsaw were conducting interesting experiments involving radio broadcasting of speech, and the Germans felt that they could learn much from the Italians as well.[18] It was also clear that this new experimental laboratory needed to concentrate on the comparison of the relevant broadcasting devices. Microphones were specifically singled out, particularly those produced by Eugen Reisz.[19] The radio periodical *Funk* gladly reported that "finally an acoustical RVS" was planned for the Berlin Academy of Music.[20]

In January 1928, four months before the opening of the RVS, Schünemann explained to the Prussian Ministry of Culture that the RVS could also create mechanical and electric sounds and, eventually, a new form of music, with the construction of electric instruments.[21] Schünemann was convinced their invention needed to play a role in the laboratory. This conclusion was not inevitable or even logical, but instead, it was a decision contingent upon the milieu of 1920s Berlin and the RVS.

In May 1928 the RVS opened in the academy's theater auditorium. The ceremony commenced with a broadcast of the first movement of Hindemith's

"Kleine Kammermusik für Windquintet" ("Small Chamber Music for a Quintet of Wind Instruments"). Leithäuser gave a lecture titled, "Die Anlage der Funkversuchsstelle mit Lichtbild- und Filmvorführungen" ("The Facility of the Radio Experimental Laboratory with Photographs and Film Presentations"). Schünemann spoke of the RVS's tasks and practical experiments. Hermann Weissenborn discussed musical learning at a distance with the radio, while Erich Fischer spoke about conducting at a distance.[22] A folk song ensued. The second movement of Mozart's Divertimento Nr. 9 (Werk 240) for oboes, horns, and bassoons was followed by his Cavatina, "Unglücksel'ge kleine Nadel," from *Le Nozze di Figaro* sung by Carola Tiedemann and accompanied by a string quintet. Johann Strauss's waltz, "The Blue Danube," arranged for a mixed choir and piano accompaniment was the broadcast finale.[23] Ironically, given that one of the main goals of the research station was to improve the transmission of radio, a harsh critic in the audience complained that the large loudspeaker used for the opening was far from ideal. One had problems differentiating between the flute, oboe, clarinet, bassoon, and horn during the performance.[24] In addition, choir music suffered from radio broadcast. In response to those complaints, the choir director Herbert Lichtenthal subsequently conducted a number of experiments in the RVS using the academy's choir. He was particularly interested in the infidelity of broadcasting consonants.[25] The RVS had its work cut out for it.

A primary objective of the RVS was to give a "scientific basis to the artistic tasks of radio."[26] It was to research the technical and musical possibilities associated with radio broadcasting and sound films, and its mission was to become a type of *Kulturträger* (bearer of culture) of these new mass media.[27] After stressing that the radio was simply unsatisfactory to trained musical ears, Schünemann listed four major areas of focus for the RVS at the May 1928 opening: the activities that go on in front of the microphone, the results behind the amplifier, the principles for the creation of a new genre of radio music (Rundfunkmusik), and pedagogical issues.[28] The RVS determined which types of music were best suited for radio, and when necessary, how they should be rearranged. New artistic material needed to be created that met the requirements of the new technology. One also needed to experiment in order to find out how microphones altered the sounds of voices and musical instruments. Finally, musicians were required to cultivate their artistic prowess on the basis of the information they gleaned from the scientific and technological experiments. Singers, speakers, and instrumentalists needed to prepare for broadcasting, far more so now than ever before. This included, above all, the musical instruction of the technical production manager. For all of this to succeed, the RVS needed to work closely with the Berlin radio station, Funk-Stunde Berlin.[29]

Funkversuchsstelle

bei der

Staatlichen akademischen Hochschule für Musik

Berlin-Charlottenburg, Fasanenstraße 1.

*

Eröffnungsfeier

am Donnerstag, den 3. Mai 1928, vormittag 11 Uhr
im Theatersaal der Hochschule.

*

Vortragsfolge:

1. **Paul Hindemith,** Kleine Kammermusik für 5 Bläser, Op. 24 Nr. 2 ⟨I. Satz⟩

Uebertragung

Erwin Frost (Flöte), **Herbert Wiese** (Oboe), **Otto Schwägerl** (Klarinette),
Fritz Huth (Horn), **Otto Pischkitl** (Fagott)
Leitung: **Prof. Arnold Frühauf**

2. Prof. Dr. **Gustav Leithäuser,** Die Anlage der Funkversuchsstelle
⟨mit Lichtbild- und Filmvorführungen⟩

3. Prof. Dr. **Georg Schünemann,** Aufgaben der Funkversuchsstelle
⟨mit praktischen Versuchen⟩

A. Fernunterricht . . . . Prof. Hermann Weissenborn

B. Ferndirigieren . . . . Dr. Erich Fischer

a) Volkslied »Ach wie ist's möglich dann«, Satz von Erich Fischer
Elena Dinicu, Charlotte Böttger, Erich A. Collin
und Rudolf Gonszarzewski

b) W. A. Mozart, Divertimento Nr. 9 ⟨Werk 240⟩
II. Satz: Andante grazioso
2 Oboen: **Herbert Wiese und Walter Löscher**
2 Hörner: **Fritz Huth und Theodor Schenk**
2 Fagotte: **Otto Pischkitl und Hans Müller**

c) W. A. Mozart, Cavatine »Unglücksel'ge kleine Nadel«
aus »Figaros Hochzeit«
Carola Tiedemann
Erich Kindscher (Violine I), **Erika Schorss** (Violine II), **Irmgard Veidt** (Bratsche)
Hans-Joachim Kittke (Violoncello), **Wilhelm Lange** (Kontrabaß)

4. **Johann Strauß,** Walzer »An der schönen blauen Donau«, bearbeitet für gemischten Chor mit Klavierbegleitung

Uebertragung

Die **Ensembleklasse** unter Leitung von **Prof. Siegfried Ochs**
Am Klavier: **Hans-Joachim Vetter**

*

Die Instrumente, die bei der Vorführung gezeigt werden, stammen aus der
Musikinstrumenten-Sammlung der Hochschule.
Die Sammlung ist geöffnet: Dienstag, Donnerstag, Sonnabend und Sonntag 11–1 ⟨13⟩ Uhr

---

Während der Vorträge bleiben die Saaltüren geschlossen.

FIGURE 4.1. Program of the opening concert for the Funkversuchsstelle (later Rundfunkversuchsstelle (RVS, or the Radio Experimental Laboratory) at the Berlin Academy of Music, May 3, 1928. *Source:* Universität der Künste (Berlin) Archive, UdK-Archiv 1b-16. Reprinted with the permission of the Hausarchiv der Universität der Künste.

While commercial enterprises such as Allgemeine Elektricitäts-Gesellschaft (General Electrical Company,or AEG), Siemens & Halske, and Telefunken contributed financially to some of the research conducted at the RVS as well as rented or donated their equipment for experimentation, the project was predominantly funded by the Prussian Ministry of Culture and the RRG: approximately two-thirds of the costs during the late 1920s and early 1930s were provided by the ministry, and the RRG, whose majority of shares were owned by the Reich's Post Office, contributed the remaining one-third.[30]

The RVS proved to be a popular attraction early on; in the fiscal year 1929 alone it welcomed approximately three thousand visitors, many from abroad, including from the United States, Great Britain, France, Japan, Italy, Switzerland, and Sweden.[31] For Berliners, however, the RVS was experiencing some growing pains. A critic lambasted an evening of lectures sponsored by the RVS. He was particularly appalled by the lack of technical knowledge of Erwin Fischer, who spoke about the measurement of sound reflection. While Hellmut Sell discussed the interesting topic of the carbon microphone's ability to pick up high frequencies, he apparently failed to compare it to other types of microphones used by radio stations. He also discussed the production of new tones and tone colors by electricity, but he allegedly was unable to describe the technological and scientific principles behind producing such sounds in a reasonably comprehensible way. H. K. von Willisen demonstrated a new form of disc recording based on a novel cutting technique. Unfortunately, he could not get the experiments to work. The reviewer warned: "The RVS must either achieve worthy scientific or practical, useful work, otherwise one will need to question its right to exist."[32]

Schünemann felt compelled to answer his critic. He spoke of von Willisen's pioneering research on cutting techniques for discs as well as Sell's carbon microphone, perfected loudspeaker, and his experiments with electric musical instruments to study the changes in their tone color. These experiments had already led to successes, namely Siegfried Borris's musical piece "Kompositionen für elektrische Stimmen" ("Compositions for Electric Voices"). He also explained how Fischer's research on room acoustics yielded a new procedure for measuring sound reflection in rooms covered in celotex insulation at the Funk-Stunde Berlin.[33] Schünemann was adamant that his RVS was performing far beyond expectations.

The RVS attracted the attention of some of the period's leading radio composers. For example, Berlin composer Max Butting mentioned the radio research facility: "Here the characteristics of broadcast music and the problem with creating a new music for radio are studied."[34] He offered a course on microphones at the RVS.[35] For Butting, perhaps the most important task of the RVS was assisting the new genre of Rundfunkmusik. Kurt Weill, the renowned German composer who set many of Bertolt Brecht's lyrics to music, welcomed the news of the RVS's creation, arguing that it should have happened much earlier.

> For a long time we had come to believe that the current state of broadcasting as a whole requires such a place [the RVS] that would first of all determine the *aesthetic* and acoustic foundations of a possible broadcasting art and, on this basis, would have an educational and constructive effect.[36]

He hoped the RVS could help achieve just that. In his view, the current broadcasting service in Germany was becoming dangerous to the art of radio, as it was based nearly exclusively on traditional forms of entertainment. If radio was to assist in the cultural liberation of the working classes, as he so desperately desired, it needed to become a new, original art, a *Rundfunkkunst*, and in essence a new type of entertainment with a new type of aesthetic. He, too, was a strong supporter of radio music (Rundfunkmusik).

> Only when conductors and directors know the full range of acoustic possibilities, only when they are clear about the tonal effects of all musical, linguistic, and noise combinations, can they begin to build a new art using their technical knowledge and experience, whose aesthetic foundations are based on technical innovation.[37]

It turns out that the RVS did offer a well-attended course on Rundfunkmusik.[38] Walter Gronostay, a young composer known for his Rundfunkmusik and film scores—particularly his work with Herbert Windt on Leni Riefenstahl's *Olympia*—plied his craft at the RVS. In 1931 Schünemann, Butting, Gronostay, and Friedrich Trautwein all attended the second conference on radio music led by Leo Kestenberg of the Prussian Ministry of Culture in Munich.[39] Rundfunkmusik was clearly critical to the RVS.

Not all radio fans, however, were encouraged by the founding of the RVS. An anonymous critic was disturbed by what he saw as the overemphasis of "electroacoustic questions."[40] He was convinced that

> technology today does not yet provide the tools to reproduce acoustically with absolute fidelity (*absolut naturgetreu*), so the technician has to make the compromise of assembling the various elements, e.g., by working with different types of speakers and microphones.[41]

Addressing that concern, music editor Ludwig Kapeller underscored the so-called comprise between art and technology that was so crucial during this period, and stressed that "it is not for nothing that the new Radio Experimental Laboratory in the Academy of Music is jointly supervised by an artist, Prof. Dr. Schünemann, and an engineer, Prof. Dr. Leithäuser."[42] The

FIGURE 4.2. The Rundfunkversuchsstelle (RVS, or the Radio Experimental Laboratory) in 1929. The laboratory worked with Siemens and Reisz microphones. *Source:* Kestenberg, ed., *Kunst und Technik.*

aforementioned critic also felt that more time and effort should be dedicated to perfecting the technique of radio drama. Music was not the only emphasis of the RVS: the spoken word was as well.

From the outset, the RVS had musicians, engineers, and scientists work together.[43]

> Areas of research being undertaken in 1929 at the RVS: soloists and musicians rehearse in front of the microphone, others listen, observe, and experiment. Acoustical trials are systemically executed with the oscillograph. Electrical tones are tried out, and all types of transmissions are studied.[44]

Fearing that the engineers and politicians were taking charge of the application of technology to music, Schünemann lamented that, "up to this point the musician has stood on the sidelines."[45] He added: "A productive collaboration of engineers, scientists, and musicians still has yet to occur, and yet we can only forge ahead if we all jointly strive for the same goal: the technical and artistic perfection of radio broadcasting and radio music."[46] The RVS was to lead the charge.

The German electrical engineering giant Siemens & Halske provided the RVS with state-of-the-art radio transmission equipment. Each room of the RVS could transmit music and speech by means of recording microphones, amplifiers, and playback devices such as the Magnetophon (also spelled Magnetophone), the first reel-to-reel tape recorder. Electrical wires ran from the amplification room housed in the attic to various other rooms, including the large and small music halls and various instructional rooms and auditoria. A double-circuit line from each room led to two tables, one with microphones and the other with loudspeakers.[47] The setup included a main amplifier for microphones and power amplifiers.[48] During the construction of the RVS in early 1928, Siemens & Halske also supplied a large loudspeaker unit, which was rented annually to the RVS.[49]

Experimenting and Tinkering

The research undertaken at the RVS was multifaceted. Some worked on improving Curt (also spelled Kurt) Stille's Fernschreiber (teletype machine), which recorded acoustical vibrations in magnetic form on a steel wire. Stille was an engineer of the Berlin radio equipment manufacturer, Vox Maschinen AG. His work culminated with the creation of a synchronized sound system for films using magnetized steel tape.[50] The RVS tested just how faithfully the device could reproduce frequencies of the highest tones: apparently, the device was impressive.[51] The engineers tested other equipment, such as Ernst Finking's Sprechapparat, a loudspeaker renowned for its quality.[52] Others investigated AEG's newly invented Magnetophone. A group within the RVS even worked on X-ray cinematography, aiming to elaborate the physiological processes involved in speaking and singing. Their work drew upon the research of the physician Dr. Viktor Gottheiner, a pioneer of the so-called science of X-ray films who attempted to capture the sound of the human voice while X-raying the speech organs.[53] Schünemann made X-ray films with Gottheiner.[54] Alfons Kreichgauer, also a student of Carl Stumpf, was a research assistant at the Institute for Psychology of the University of Berlin and taught musical acoustics at the Berlin Academy until being replaced by Trautwein.[55] In early 1929 Kreichgauer was employed by the RVS to test loudspeakers and Reisz microphones.[56] Scientists and engineers at the RVS also attempted to synchronize music to silent films. They employed a technique developed by Tobis Film, an early German film production and distribution company, for using light to record sound on film. Working with Western Electric's Vitaphone, they sought to improve the process whereby sound was recorded and played in conjunction with filming.[57] Trautwein worked on a procedure for the reception and reproduction of pictures and sounds: the differences in tone volume and picture brightness could be amplified, meaning that soft tones would become softer, loud tones louder. Bright points on a screen would become brighter and dark points darker. Clearly this new principle of amplification was relevant to television as well as improved radio transmission.[58]

Composers also tinkered with the creation of new sounds and constructed new musical genres with their own aesthetic characteristics at the RVS. The experimental technique of changing the rotational velocity of a phonograph was employed by Hindemith, who was professor for musical composition at the Berlin Academy of Music and a frequent visitor of the RVS, where he taught a course on music in films during the academic year 1929–30.[59] He referred to the small room in the attic of the instruction building of the academy on Fasanenstrasse as the "Electric Studio Berlin."[60] His experiments of *grammophonplatteneigene Stücke* (pieces specifically for gramophone records)—the Austrian composer Ernst Toch referred to these as *Grammophonmusik*, or what we would today call turntablism, whereby the phonograph itself becomes a musical instrument—were composed solely on gramophone records. They debuted at the Neue Musik Berlin 1930 festival.[61] Hindemith's performance featured three single-sided 78s with the date "30 June 1930" and "R. V." stamped on them, meaning that they were made at the RVS, where engineers and physiologists had been employing a technique of increasing the velocity of gramophones in order to study formants.[62] One record was labeled "Gesang über 4 Oktaven" ("Song Greater than Four Octaves"), upon which a brief melody and its variations were sung. It featured two instruments: the viola and the xylophone. Changing the speed of the record changed both the pitch and the tone color of the viola, which at times resembled a violin and a cello, and the vocal sounds began to resemble those produced by musical instruments.[63] Toch experimented with vowel sounds and consonants. A review of the Neue Musik Berlin 1930 festival by *World-Radio* commented on Hindemith and Toch's "experiments."

> The experiments with new instruments comprised music for gramophone: for instance (by Hindemith), vocal pieces for "impossible" registers as when the machine plays very fast or very slow, or instruments with high or low harmonics "filtered out." Toch composed several brilliantly effective pieces made out of vowel rhythms for vocal chorus.[64]

That metamorphosis of the tone color of the vocal sounds into instruments clearly struck a chord with Schünemann:

> If vowels are sung and are raised in pitch, curiously strange sounds ring out; and if they are combined with consonants in the manner of solfège syllables, a nearly instrumental sound arises. Ernst Toch made such recordings in the Rundfunkversuchsstelle for Neue Musik Berlin 1930 festival.[65]

Toch's "Gesprochene Musik," an example of *Grammophonmusik*, was composed exclusively for a speaking choir.[66]

The RVS and the Fidelity of Radio Broadcasting Equipment

One major responsibility of the RVS was the testing of equipment relevant to sound production, amplification, broadcasting, and recording. According to radio engineers as well as Schünemann, the key to improving the fidelity of radio broadcasts was to reproduce the formant region, or the specific frequency band encompassing the overtones that characterize the sound. One needed to find a physical and technological solution to render the formants of the broadcast music the same as those of the original instruments. One, of course, also needed to study room acoustics.[67] Recall that Karl Willy Wagner commented on how the transmission of music was a far greater challenge for radio engineers than the broadcasting of the voice.[68] Musical aesthetics dictated technical rigor. As we shall see in the next chapter, formants played a crucial role in Trautwein's invention of the trautonium.

Schünemann was familiar with the numerous problems that plagued the transmission of various musical instruments with different tone colors. The improvement of the technique for the fidelitous broadcasting of the various instrumental tone colors and human voices was paramount.[69] The RVS was dedicated not only to rendering a subjective judgment of the quality of reproduced speech or music but also to generating an "objective picture" of the reproduced sounds. To do so, they used oscillographs and harmonic analyzers supplied by Siemens & Halske.[70]

Schünemann insisted that the RVS set up a systematic series of experiments to study the modifications of musical notes caused by the microphone and amplifier for the instruments of an orchestra, ranging from the contrabass to the piccolo. These experiments were to be conducted according to Carl Stumpf's interference method with an oscillograph. Oscillograms of the overtones of various instruments (i.e., their natural sound, or *Naturklang*) were compared with the oscillograms of the broadcast sound.[71] Interference tubes[72] could cancel out overtones, thereby generating simple tones that produce pure sine waves; they were critical to the synthesis of vowels, speech sounds (*Sprachlaute*—or distinctive units of speech), and sounds produced by musical instruments.[73] Schünemann expressed his hope that Stumpf's Institute of Psychology would lend the RVS the necessary equipment for interference studies as he wished to build upon this research tradition in order to analyze and synthesize sounds.[74] He was convinced that these studies would inform musicians of the interactions of various sounds with respect to room acoustics and the numerous sound combinations of musical instruments.[75] In addition to the studies on the analysis and synthesis of sound based on resonance and interference, experiments dealing with the dynamics, rhythm, and agogics needed to be conducted.[76] Formants of both human speech and musical instruments were to be investigated in order to render the sounds generated by the broadcast voices and musical instruments equivalent to their original counterparts.[77]

> We shall determine the change in the structure of the overtones caused by the microphone for each instrument. In addition to evaluating the interference methods [of Carl Stumpf], we shall analyze the curves produced by oscillographs and be able to use the tuning-fork method for the initial findings. The material should be ordered systematically and serve as a basis for further experimentation. We shall then be in a position not only to test the microphones musically and to investigate them scientifically, but shall also be able to suppress distortions by either strengthening the formants from the outset or by overemphasizing all overtones, which are obstructed by the microphone. Naturally one can prompt the player, through technical means, to brighten or temper the sound in this or that direction. And this brings us back to learning how to deal with the microphone. These acoustical-scientific experiments can only become practically meaningful if the musician makes available his rich store of instrumental and technical experiences. In addition to the characteristic formant regions, secondary effects—such as blowing and bowed sounds, articulation, and dynamics—also have an influence on the character of the sound.[78]

For example, he contrasted A4 (435 Hz at the time) blown by a trumpeter with the broadcast of that note. There was a clear deviation in the number and strength of the overtones in the graphs.[79] He also included old instruments in his analyses. That aforementioned pesty critic wondered why in the world Schünemann was interested in using oscillographs to analyze sounds of old instruments.

> We ask ourselves, what is the purpose of writing down with the oscillograph how the flute of Frederick the Great sounds on the lips of Prof. Schünemann? What is the purpose of audio recordings of instruments from the past centuries . . . ?[80]

In addition to the technological difficulties associated with broadcasting, there were psychological obstacles to be negotiated. Schünemann stressed how the conditions of recording were quite different from a normal performance. The presence of the microphone could be intrusive. The recording rooms were usually rather small and bedecked in heavy, sound-absorbing curtains. The atmosphere affected the coloring of the performance: since the surroundings were so "unnatural," the musician often "overplayed" and "exaggerated" passages, and the resulting music seemed "contrived."[81] Musicians certainly needed to instruct their students on how to adjust their playing technique accordingly.[82] As Jonathan Sterne has so persuasively argued, the so-called original sound being broadcast was clearly just as contrived and artificial as the sound that emanated from the radio and entered into the ears of the listeners. Both were predicated on human-machine interactions and a network of skills and practices. That is to say that there was nothing natural about the original sound.[83] So the *Natur* in *Naturgetreue* had a rather hollow ring. Or, the "original"—to borrow Adorno's word—was just as artificially constructed as its reproduction.

Student performances were sources of experimentation for the scientists and engineers: the vibrato of cellists and violinists, the pedaling of pianists, the soft phrasings, the dynamic contrasts, the changing of the musicians' locations in a large ensemble, the various registers of the wind instruments, and the acoustical formant regions all offered opportunities for improving the design of the microphone.[84] Musicians welcomed the assistance of scientists and engineers. This was an area where they, physiologists, physicists, and engineers at the RVS could enhance recording, broadcasting, and musical pedagogy.[85]

In 1931, Schünemann informed the Prussian Ministry of Culture of the RVS's areas of focus. First was testing the artistic prerequisites for the successful transmission of music, specifically radio, records, sound films, and electric music, the last of which had been firmly ensconced in the RVS from 1930 on. The second goal was establishing courses of study on broadcasting speech and music for lecturers, singers, instrumentalists, and composers with a view to teach all of them, among other things, how to use a microphone properly during radio broadcasts, as the broadcasting studio was a different environment than a concert hall or a theater stage.[86] The RVS offered instruction on radio speaking and lecturing; speech and gestures for sound movies; educational radio; singing and choral work for radio; radio composition and scoring; Gebrauchsmusik; film music; radio music; and electroacoustics.[87]

Microphones, Loudspeakers, and Amplifiers

Since the RVS was established during the rise of radio, its trajectory sheds light on the history of broadcasting equipment. It was an important independent laboratory to test microphones, loudspeakers, and amplifiers.[88] Schünemann stressed the importance of microphone research at the RVS.[89] Activity reports of the RVS instructors are testament to the amount of microphone training in the Berlin Academy's curriculum.[90] Berlin film director and radio reporter Alfred Braun taught his students how to use the microphone for singing and speaking, particularly for radio dramas.[91] Of particular importance was the precise location of the microphone. Karl Würzburger and Karl Graef from Deutsche Welle of the Königs Wusterhausen radio station co-taught a course (popular among students) that discussed the psychological conditions of the effect of microphones on performances.[92] Bruno Seidler-Winkler taught a related course on the types of instruments and voices best suited for radio.[93] For broadcasting purposes, the RVS possessed a Siemens & Halske Protos microphone, a Riegger condenser microphone, and a Klangfilm-GmbH condenser microphone. It also had Siemens & Halske ribbon microphones connected to loudspeakers in the various instructional rooms.[94]

As part of its mission, the RVS tested microphones sent from various companies.[95] Yet the goal was not to promote one brand over another. The RVS would not offer certificates of performance; instead, it would report the findings of its investigations to the devices' owners and manufactures.[96]

The RVS often experimented with Reisz microphones, considered by many to be the best for radio broadcasting during this period.[97] In August 1928, Eugen Reisz supplied the RVS with two different types of microphones, Types 104 and 105. Having left Siemens & Halske in 1925 to establish his own Berlin-based company, he was now a competitor of the renowned electrical engineering company and was keen to hear how well his microphones compared to his former employer's. He had heard a lecture by Schünemann in Göttingen in which he had thought that Schünemann had cited erroneous measurements of both companies' microphones. According to Reisz, measurements made by the Radio Corporation of America (RCA), Western Union of New York, and the Marconi Company in London demonstrated that the frequency range for his microphones was flat from 30 to 8000 Hz, meaning that there was no linear distortion in that range.[98] He insisted that

> it is completely out of the question that at high frequencies, S. &. H's [Siemens & Halske's] condenser microphone is superior. If, according to the oscillograms, the overtones of string instruments appear more clearly in the condenser microphones of S. & H. [than in mine], this does not prove to me that the S. &. H. microphone works properly. As far as I know, the occurrence of these harmonics is due to resonance of this particular microphone.[99]

According to Reisz, Siemens & Halske's microphone suffered from nonlinear distortions.

In December 1930 Schünemann and the head engineer of the RVS, Friedrich Wilhelm Grunel, spoke to Reisz in his laboratory about gramophone recordings with a wax stylus that possessed a flat frequency response curve from

60 to 7000 Hz. They were particularly impressed by the "special microphone," which yielded "quite extraordinary recording results."[100] The RVS also investigated the acoustics of the academy's performance rooms with the broadcast of a piano using a Reisz microphone and a loudspeaker. They noted the fullness of sound and the fidelity of the lower piano pitches.[101]

Telefunken, AEG, and Reisz provided the RVS with loudspeakers.[102] Additionally, the RVS had a number of Rice-Kellogg loudspeakers at its disposal.[103] Due to a patent exchange agreement with General Electric, which developed Rice-Kellogg loudspeakers, AEG was permitted to sell these loudspeakers in Germany, as the international patents were managed by the German company Telefunken (a subsidiary of AEG at the time).[104] Marketed by RCA and by AEG in Germany, it would become one of the most popular loudspeakers of the period. During the early 1930s, the RVS also had Körting loudspeakers, which were built upon an entire wall with sound baffles. Founded in Leipzig in 1889, Körting & Mathiesen AG had originally produced arc lamps for street lighting. In 1923 they began to manufacture transformers, power supply apparatus, and choke coils for radios. Two years later, two employees, Oswald Ritter and Wilhem Dietz, created a company that sold dynamic loudspeakers under the brand name of "Körting."[105]

The RVS's teaching rooms and auditoria had a large number of loudspeakers produced by Siemens & Halske. They included the electrodynamic loudspeaker, Protos (Siemens 072), invented in 1926 and used in radio immediately thereafter. Its membrane consisted of a folded piece of pertinax, which oscillated in a magnetic field thereby creating sound waves in the surrounding air. Its selling points included "fullness of tone", "a reproduction true to nature", and "minimal use of energy."[106] An advertising brochure from Siemens & Halske of their Protos stressed its ability to reproduce equally well the lower and upper tones. The lower tones possessed a true richness, while the upper tones of the various musical instruments recreated the unique tone colors. Pianos were always a challenge for loudspeakers since the fullness of the instrument was predicated upon its resonance, which was difficult for a loudspeaker to pick up. Often the overtones of the higher pitches would be lost, thereby destroying the reproduction. These were not problems for the Protos.[107]

Another type of Siemens & Halske loudspeaker found at the RVS was the folding (portable) speakers (Siemens-Faltlautsprecher). This loudspeaker possessed an electrodynamically powered membrane; therefore, it was more powerful than the Protos. A metal conductor, which moved freely between the poles of a strong electromagnet, was attached to the seam of the membrane. The amplified current of the microphone flowed through the conductor, causing the membrane to swing in rhythm with the sound waves. This in turn resulted in the vibration of the surrounding air. Siemens & Halske's moving coil loudspeaker (Riesenblatthaller), invented by Hans Riegger in 1923 and used in public address systems in the 1930s, could be found in the RVS's large music auditorium. The Riesenblatthaller was another type of moving-coil, dynamic loudspeaker. It possessed a corrugated aluminum membrane, which ensured that the forces were distributed equally. The conductor was housed between the poles of powerful electromagnets. Because the membrane was so large, sizeable columns of air with considerable amounts of energy could be brought to vibrate. The Riesenblatthaller was suitable for large auditoria and outdoor events.[108]

The investigation of loudspeakers was of considerable importance to the RVS, second only to the institute's investigation of the microphone.[109] They tested loudspeakers manufactured by small companies, such as Lenzola-Lautsprecher-Fabrik and Tempel-Lautsprecher-Fabrik, both in Krefeld; Finking-Reyton of Leipzig; Otto Knöllrer of Neu-Ruppin; and Pama Paper Maché Company of Munich.[110] The RVS engineers were also interested in a newly designed "spectral sound membrane loudspeaker," which was an electrodynamic loudspeaker invented by Berlin engineer Hans Tschirner.[111] They discussed whether certain combination tones—such as those generated when one listened to two instruments (for example, an oboe and flute) played simultaneously through a loudspeaker—were a physiological product of the ear (meaning that they were "subjective") or the result of the loudspeaker resonating with the original sound source (thereby creating "objective" combination tones).[112] Trautwein insisted that the ears of his assistant, the trautonium virtuoso Oskar Sala, were superior to the oscillograms. When an AEG loudspeaker was judged to be flawless according to frequency measurements, as indicated by the oscillograms, Sala claimed to have distinctly heard a dissonant *Hallformant* (resonance formant).[113] The human could not totally be erased by scientific analyses.

The RVS tested amplifiers as well: it had three types of amplifiers that operated with vacuum tubes: a microphone main amplifier, an output amplifier, and a power amplifier.[114] Various vacuum tubes used for amplification were provided to the RVS by Siemens & Halske, Telefunken, AEG, and a number of electronics companies in Berlin specializing in radio parts including Leopold Kling, Walter Arlt, Hansa Funk Berlin, Tobis-Klangfilm, Nora and Lorenz-Radio Vertriebs-Gesellschaft.[115] Braun tubes, better known as cathode-ray tubes outside of Germany, were of particular interest.[116] RVS testing from October to December 1929 revealed that Siemens & Halske's microamplifiers were not free from linear distortion.[117]

By 1930, the RVS was already enjoying an outstanding reputation for its work on improving radio, particularly the requisite technical equipment.[118] While adjudicating these apparatuses, physicists and engineers often consulted the works of Erwin Meyer, Paul Just, and Martin Grützmacher of the Reich's Ministry of the Post Office.[119] Meyer's scholarship was particularly important. Schünemann met with Meyer and Wagner of the HHI to discuss a collaboration between the two institutes.[120] Moreover, like Wagner, Meyer sat on the board of directors of the RVS.[121] In 1930 Meyer taught a course on electroacoustics at the RVS. The course dealt with issues of physical and physiological acoustics related to sound recording and sound playback devices. It also discussed the electrical parts of sound transmission and recording equipment for radio stations, such as amplifiers, radio receivers, and transmitters as well as the technologies relevant to gramophones and sound films. Part of its curriculum included occasional experiments at the HHI. Apparently, it was not well attended.[122]

Friedrich Trautwein and Oskar Sala at the RVS

By far the most important project carried out at the RVS, for the purposes of this book, was Friedrich Adolf Trautwein's invention of the trautonium. Trautwein was born in Würzburg, Bavaria, on August 11, 1888, to a pastor, Adolf Trautwein, and a housewife, Babette (née Gerner). As a young child, he learned to play the organ for his local church. He finished his *Abitur*, or the German qualification obtained at the end of secondary education, in 1906 at the Heidelberg humanistic gymnasium, where he was the founding director of a student orchestra. His educational trajectory to that point was typical of the Bildungsbürgertum. After his studies in electrical engineering at the Technical University of Karlsruhe, from 1906 to 1908, he interrupted his studies to become an apprentice to numerous post offices from August 1908 to August 1909, working on technical communications. In the winter semester of 1908–9, until the summer semester of 1911, he studied law at the Friedrich Wilhelm University of Berlin, finishing his legal exams in the fall of 1911, which qualified him for a high-level legal internship in the Reich's Ministry of the Post Office. While he was working on his legal degree, he was also studying electrical engineering and technical physics at the Technical University of Berlin. During the summer semester of 1912 and winter semester of 1912–13, he switched to studying physics at the University of Heidelberg.[123] His studies were interrupted by the war.

During World War I, Trautwein was a volunteer radio operator, becoming a lieutenant of the Telegraph Battalion Nr. 4 of Karlsruhe.[124] After the war, in 1919, he finished the examination to become a postal assessor, and in June 1921 he received a doctorate of engineering sciences (Dr. Ing.) in electrical engineering at the Technical University of Karlsruhe under the supervision of Herbert Hausrath, professor of communication and signal engineering

(*Schwachstromtechnik*). His dissertation was published in three parts, two in 1920 and one in 1921. The articles detailed the application of vacuum tubes in electrical metrology and the measurement of signal attenuation at high frequencies.[125] He demonstrated that a number of metrological problems could be solved via the rectification effect (or the conversion of AC to DC) of vacuum tubes.

In 1921 he accepted a senior civil service position in wireless telephony at the Reich's Ministry of the Post Office in Berlin. Two years later, he led the Post Office's technical section that was devoted to the creation of German's first radio station, Radio-Stunde Berlin—later called the Funk-Stunde Berlin—that began broadcasting on October 29, 1923, as discussed in chapter 1. Six days before that first broadcast, he assembled the audio frequency part of the station with amplifiers.[126] In essence, he was a part of German public radio from the beginning.[127]

In 1923 he left the public sector and entered into the electrical engineering industry, working as head engineer for the radio transmission department of E. F. Huth Berlin. After the company was dissolved, he was employed as head engineer for Radio Company Erich & Grätz and Radio-Loewe, both in Berlin.[128] During the early 1920s he was particularly interested in modulation,[129] improving the recording qualities of microphones, and improving the transmission performance of radio, arguing that a successful radio broadcast was predicated on both the frequency and the amplitude of transmission.[130] This was yet another definition of fidelity. In 1925 he published a work on wireless telephony and telegraphy written for the educated layperson.[131]

Meanwhile, the reputation of engineers declined in the public's eyes, as Germany lost the so-called technology-driven war.[132] Engineers actively sought governmental posts, hoping to slash the monopoly held by lawyers in those positions. Engineers insisted that they were needed to reestablish the German economy after the Reich's humiliating defeat. In late March 1919, they scored a major victory, as the Prussian ministry appointed engineer Hans Bredow, the so-called father of German radio, as minister of the Reich's Post Office.[133] Alas, such victories were rare. The Verein Deutscher Ingenieure's report of 1932 complained that governmental officials of the Weimar Republic had rendered engineers irrelevant. One can begin to see how many happily embraced the new future the Nazis promised them.[134]

There was a drastic increase in the number of nonacademic engineers—the darlings of industrial managers because of their cheap labor—at the turn of the century. The Wilhelmine government did its best to bolster *Maschinenbauschulen* (Machine Building Schools), spending more money on them than on the technical universities, where academic engineers were trained.[135] Professional engineers had established a new association in 1909, called the

Verband Deutscher Diplom-Ingenieure(VDDI), to create distance between themselves and nonacademic engineers and technicians. The association of professional engineers came under the leadership of Alois Riedler, the rector of the Technical University of Berlin.[136]

The VDDI was politically conservative: they generally stayed away from the Social Democratic Party.[137] Their leaders embraced "an organic and communal ideology" that was romantic and sought to portray academic engineers as the nation's heroes.[138] The desperate (and ultimately unsuccessful) attempts by academic engineers to fashion themselves as the educated elite, on par with the Bildungsbürgertum, resulted in their anticapitalist sentiments and their hostility to the old political and social order. This facilitated their support of ring-wing ideologies, or what Jeffrey Herf has called "reactionary modernism."[139]

Trautwein's trajectory—a pastor father, a law degree—was typical of the educated upper middle class. His Dr. Ing., however, was not a typical degree of that social class. Clearly, he belonged to the class of highly educated engineers, and he wished to be recognized as a Bildungsbürgertumer.

As he would later recall, he was of the opinion that the development of electric musical instruments would have been unthinkable without radio.[140] While researching radio transmission, he became interested in the creation of electric sounds and electric music.[141] In 1922 he was granted his first patent, for generating vibrations in an electric circuit and using vacuum tubes to amplify the signal; the design was based on a feedback circuit, or one in which the output is routed back as an input to boost amplification.[142]

In 1924 he obtained his second patent, for the use of electric circuits for the production of musical tones possessing certain timbres. Included among the patent claims was a process by which the overtones necessary for the synthesis of sound were first sifted out of a harmonic mixture by means of resonant circuits and then mixed together in definite ratios of amplitudes.[143] In modern terms, Trautwein had created a resonant filter that could produce various tone colors from sounds rich in overtones generated by glow discharge tubes.[144] Trautwein believed that a resonant filter, which eliminated some overtones while allowing others to pass through, could be used to select for the harmonics of each tempered semitone.[145] His technique was based on the use of AC generators, glow discharge tubes, capacitors, resistors, and inductors to imitate human vowel sounds and the tone colors of various musical instruments.[146] This second patent proved critical to his development of the trautonium.

Sometime in 1930, Trautwein was appointed as lecturer in musical acoustics at the RVS of the Berlin Academy, eventually replacing Carl Stumpf's student Alfons Kreichgauer, who last taught the course in July 1930. Shortly thereafter, Trautwein took over the instruction in electric music.[147] In early 1930, as Oskar Sala recalled nearly three decades later, he met Trautwein while the engineer was spending much of his free time during the evenings tinkering with the creation of sounds from electric circuits at the RVS.[148] Trautwein pursued numerous projects relevant to electric music, physiology, and radio and recording engineering during his tenure there. As he himself admitted, his research in electric music "first and foremost covers the problem of the electrical formation of sound . . ."[149] He added: "Second, my research deals with the design of a playing technique that permits all artistic forms of expression and relieves the artist of any unnecessary mechanical work."[150] He not only wanted to reproduce the tone colors of previously existing musical instruments, but also to generate new, "and perhaps musically more valuable sounds."[151] Finally, as part of his teaching duties, Trautwein designed a radio-technical practicum, which instructed students on how to measure and conduct experiments using instruments relevant to radio and sound movies.[152]

In an essay he sent to Schünemann titled "On Acoustical Problems from the Frontiers (*Grenzgebieten*) of Music, Physics, and Physiology," Trautwein shed light on the cross-disciplinary problems he was tackling at the RVS by discussing the current state of broadcasting and sound recording.[153]

> The development of broadcasting and sound recording has created a number of problems, which fall into the domains of music as well as physics and physiology, the solution of which is therefore conveniently taken over jointly by representatives of these fields of research. The technical needs that give rise to the problems also exist in concert music and musical instrument technology . . . [154]

The distortions caused by amplification needed to be corrected.

> The technique of amplification added yet another requirement [of radio broadcasting], namely the preservation of strictly linear relationships in all processes of energy transformation. The nonlinear distortions were already present in the microphones and telephones used in telephony, but were less distracting, particularly if one was only dealing with voice transmission. In the case of amplifiers, however, the nonlinear distortions are particularly strong if one overloads the tubes . . . The ear is very sensitive to such [nonlinear] distortions; therefore, for amplifiers one must basically rule out the use of the nonlinear characteristic part, since the usable part would be so small that it would have no economic advantage. For microphones and loudspeakers, the largest distortion has been established and methods for its measurement have been developed. This determination does not necessarily require a musically trained ear, because the differences are so prominent. With respect to the nonlinear distortions, the question arises about the permissible differences in amplitude. The range, which remains

> between the noise level and the maximum controllable amplitude, is not greater than 1:50 with a normal technical effort, while a musical work requires amplitude differences of about 1:1,000.[155]

Trautwein underscored how the collaborations taking place at the RVS could address those challenges:

> Here is a problem that requires the collaboration of musicians and engineers: the musical judgment of the disadvantages of amplitude restriction, the adaptation of the [musical] composition to this constraint, and the investigation of the laws governing amplitude restriction.[156]

He also was interested in pursuing research on acoustical transients—or ephemeral, high-amplitude sounds that occur at the beginning of a waveform, which, as depicted by oscillograms, differed radically depending on the vowel sound or musical instrument and which he felt contributed to tone color.[157] According to Trautwein, physiologists could provide vital assistance in the areas of electro-technical distortion and fidelity of acoustical transmission.

Another problem that needed to be addressed was the phase relationships between simultaneously occurring frequencies. Electroacoustic transmissions, which were based on correct and precise measurements both of the frequency band and linearity and therefore in theory should have been aurally pleasing, in actuality were sometimes unsatisfactory to the musical ear. As a result, Trautwein concluded that it "can therefore be assumed that the ear takes on further characteristics of the tone [*Klangbild*] than frequency and amplitude."[158] The physiologist needed to investigate that aspect of hearing. He continued:

> Recently, it has been recognized that, in particular, the oscillation and equalization processes of the sound vibrations are important for the physiological sensation of tone. Obviously, one pays special attention to these processes when listening . . . At the Physics Conference in Bad Elster in 1931, these questions were dealt with extensively. Particularly interesting are the reports of [the physicist Hermann] Backhaus, who determined the duration of acoustical transients by means of numerous oscillograms for speech sounds and sounds of musical instruments. The duration of the transient process is very different. It is particularly long with flute sounds and short with vowels like e and i. This explains why a flute tone, which is nearly sinusoidal in the steady-state, gives a completely different sound than that of an electronic audio oscillator.[159]

This interdisciplinarity of the RVS's research program was underscored by Schünemann in his edited series. He planned to publish a collection of written works where the research of the RVS could be made known to educated laypersons.[160] The first (and only) book in the planned series was Trautwein's *Elektrische Musik* of 1930.[161] It was one of the earliest pieces in Germany dedicated to electric musical instruments. Trautwein started off by situating this branch of musical instruments in the scientific context of broadcasting, and he drew informative parallels between electric musical instruments and apparatuses used in radio broadcasts and in electrical measurements.

In an interview from that same year advertising his new instrument, Trautwein explained:

> While electroacoustics has occupied itself in the last few years primarily with the problems of reproduction, I would like to provide new expressive possibilities for the creative musician. Mechanical music [i.e., the gramophone] has not enriched art as such, but for the most part only disseminated it. Above all, I hope through my work to serve creative art and to contribute to the reconciliation of two falsely opposed branches of the human spirit: art and technology.[162]

He echoed that sentiment over two decades later: "Electronics means something very different than mechanization to an active musician. Quite the contrary, electronics leads to the highest artistic intensification. It can solve the great cultural task of rescuing music from its current passivity, if artists, art educators, and engineers stand together."[163]

In Trautwein's view, the goals of electric music, as opposed to mechanical music, were threefold. First, as is seen from the quote above, it was to serve musicians interested in exploring the possibilities afforded by the new technology in generating novel musical sounds and timbres. Second, it was to provide musicians with the highest degree of technical perfection. Finally, it should guarantee that each new technological discovery leads to new musical advances.[164]

While Trautwein could play his trautonium, the leading virtuoso of the instrument was undoubtedly Oskar Sala. Sala was born to an optometrist and a singer on July 18, 1910, in the small, industrialized city of Greiz in the German state of Thuringia, between the city of Leipzig and the border of what is now the Czech Republic. He was a gifted pianist as a child, and in March 1929 celebrated the successful completion of his *Abitur* and went to Berlin in 1929 to study piano.[165] Yet he soon shifted his studies to musical composition. He studied under Hindemith, who in 1930 introduced him to Trautwein, as the composer wanted his students to see the new experiments being conducted at the RVS.[166] As a viola player, Hindemith had been particularly intrigued by the trautonium because it was based on a metal strip that resembled a string rather than a keyboard.[167] That meeting sparked a collaboration. Trautwein and Sala quickly went to work designing the trautonium. Sala concurred with

Trautwein that the greatest achievement of so-called electro-synthesis was that it gave musicians a solo instrument that could fulfill all of their artistic aims.[168]

Berlin radio during the 1920s was the catalyst for coordination between a number of disciplines with their intellectual commitments, toolkits, skills, and practices. It is a story that reflects the vibrant communities that actively engaged with one another to improve radio fidelity. Given the economic challenges of the late 1920s and early 1930s, it is striking that the Prussian Ministry of Culture felt compelled to continue funding the RVS. Radio's role in German culture was deemed too important to succumb to its flaws, regardless of the amount of money needed to improve it. While other companies, such as RCA and Bell Telephone Laboratories, were asking similar questions and solving them along similar lines, Berlin's RVS was unique in that musicians and leading composers worked together with engineers and scientists in a laboratory housed in a musical academy. Music, not speech, took center stage. As a result, and as discussed in the ensuing chapter, Germans were able to construct a new, sophisticated musical aesthetic based on a new genre of musical instruments. In the space and with the expertise provided by the RVS, Friedrich Trautwein was coming to terms with formants; artificially (i.e., electrically) creating vowel and musical-instrument sounds; and inventing the trautonium.

5

The Original Trautonium

ONE FASCINATING technical and cultural product of Weimar radio was electric music.[1] It certainly would not be an exaggeration to say that electric music's origins reside in radio. Much like radio, electric music was experimental. Just how popular was electric music during the Weimar Republic? It turns out, the proverbial jury was still out, as electric music received mixed reviews from critics. Much of the harsher criticisms centered on the role of soulless machines playing music and how this shackled the human spirit. Other opponents simply did not like the futuristic sounds.[2] Others still feared the replacement of traditional instruments with these new instruments, a concern that a number of the inventors vehemently denied. Those who were more impressed with the new genre were lured by its ability to increase the palette of sounds and tone colors.[3] The engineer A. Lion felt that musicians and the lay public "were standing before a miracle" and praised the "brilliant utilization of technical means for musical purposes."[4] Music critic Arno Huth insisted that the electric current used to generate the tones did not "kill aesthetic sensitivity."[5] On the contrary, electric musical instruments permitted aesthetic sensibility to act more directly than ever before. Lion concurred: electricity enabled the "intimate connection of this sound material with the individuality of the performing artist."[6] Some welcomed the new musical instruments as representations of modernity. Others condemned them for precisely that reason.

As we saw in the previous chapter, the trautonium was invented at the Rundfunkversuchsstelle (Radio Experimental Laboratory, or RVS). The skills, techniques, and practices employed to build the trautonium were also used in radio stations. Similarly, the trautonium's components, such as LC circuits, high-low pass filters, vacuum tubes, loudspeakers, thyratrons, and glow-discharge tubes, were also found in those burgeoning broadcast stations.[7] The RVS's oscillographs were used to compare and contrast the sound waves with those generated by acoustical instruments and vowels. Given his industry connections, Friedrich Trautwein was able to procure an oscillograph, which he and Oskar Sala used for the invention.[8] The trautonium was the culmination of cutting-edge research in a discipline that radio created: electroacoustics. The trautonium had two critical attributes: its ability to alter its tone color to imitate conventional acoustical instruments, and its creation of new, futuristic sounds and timbres. The first four chapters explain the context in which the trautonium was invented. The trautonium ties together the themes of timbre, musical aesthetics, formant research, radio-broadcasting fidelity, and political debates about modernity.

One might wish to see the trautonium as a mere precursor to synthesizers, such as Harry F. Olson and Herbert Bekar's Mark I and II of RCA during the 1950s, or the Moog synthesizer of the 1960s and '70s. Sala vehemently denied the similarities between the synthesizer and his mixture trautonium for two reasons.[9] First, he clearly had a financial interest in disambiguating the trautonium from the newer synthesizers. Second, as discussed in chapter 8, he insisted that what he did as a composer and performer could only be executed by a human interacting with an instrument: it could not be replicated by a synthesizer. He continually underscored that the human performer was a critical interpreter of a musical piece, and he did not wish to relinquish a performer's authorship to a machine. I personally do not find asking precursor questions all that helpful historically. While there are a number of important similarities between the trautonium and the more modern synthesizers, I would argue that the trautonium is much more than a mere proto-synthesizer. As we shall see, the trautonium was relevant to physiological theories of vowel formation and was inextricably linked to notions of what constituted sound, tonality, the human voice, and timbre.

Early Electric Musical Instruments

While instruments that generated and distributed music electrically had existed previously, such as Thaddeus Cahill's telharmonium or dynamophone patented back in the 1890s, the late 1920s and early 1930s were when electric musical instruments captured the public's attention and imagination. Perhaps the most famous electric musical instrument was the theremin, invented by Soviet physicist and cellist Lev Sergeyevich Termen (Anglicized to Leon Theremin).[10] Invented around 1920, the theremin's otherworldly and supernatural sound was generated by two battery-powered radio frequency oscillators. One oscillator remained at its particular pitch while the other was variable and determined by the distance of the performer's hand from the instrument's antenna. The closer the hand was to the antenna, the higher the pitch. The instrument enjoyed a range of six octaves. A second antenna, which was oriented perpendicular to the first, controlled the volume. The performer did not actually touch the instrument; rather, they extended their hands in the electric field. Based on the principle of heterodyning oscillators, a radio signal processing technique, new frequencies were generated by mixing two frequencies. The theremin was all the rage during the late 1920s, particularly in Berlin.

Jörg Mager, considered by many to be the "father of German electric music," was an engineer, organist, and inventor of the spherophone.[11] His work exemplified the role that German radio played in the development of electric musical instruments. In his 1924 work *Eine neue Epoche der Musik durch Radio* (*A New Age in Music through Radio*), Mager stressed the importance of radio in the production of music rather than just as a transmitter of it.[12] He wrote, a tad optimistically, in the periodical *Der deutsche Rundfunk* (*German Radio*):

> The music of the future will be attained by radio instruments! Of course, not with radio transmission, but rather direct generation of musical tones by means of cathode [ray tube] instruments! Indeed, cathode music will be far superior to previous music, in that it can generate a much finer, more highly developed, richly colored music than all our known musical instruments.[13]

Being committed to microtonality, he wished to build an instrument that could generate infinitely variable pitches. In 1921 he constructed an "electrophone," later to be called a "spherophone," which employed vacuum tubes.[14] The simple device consisted of a semicircular metal plate and a hand crank positioned above it. A button on the handle connected the circuit and resulted in the device sounding a pitch, which was controlled by the crank. As the performer turned the crank, the capacitance of the circuit was changed, thereby raising or lowering the pitch. The instrument transitioned between tones with a glissando effect: it was therefore microtonal. Like the theremin, the physical principle behind the instrument was heterodyning.

French cellist and radio engineer Maurice Martenot invented and patented the ondes Martenot, or Martenot waves (musical waves) in 1928. Martenot wished to reproduce the tones generated by radio oscillators used by the military during World War I. Used in Olivier Messiaen's "Fête des belles eaux" for six ondes, "Trois petites liturgies de la présence divine," "Saint-François d'Assise," and "Turangalîla-Symphonie," the instrument featured a metal ring that was worn on the right index finger. By gliding the ring along a wire, tones similar to the theremin were generated. A keyboard was added in subsequent designs. The tone color could be altered by moving the left hand. A number of other works for the ondes Martenot were written by leading composers, including Arthur Honegger and Edgard Varèse.[15]

Other electric instruments of the period included the Neo-Bechstein grand piano, which was what we would now call an electro-mechanical instrument; Vierling's electrochord, which was similar to the Neo-Bechstein piano; the hellertion, built by electrical engineer Peter Lertes and pianist Bruno Hellberger, which generated pitches with a ribbon controller that, when depressed. came into contact with an audio oscillator; and, of course, Friedrich Trautwein's trautonium, the most famous of the German electric musical instruments. It is not surprising that these instruments, in their nature as byproducts of the advances in radio equipment, were featured in the annual radio exhibitions taking place throughout Germany.[16] For example, an entire exhibition, titled "Electrical Orchestra," was dedicated to electric instruments at the Radio Exhibition of 1932 in Berlin.[17] The lives of many of these instruments were ephemeral. The trautonium, however, thrived throughout the 1930s until the present and in various cultural, technological, and scientific venues.

Synthesis and Human Hearing

As Trautwein explained in October 1951, the origins of the trautonium dated back to 1921, when the Reich's Ministry of the Post Office tasked him with building a radio station in Berlin. At said radio station, he would begin conducting research that would culminate in his aforementioned patents. Since harmonic analysis produced highly accurate depictions of the relative amplitudes of an instrument's overtones, one might assume that the process is reversible. That is to say that one might simply start out with the constituent overtones with their correct amplitudes and have them blend together to produce the complex wave characteristic of the vowel sound or the sound of a musical instrument. Alas, additive synthesis could never reproduce the desired tone color.

A brief explanation of the properties of the human ear is relevant here. The human ear does not hear linearly. Our hearing is logarithmic; therefore, one cannot simply reverse engineer a sound from a graph of its constituent overtones and their relative amplitudes generated by oscillators as broken down by harmonic analyzers.[18] The scientific apparatus "hears" the sound differently than we do. As the famous Bell Telephone Laboratories acoustician Harvey Fletcher pointed out: "Apparatus and methods are available for making accurate measurements of intensity, frequency, and overtone structure but not for measuring any of the *subjective* characteristics, loudness, pitch, or timbre."[19] Our hearing is much more sensitive in the range of 3000 to 4000 Hz. For example, it was known as early as 1925 that in order to produce a barely audible tone of 16 Hz, one needs to supply an energy level that is one million times greater than to produce a barely audible tone at 512 Hz. Similarly, to generate a barely audible tone of 8192 Hz one needs ten thousand times more energy than that barely audible tone at 512 Hz.[20] As Trautwein himself realized early on in his research on the creation of tone colors, "Due to the logarithmic sensitivity of the ear, even large differences in amplitude are hardly noticeable."[21] The lack of linearity also affects human detection of the overtones and their amplitudes, and therefore the perception of timbre.[22]

There are also examples of the human ear perceiving something that does not exist independently of it, which is to say, it does not physically exit. If two pitches are played simultaneously, our ear can be convinced of a third tone. Sometimes that tone is picked up by a measuring device. Other times, it is not. In the latter case, the phenomenon is known as a subjective (or ghost) tone, as the tone is a product of the human ear. There are also cases in which the ear hears a fundamental tone, which is actually absent, if the dominating overtones occur in the requisite amplitude ratios.

Another reason why physicists and electrical engineers could not simply synthesize tones based on harmonic analysis: timbre is a dynamic process, i.e., it changes over time, and ears detect that change. Trautwein himself admitted that additive synthesis had been attempted many times, and each time was a failure.

> If one now sets the task of creating artificial syntheses of such sounds from the results of the [harmonic] analysis, one will find that a combination of harmonic partials does not yield satisfactory results. Attempts have often been made to let as many tone generators of single-wave tones sound together as overtones in their correct amplitude ratios, which were determined by analysis. Such attempts at imitation did not produce any satisfactory results. The audible impression of many tone generators does not completely melt into a uniform tone, as if there were only one tone generator. Rather, the ear generally recognizes the multitude of sound sources.[23]

He and others suggested that "the ear, confined to resonance effects, pays particular attention to transient processes (*Ausgleichsvorgängen*) for recognizing vocalizations and timbres."[24] According to Trautwein, Backhaus, who studied the harmonics of notes of several violins, criticized Helmholtz, Stumpf, and Wagner for limiting their analyses of the sounds generated by musical instruments to static depictions rather than offering a "*mechanism of the production of sound*."[25]

Backhaus claimed that physicists and engineers could not build electric musical instruments that could reproduce the tone colors of musical instruments because of the inability to recreate the onset of the tone (the acoustical or attack transients) as well as the subsequent modifications of the tone color, from the decay to the sustaining of the tone. Trautwein responded to this claim in a lecture at the Society for Technical Physics in September 1930 in Königsberg (now Kaliningrad, Russia) by stating that his trautonium reproduced precisely that metamorphosis.[26] What is fascinating about Trautwein's research on tone color is that he was actually working on what we now call the envelope of sound.[27] He recreated electronically what is now referred to as the attack, decay, and release of the acoustic envelope. The sustaining phase generally refers to the volume. Each of those phases contain different timbres, and if one wished to mimic an instrument, one needed to reproduce those timbres at the correct time in the envelope. It seems that the trautonium is the earliest instrument built with the acoustical envelope in mind.

The Instrument

Back in 1930, there were two ways to generate electric vibrations that were rich in overtones. One was to create sinusoidal vibrations by means of tube transmitters or alternating current generators. The second was to produce distortions of sinusoidal curves, similar to those produced by frequency multiplication. This could be done through the use of extremely distorted vibrations produced by glow-discharge tubes filled with neon. Trautwein chose the second method, as many more changes in timbre were possible using a glow-discharge tube than using sinusoidal vibrations of AC generators or tube transmitters.[28]

Trautwein was convinced that the replication of musical and language sounds could only be satisfactorily achieved if one considered the physical notion of impulse (or shock) excitation (*Stoßerregung*) rather than reverse engineer the tone from its analytically ascertained components.[29] He writes: "Electrical impulses of short individual duration and periodic succession can be generated by discharges in inert gas tubes."[30]

The original trautonium comprised a wire placed over a long metallic rail, which produced a sound when it was pressed.[31] A glow-discharge (neon) tube circuit[32] generated sawtooth vibrations (*Kippschwingungen*) whose frequencies primarily depended on the time constant of the resistance used in the metal strip and the capacitor used in the circuit.[33] Trautwein held the capacitance constant with a varying resistance, thereby changing the frequency of the oscillation: varying the time constant that resulted from the changes in the resistance altered the frequency. The frequency was also dependent upon the voltage. The vibrations produced by the glow-tube circuit were then sent through resonant filter circuits. These circuits blocked out certain frequencies and allowed others to pass through. The timbres the instrument could generate were created by adding together those filtered frequencies, or what is referred to as subtractive synthesis.

The wire was user-friendly. The pressure of the forefinger caused the wire to come in contact with the resistance strip, resulting in a tone, the volume of which depended on the pressure of the finger.[34] A foot pedal was added to

FIGURE 5.1. Friedrich Trautwein playing his trautonium, 1930. *Source:* Kestenberg, ed., *Kunst und Technik.*

increase the volume more substantially. Musical intervals needed to correspond to the same length on the wire. This could be achieved by varying the resistance distribution. On the recommendation of Schünemann, the trautonium was independent of a particular temperament and therefore could be used to perform microtonal pieces or ones not limited to the traditional twelve semitones of the octave; hence, a wire was used instead of a keyboard. It was able to produce vibratos and glissandos, just as a number of string instruments could.[35] One changed the formants on the trautonium by using rotary capacitors or potentiometers located on the instrument's front panel. Additional capacitors enabled the performer to smoothly adjust the numerous formants thereby changing the timbre to recreate numerous orchestral instruments. The performer simply needed to flip a switch or turn a dial in order to change the timbre.[36]

Radio's influence on the trautonium was clear. Trautwein concluded his *Elektrische Musik*: "The same principles apply to the construction and design of the trautonium as to the construction of radio receivers . . . One can see that the design and shape are based entirely on the radio receiver."[37] He even recommended the use of the trautonium as a radio receiver.[38] The imperfections of the radio equipment needed to be ameliorated. Another key similarity was the

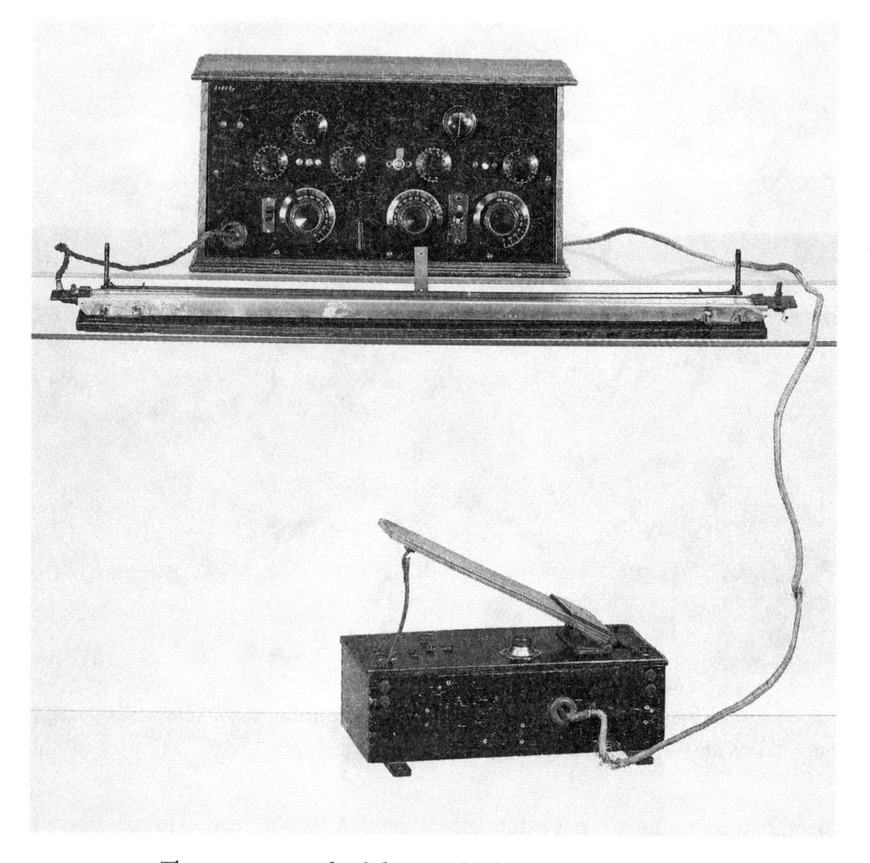

FIGURE 5.2. The trautonium, built by Friedrich Trautwein with the assistance of Oskar Sala, 1930. Reproduced with the permission of the Deutsches Museum (Munich) Archives.

absence of a microphone. As we saw in the previous chapter, microphones were problematic. When it came to the invention of musical instruments, Trautwein felt it best to circumvent microphones all together. Microphones were rendered superfluous with the trautonium, as the alternating current could be used directly for transmitter modulation. This is to say that the trautonium played directly into the modulator circuit.[39] Engineer Joachim Winkelmann explained how the trautonium could be used as a radio receiver and gramophone amplifier.[40] An advertisement for the instrument queried its readers:

> Do you know this new radio—universal—musical instrument? The trautonium, an invention of Dr. Trautwein, is a universal musical instrument based on the principles of the radio, on which you can play or play back all known

> musical instruments from the timpani to the flute yourself. In addition, the trautonium is a radio receiver and gramophone amplifier at the same time.[41]

While the original trautonium was monophonic—i.e., it could only play one part—Trautwein envisaged a multipart version from the beginning.[42] A multipart performance by one player could be achieved by laying several wires next to one another. He also conceived of a multipart trautonium wherein one played the instrument as if it were an organ, albeit in various musical intervals other than semitones. The circuit could be constructed such that ten voices were provided, one voice for each finger.[43]

Through his collaboration with Schünemann and other musicians at the RVS, Trautwein built an electric musical instrument that could produce various sounds, including but not limited to musical sounds. Reminiscing many years later about his initial collaboration with Trautwein, composer and trautonium virtuoso Sala recounted the story of how one day they began experimenting with an apparatus that would eventually become the trautonium. After initially generating a "peeping sound," they were able to generate vowel sounds:

> To get hold of such a bar, to attach a strip to it and to switch on a small glow tube behind it and then it needs an amplifying tube, of course, and when you played it like that, it made a peep and it just produced a sound, and that was at least a chance for some demonstration . . . And we [Trautwein and Sala] tried out many beautiful things there. He once got there with a thick transformer and said, I've brought this along, now let's connect it up and then we'll add a capacitor, just a small one, and all of a sudden we were amazed to hear: u, o, a, o, u, oa, oa. We were speechless, he [Trautwein], too, of course. Naturally, he had thought about it. That it ought to work. But then when it does work, you exclaim "wow!" You might say that was the revelation for every one of us. We now had a range of tone colors. . . . [44]

Sala repeated his account of the early trautonium's vowel synthesis, which actually preceded the generation of musical sounds. It sounded "as if someone was speaking vowels with a deep or middle-range voice, in various pitches. Suddenly we heard 'vau vau' and 'miaow.'"[45] Immediately thereafter, Hindemith joined Trautwein and Sala, and they were able to generate musical notes with the electrical contraption. Sala was struck by the similarity between certain vowel sounds and those generated by certain musical instruments, for example, the vowel U and a bassoon, O and a clarinet, A and an English horn, E and the oboe, several shades of the vowel A and the saxophone, the mixture of U and E and the cello, and the mixture of A and I and the violin.[46]

Many years later, Sala recalled Trautwein's response when the trautonium successfully produced vowel sounds.

FIGURE 5.3. Friedrich Trautwein (standing on the left), Paul Hindemith (standing on the right), Oskar Sala (seated at the trautonium), and an unknown violinist in the Rundfunkversuchsstelle (RVS, or the Radio Experimental Laboratory), around 1930. Reproduced with the permission of the Deutsches Museum (Munich) Archives, PT_14055.

> We electronically imitated the natural generation of vowels. Our glottis in the larynx provides us with the highly variable sawtooth vibrations, and the oral cavity corresponds to the electrical resonance circuit composed of a coil and capacitor. According to Carl Stumpf's formant theory, I calculated the values for self-induction and capacitance for the necessary resonance frequencies.[47]

Alas, Stumpf's theory in the end would prove impractical for the workings of the trautonium, as the transformer needed to be extremely large and the connected wires particularly thick. As Sala conceded:

> The vowel sounds were just a byproduct for us, of course. But now we had timbres, because every position of the capacitor delivered a different one, from u-like to i-like and beyond—even more that are outside the language range.[48]

Schünemann wrote in 1931 that the trautonium was "the richest monophonic instrument of our times." It superseded all other electric musical instruments, and in his opinion "it is the only true musical instrument that one needs to play in order to generate the change in tone color and the variation of the human voice."[49] Trautwein also underscored the importance of the trautonium to alter its tone color:

> Because of its impressive expressive abilities, the trautonium would deserve to be regarded as a remarkable advance in musical instrument construction, even if it were limited to one of its beautiful timbres. Music has long had a need for many and varied timbres, but up until now there has not been an instrument that combines a large number of timbres and allows for a wide range of gradual shifts in timbres. So far, only the "either-or" of the timbre was possible. Before electromusic, gradual changes in timbre were known only to a limited extent.[50]

In his influential work, *Elektrische Musik* of 1933, electrical engineer Peter Lertes, the co-inventor of two electric musical instruments—namely the hellertion and the heliophon—argued that the various timbres produced by the trautonium were superior to all other electric instruments of the period.[51] On January 27, 1933, Schünemann wrote to Emil Mayer, director of Telefunken, concurring with Lertes: musicians agreed that the trautonium was the most versatile and interesting of the electric musical instruments.[52] Schünemann assisted in increasing the instrument's publicity, writing that "one of all musicians' long-standing dreams has now been fulfilled: we have an instrument that corresponds to all of our musical wishes, as it can be employed and changed in numerous ways thus uniting the advantages of numerous musical instruments."[53] Walter Germann, Telefunken's engineer who collaborated with Trautwein on the building of a later version of the instrument, the *Volkstrautonium*—discussed in detail in the next chapter—also commented on the various timbres of trautoniums in general, and the Volkstrautonium in particular: "The player has completely at his control the variation of the timbre of the instrument within wide boundaries. He can, for example, give the tones of the instrument the timbre of a violin, 'flute-cello' or bassoon. Herein lies the great advantage of the trautonium over all other available musical instruments."[54]

While Telefunken heavily subsidized the Volkstrautonium, the earliest trautoniums were financially supported by the Prussian Ministry of Culture with assistance from Telefunken and AEG, specifically its Klangfilm GmbH, where AEG engineers collaborated with colleagues from Siemens & Halske AG on sound-film technology from 1930 to 1932.[55] Such subsidies were necessary since the resources needed were substantial. Musician Leo Kestenberg, an influential member of the Prussian Ministry of Culture and RVS board member, defended Trautwein's research costs incurred by the RVS. He stated that Trautwein's research fit squarely within the framework of the RVS and that musicians could generate some valuable ideas from the experiments.[56]

A number of composers, including Igor Stravinsky, Ernst Krenek, Paul Hindemith, and Alois Hába, told Schünemann that they hoped to incorporate the trautonium in their compositions.[57] Hába, for example, was most likely interested in using the trautonium for his microtonal compositions. Even Arnold Schoenberg stopped by the RVS to have a look at the trautonium. He encouraged Sala to increase the range of the instrument to match the piano's.[58]

Formants and the Trautonium

Trautwein emphasized that the trautonium was the instantiation of his physico-physiological work on Hallformanten, or resonance formants (also translated as reverberating formants or tone-formers), which were based on Ludimar Hermann's theory of formant formation discussed in chapter 3.

> My Hallformanten hypothesis is very similar to the Hermann's pulse excitation theory (*Stoßerregungstheorie*). The idea of self-oscillating structures, which sometimes have numerous natural frequencies, provides such a simple and clear understanding of the mechanism of sound generation that it would be worthwhile to base further physical and musical investigations on it.[59]

As we saw in the previous chapter, Trautwein's research on formants at the RVS was primarily dedicated to researching the electrical production and transmission of musical sounds. Formants, or the frequency bands of the sound spectrum where the overtones determine the tone color, are the tangible link in forming a theory of tone color, speech sounds, and electric musical instruments. Recall that Helmholtz defined tone color in terms of the ratios of the amplitudes of the harmonics of the fundamental pitch. Hermann, on the other hand, believed that the characteristic tone of each vowel sound did not need to be harmonic with respect to the fundamental tone: inharmonic overtones also played a role. Intermittent pulses of air produced by the glottis, exiting the larynx and then entering the oral cavity, initiate the generation of eigentones—the characteristic tones produced by a vibrating body. In this case, the tones are acoustical transients that decay and are subsequently damped. This process, referred to as *Stoßerregung*, or impulse (or shock) excitation, uses an ephemeral oscillatory discharge in a primary circuit just long enough so that the greatest amount of energy is transferred from the primary to the secondary circuit: a damped alternating current is generated in which the duration of the impressed voltage is short when compared with the duration of the generated current. This method had been used for German wireless telegraphy by Telefunken for over fifteen years.[60] The resulting damped frequencies were the formants. Trautwein referred to them as Hallformanten. Sala believed that "the overtone theory [of Helmholtz] then fell out of favor with us. Their formulas were utterly useless for the practical construction [of the trautonium]."[61] Siding with Hermann, Trautwein also thought that inharmonic frequencies were important to timbre.[62]

> Almost two years ago [in 1930], I expressed the assumption that the formants are essentially decisive for the impression of physiological sound. The formant concept has already been established by Helmholtz, but in the sense of preferred harmonic regions of the continuous vibration state. In order to differentiate my idea from that of Helmholtz, I chose the term "resonance formants." This expression is from the musical observation of vowels and instrumental tones that clearly correspond to the mostly inharmonious formants . . . [63]

While Helmholtz was correct in identifying the importance of overtones to timbre, according to Trautwein, his theory corresponded more to a mathematical presentation of a particular musical tone by means of a Fourier series rather than the actual physical processes that occur in musical instruments, particularly the trautonium. Hermann's theory was more applicable to musical instrument making. Helmholtz's theory also led to the misconception that the same ratios of the fundamental tone and overtones result in the same characteristic timbre even with different fundamental tones.[64] Hence, Hermann's theory was critical to the way in which Trautwein conceptualized the physical theory of and carried out the electrical work relevant to his invention.

The main circuit for the "physico-physiological fundamental experiment" formed the basis of his trautonium.[65] Referring to the Figure 5.4, a resonant circuit S, whose natural frequency (eigenfrequency) lay between 400 and 4000 Hz, was triggered by an interrupted alternating current, for example, by an interrupter consisting of a glow tube (1), a resistor (2) of the order of 1 megohm, and a capacitor (3) of a few thousand centimeters. The capacitor (3) could either be connected in parallel to the glow tube (1) or to the resistor (2). The current impulses were transmitted to the circuit S through an amplifier tube. In order to lower the damping of the circuit and make it adjustable, the circuit was fed back into the grid of the tube with the aid of a capacitor (5). The resistor (6) of about 1 megohm was used to keep the coupling of the circuit S with the interrupter formed by the feedback capacitor (5) as low as possible. The damping resistor (7) altered the formants thereby changing the timbre.

The oscillations of the resonant circuit S were fed to the loudspeaker via an amplifier. The frequency of the interrupter was set by selecting a large resistor (2) and a large capacitor (3) such that the pitch was just below the auditory range. One clearly heard a waveform fading away, similar to the periodic

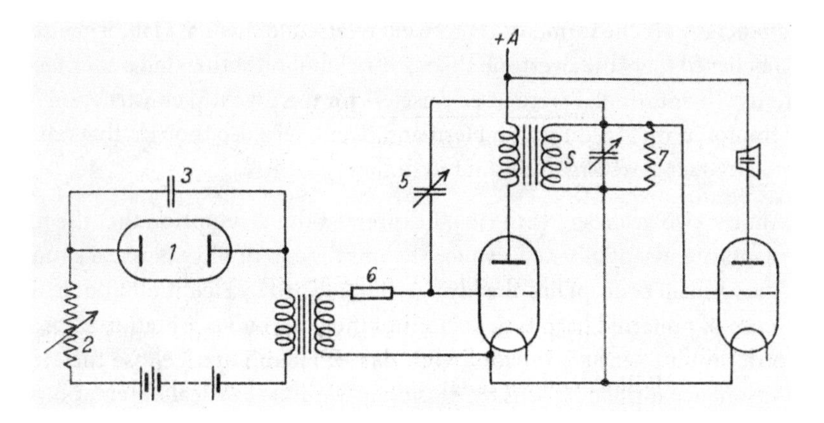

FIGURE 5.4. The circuit for Trautwein's "physico-physiological fundamental experiment." *Source:* Trautwein, *Elektrische Musik,* 18.

striking of a bell. The decay time of the wave trains could be adjusted with the help of feedback, and, depending on the setting, one could get slowly decaying bell tones similar to sounds produced by a xylophone. Vibrations that faded quickly were reminiscent of the drums in a jazz band. If one increased the excitation frequency in the auditory range, starting with impulses that decay relatively quickly by reducing resistance (2) and capacitance (3), an interesting physiological phenomenon occurred. The ear no longer heard the eigentone of the resonant circuit S, but it now perceived the excitation frequency as the fundamental tone of a pitch with a particular tone color. Depending on the choice of the frequency of the Hallformant, the ear perceived a different timbre. A low Hallformant produced a dull, bassoon-like timbre. A Hallformant in a medium pitch range created timbres similar to the clarinet, while a high-end Hallformant corresponded to sharp tones resembling a trumpet's sound. As expected, the impression of a pronounced tone color disappeared at high pitches due to the low degree of sensitivity of the human ear. The tone-coloring effect of formants at very high pitches was also very low. Trautwein's experiments were therefore particularly effective with the fundamental tones of the middle and lower registers. The glow-tube breaker oscillation generated mostly violin- or cello-like timbres even without the aid of the resonant circuit S, most likely due to the presence of formants (that determine the violin's timbre) in the resonating membrane of the loudspeaker.[66] When Trautwein was able to produce numerous vowel sounds during his early experimentation with the trautonium in 1930, he felt as if his theory of vowel production had been vindicated: "We have electrically imitated the natural production of vowels."[67]

Unlike L. Hermann, Trautwein applied the impulse excitation theory to musical instruments, not just vowel sounds.

> Musical timbre is mainly caused by one or more Hallformanten, which are mixed with the fundamental tone. Hallformanten are damped trains of oscillations possessing a particular frequency. They are always higher than the frequency of the fundamental, and which can be inharmonic with respect to the fundament. The Hallformanten always fade away during the course of the fundamental period or are extinguished by the beginning of the next period. . . . The Hallformanten are usually triggered by one or more discontinuities during each period of the fundamental oscillation.[68]

Trautwein did not think that formants originated via resonance of overtones enhanced in the oral cavity as Helmholtz had claimed some seventy years earlier. Rather he believed that the impulse excitation frequency generated by the glottis, which served as a valve for the air traveling from the lungs to the mouth, initiated acoustical (or attack) transients (or the eigentones) in the oral cavity that decayed. He stressed that the physical process of the production of sound by musical instruments was "of a very related nature" (*ganz verwandeter Natur*) to L. Hermann's observations of the processes of the human vocal apparatus.[69] For example, in the case of musical instruments, resonators (which were sent into motion by impulse excitations) consisted of sounding bodies: wooden bodies in the case of string instruments and tubular wooden or metallic structures in the case of wind instruments.[70] The effects of the glottis, the chambers of the oral cavity, and the material and shape of instruments could be artificially generated by coupling electric circuits.[71]

> The scientific significance lies in the physico-physiological impression of the synthetically generated sounds [compared] with the timbres of numerous musical instruments and speech sounds. This suggests that the physical processes are related in many cases.[72]

In short, he was convinced that "a large group of acoustical phenomena can find their physical explanation through the coincidence of impulse (or shock) excitation and the phenomenon of resonance of coupled systems."[73] Some early reviews of the instrument discussed this critical scientific theory that formed the basis of the trautonium.[74] As Tresch and Dolan have so compellingly shown, a look at scientific and musical instruments shows us the historically contingent relationships between science and music throughout the centuries.[75] In this case, we have an instrument that belongs to both knowledge domains.

Referring to Figure 5.6, row c shows the basic pattern of an oscillation that was generated according to the formant theory in a Fourier series. Rows a and b are the oscillation's composite parts. The excitation frequency has a

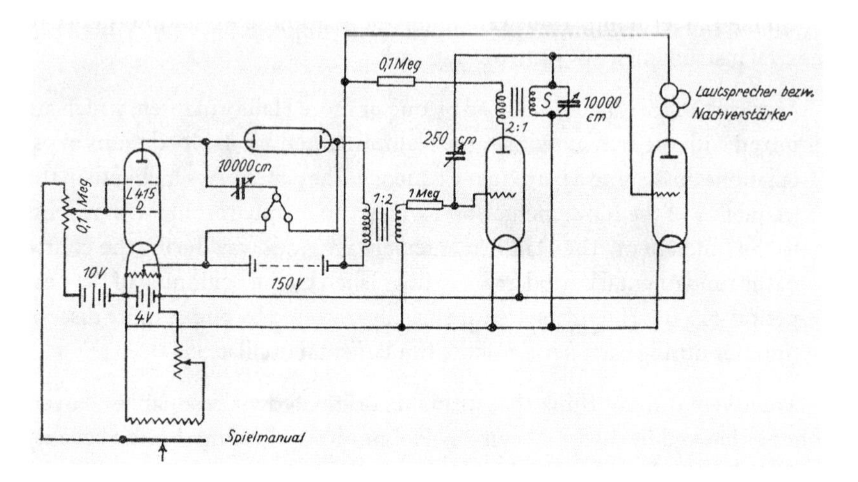

FIGURE 5.5. The circuit diagram for a monophonic trautonium.
Source: Trautwein, *Elektrische Musik*, 29.

discontinuity in each period. This discontinuity affects a structure that is capable of self-oscillation in such a way that the impulse excitation gives rise to a damped vibration train whose decay time is determined by the reduction in the damping of the structure in row b. The process is similar to the oscillations that occur when an electrical oscillating circuit is excited by interrupters.

The oscillation process in row c is generated by a galvanic mixture of interrupter excitation oscillations and Hallformanten. In many cases—such as speech processes and acoustic musical instruments as well as electrical synthesis—it is not the interrupting oscillation process that triggers the Hallformant directly, but rather the Hallformant is formed through the mediation of magnetic, capacitive, or elastic coupling. The secondary process of an interrupted oscillation has the form depicted in row d. Assuming such an exciter oscillation, an oscillation pattern (or mode), depicted in row e, is created that is similar to the damped vibrations well known in radio technology. This mode of oscillation can be viewed as a basic type of musical timbre. In the case of the oscillation modes of acoustical instruments, one often needs to assume a mixture of primary and secondary forms of impulse excitation. Trautwein demonstrated pictorially the link between synthetically generated sounds, musical timbres, and speech sounds. He turned to oscillograms generated by oscillographs to confirm the analogy.[76]

Specifically, he compared the oscillograms of the vibrations of his trautonium with those of a bowed violin taken by O. Krigar-Menzel and A. Raps[77] and the oscillograms of speech sounds provided by Ferdinand Trendelenburg, acoustical engineer at Siemens & Halske. Their similarities clearly illustrated the point of discontinuity and the beginning of the damped vibration characteristic

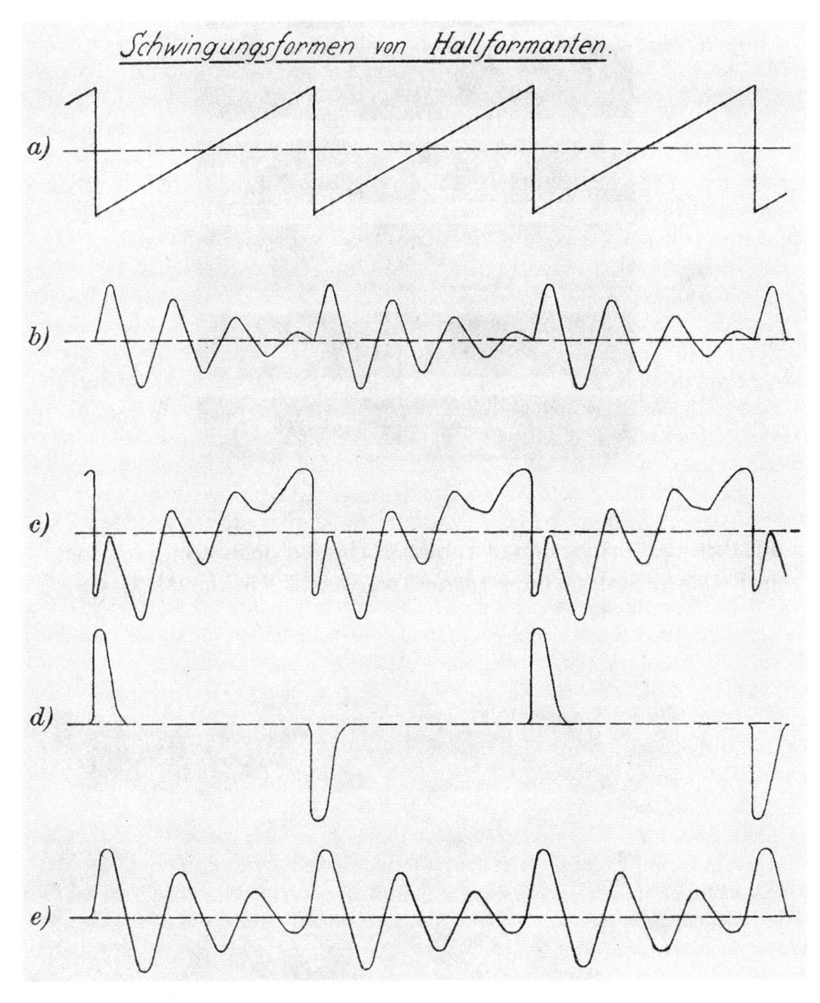

FIGURE 5.6. Oscillation modes (or patterns) of the *Hallformanten*, or resonance formants. *Source:* Trautwein, *Elektrische Musik*, 14. Rows a and b depict the oscillation's composite parts. Row c represents the basic pattern of an oscillation generated according to the formant theory in a Fourier series. The secondary process of an interrupted oscillation has the pattern depicted in row d. Row e illustrates an oscillation pattern similar to those of the damped oscillations well known in high-frequency (radio) technology.

of Trautwein's Hallformant.[78] His analysis of the oscillograms confirmed that speech sounds, tone colors produced by musical instruments, and galvanic circuits were all characterized by formants.[79] Situating his work firmly within the tradition of L. Hermann's formant research, Trautwein was able to create new timbres.[80] Much like his colleagues at the RVS, he drew upon the previous work

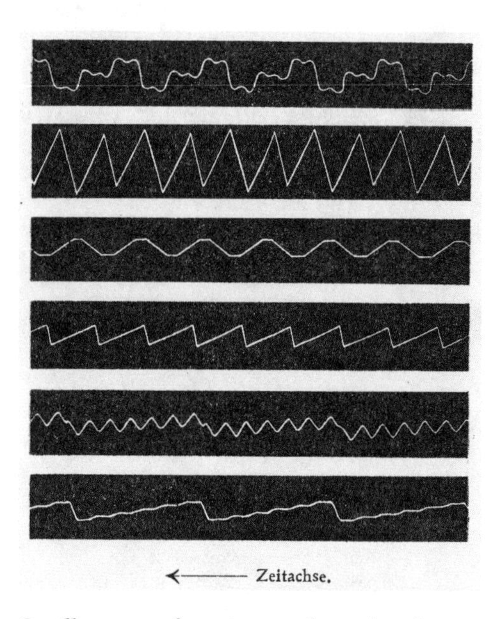

FIGURE 5.7. Oscillograms of a point on a bowed violin string produced by Krigar-Menzel and Raps. *Source:* Trautwein, *Elektrische Musik*, 15.

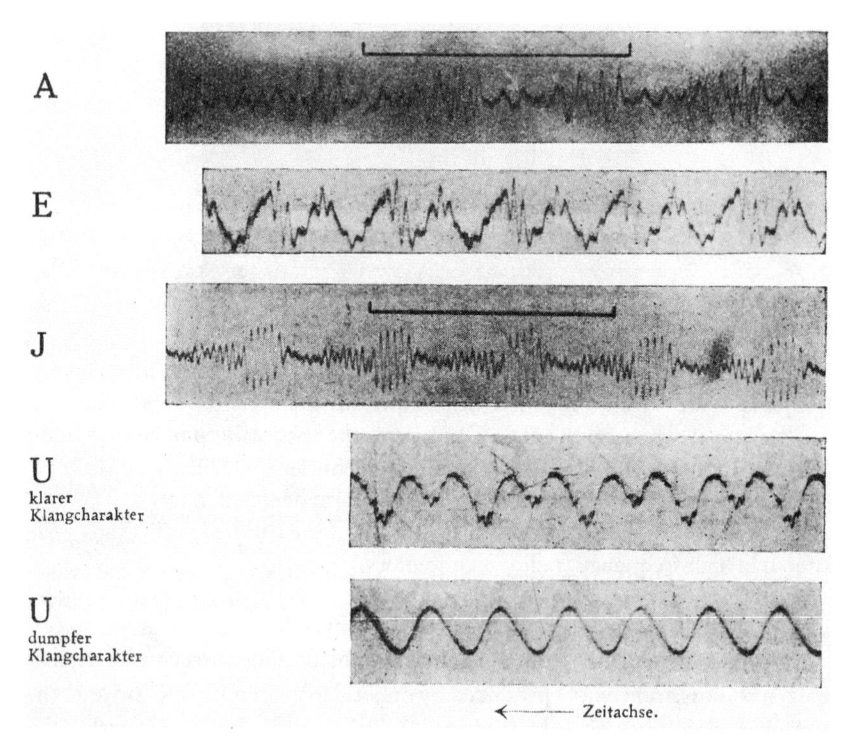

FIGURE 5.8. Oscillograms of vowels A, E, I, clear U, and dull U produced by F. Trendelenburg. *Source:* Trautwein, *Elektrische Musik*, 16.

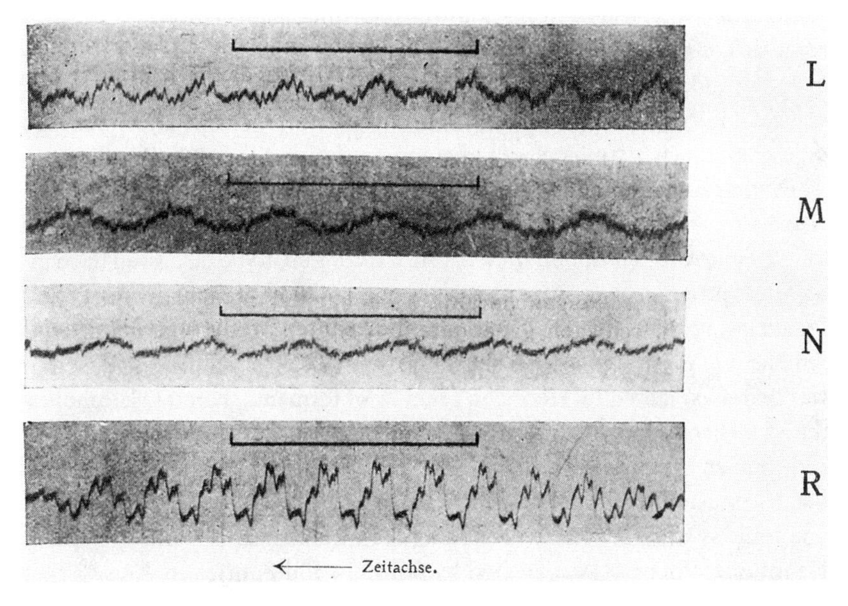

FIGURE 5.9. Oscillograms of voiced consonants L, M, N, and R produced by F. Trendelenburg. *Source:* Trautwein, *Elektrische Musik*, 17.

of not only Helmholtz and Hermann, but also Carl Stumpf, Dayton Clarence Miller, and Karl Willy Wagner on generating partials. He also relied on Erwin Meyer's work on the use of electroacoustic analytical techniques to investigate the sound spectra generated by musical instruments.[81] He credited his contribution to the theory of formants to his collaboration with musicians, particularly Schünemann and Sala.[82]

Trautwein based his electric musical instrument on the theories explaining the physics and physiology of human speech: "The trautonium is an electrical analogy of the mechanism of sound creation of musical instruments and the human vocal organ."[83] He was reported as insisting back in 1930 that "we're imitating the way natural vowel sounds are produced. The glottis is a flexible multivibrator, and the oral cavity corresponds to the resonant circuit consisting of the coil and the capacitor."[84] Psychologist and musicologist Carl Stumpf, who spent much time and effort thinking about timbre, visited the RVS and was reportedly shocked by the trautonium's creation of vowels and animal sounds: "But that can't be! You can't possibly incorporate so many overtones in such a small box [the trautonium]."[85] Electric circuits could obviously generate more overtones than tuning forks.

In September 1937 Trautwein delivered a lecture at the General Voice Congress in Paris reiterating the importance of physiology to his research. He summarized what had been well known at the time, namely that the frequency ranges of consonants were considerably higher in pitch than the frequency ranges of vowels and that consonants possessed inharmonic frequencies. Furthermore, vowels, consonants, and musical instruments could be characterized through different frequency bands as well: these bands were the formants.[86]

Even as late as 1949 Sala was still convinced that L. Hermann and Trautwein's views of formants and tone color production were correct. He claimed that his music for *Faust I* on the concert trautonium, to be discussed in chapter 7, was proof positive of Hermann's impulse excitation theory of tone color formation and a total refutation of Helmholtz's resonance theory of timbre.[87] Similarly, in 1955, Trautwein still thought that the physical process of sound was better explained by Hermann's theory of formants than by Helmholtz's theory of harmonics.[88]

In the end, Trautwein's *Hallformantentheorie* did not hold up to scrutiny. Electric musical instrument maker and engineer Oskar Vierling showed that one could synthesize sounds (including Trautwein's Hallformanten) by using Helmholtz's theory (as corrected by Stumpf) and Fourier analysis.[89] That Hallformanten could also be created by a noise generator was also a problem: in this case, Hallformanten were formed not as a result of a voltage impulse, but rather from resonance from a continuous sound spectrum.[90] Similarly, Ferdinand Trendelenburg declared a year earlier that Helmholtz's theory held up to numerous physical observations.[91]

Trautwein stressed that his experiments on the electrical synthesis of vocal sounds and musical instruments at the RVS were of epistemological worth since they confirmed earlier hypotheses on the nature of how speech sounds and musical sounds were generated. They offered precise measurements for physical and physiological theories of speech and provided valuable clues for the pronunciation of vocals while singing.[92] In addition, his trautonium was intended to produce new musical possibilities, to change tone colors at a rapid pace, to create new sounds and tone colors, and to expand of the boundaries of musical dynamics.[93]

The Trautonium's Debut

On the evening of June 20, 1930, as part of the Neue Musik Berlin 1930 festival, three trautoniums premiered at the Berlin Academy of Music.[94] Trautwein provided the introductory lecture, "Technische Grundlagen elektrischer Musik" ("Technical Principles of Electric Music"), followed by the playing of "Sieben Stücke für drei Trautonien" ("Seven Trio for Three Trautoniums") composed by Hindemith, who played the upper voice part. Sala played the middle voice part, and Berlin Academy professor of piano Rudolf Schmidt

FIGURE 5.10. Paul Hindemith piece, "Langsam," from "Sieben Stücke für Drei Trautonien" ("Seven Pieces for Three Trautoniums"), premiering at the Neue Musik Berlin 1930 festival on June 20 at the Rundfunkversuchsstelle of the Berlin Academy of Music. "Experiment of a composition for Dr. Trautwein's electric music instrument. Paul Hindemith." *Source:* Trautwein, *Elektrische Musik*, 38–39.

played the bass voice part. Each trautonium had a slightly different tone color, and several loudspeakers were placed above a chandelier in the middle of the hall to generate a "sophisticated blending" of the three parts.[95] Hindemith viewed his compositions as experiments in line with the ideology of the Neue Sachlichkeit. Apparently, Trautwein only briefly mentioned at the festival the trautonium's ability to reproduce vowel sounds.[96]

As mentioned in the introduction, the reviews of the June 1930 concert were mixed. Engineer A. Lion was impressed by the range of the pitches and their dynamics. Of particular interest were tone colors, not only those of which recreated the more traditional instruments, but also new ones.[97] This is an important point. From the beginning, Trautwein and Sala insisted that the instrument could create "unlimited possibilities in the choice and combination of new tone colors."[98] Trautwein continued:

> Our view up to now was based on the few instruments which, through mechanical construction, were picked out of the infinite range of timbres.

> The electro-synthesis gives the possibility to generate all conceivable timbres in a continuous transformation, not only to match the characters of all instruments, but also form completely new, specific color values (for example, changing a tone from one timbre to the other).[99]

Sala agreed, declaring that this instrument should not be viewed as a surrogate for another musical instrument. Rather, the goal was "to realize the impossible and to improve upon imperfection."[100] Creativity was therefore a vital characteristic.

Lothar Band, editor of the journal *Funk*, commented on the instrument's ability to imitate numerous timbres. He was also struck by how one could smoothly and effortlessly transition from one timbre to another and was impressed by the range of tone colors that could be reproduced, specifically mentioning the tuba, clarinet, English horn, oboe, flute, and organ.[101] Band's review appeared on the front page of *Funk*, pointing to the importance of the instrument. While the trautonium came closest to recreating the tone color of the family of woodwinds and the main register of an organ, Band felt that the generated tone colors were not precisely identical to the original instruments. The trautonium's tones possessed their "own, special" timbres.[102] A reviewer for *World-Radio* was equally enthusiastic:

> Of outstanding interest was the demonstration of Electrical Music by Dr. Friedrich Trautwein, of the Broadcasting Experimental Station. In this instrument the tone was produced by electrical means. Its wealth and variety was [sic] astonishing, and it is not too much to say that the music lover will shortly have at his disposal an instrument which he can play himself with a minimum of practice and with unlimited scope of range and tone colour.[103]

Walter Abendroth, a Romantic and conservative German composer, on the other hand, was disappointed, calling the performance "unsatisfactorily constructed." He said Trautwein's lecture was "unintelligibly spoken with poor illustrations and insufficiently illustrated."[104] He labeled the instrument a "laboratory joke" (*Laboratoriumswitz*) not worthy of a public performance.[105] Fritz Stege, who would go on to become an important music critic during the Third Reich, agreed, writing that Trautwein's lecture was boring and that the inferior illustrations did not help the situation. Hindemith's playing of his original compositions was also deemed unimpressive, although Stege did not know if this was a result of the composition or the instruments themselves. Nevertheless, Stege thought that the trautonium was "without a doubt astonishing" and was superior to the earlier electric instruments invented by Theremin and Mager.[106] He was also impressed by the instrument's ability to vocalize vowel sounds. A reviewer for the *Deutsche Allgemeine Zeitung* was also disappointed, noting that few audience members, of which there were apparently no electrical engineers or technicians, could understand Trautwein's introduction. As a result, the concert was "flawed."[107] Given the biting sarcasm of the Berliners, a rather funny joke was circulating among musicians in the capital: "Why is it called a trautonium? Because a certain amount of daring [*Traute*] was necessary to play the thing."[108] Technical difficulties and discordant tones plagued the early performances. Taking a different stance, music critic Karl Holl wrote in the *Frankfurter Zeitung* that the age of electroacoustic instruments had arrived. Echoing that sentiment, Berlin newspaper *Morgenpost* referred to the trautonium as "the sensation of the Neue Musik Berlin."[109] An article in *Melos*, a Berlin musical periodical, commented on Trautwein's ability to produce a metamorphosing timbre for a single tone. His instrument was an enrichment to electric music.[110]

In June 1931 Hindemith composed a piece for trautonium with string-instrument accompaniment, "Konzert für Trautonium und Streichorchester" ("Concerto for Trautonium and String Orchestra"). It debuted at the Munich Radio Music Conference in July and was directed by Hindemith himself. Subsequently, the piece was played at a concert in the renowned Singakademie in Berlin under the direction of Hans Rosbaud for the Funk-Stunde Berlin.[111] The instrument had not changed all that much from the one that debuted a year earlier. The reviews were once again mixed. One skeptical reviewer noted that the instrument was

> not dependable. The tone was often unclean and, despite the variability of sounds produced, hard and rigid. The matter is certainly not hopeless, but at the moment it is not yet ripe for artistic achievements . . . Oskar Sala played, turned, and screwed the keys and knobs of the trautonium, as a knowledgeable radio electrician must do, certainly putting the invention in the best electrical light.[112]

Another review of the same performance of Hindemith's piece had a different take. "The instrument has gained significantly in intonational purity and softness of sound and now produces all sorts of flute-like, saxophone-like, tuba-like, and even string-like colors."[113] Aesthetic issues were imbricated with technological issues.

The instrument slowly gained in popularity worldwide. Originally, the instrument appealed to string players, as they possessed the requisite performance skills. As time went on, later versions created by Sala—discussed in the next chapter—were more suited to keyboard players, particularly organists, as intervals between keys were marked by auxiliary keys, which one depressed in order to make contact with the wire. Later versions also possessed foot pedals.

Joachim Winkelmann wrote a how-to manual for radio hobbyists on building a trautonium on their own.[114] Similar instructions on building a trautonium were featured in leading German radio periodicals, including *Funk* and *Funk Bastler*.[115] American composer Henry Cowell, a pianist and co-inventor of the rhythmicon with Theremin, was interested in purchasing a trautonium for experimental use and for concerts at Manhattan's New School of Social Research.[116] In 1933 *Radio-Craft* introduced the trautonium to amateur radio fans in the United States and explained how to build one.[117] Luxemburg-American inventor and science fiction author Hugo Gernsback, the founder of the Wireless Association of America, kept Americans abreast of electric musical instruments that were created from technologies generated by the radio broadcasting industry.[118] After witnessing a performance of the trautonium in Berlin in the summer of 1932, Gernsback optimistically remarked:

> The Germans have made great headway in electronic music, and many of their instruments have already been exported to this country. I understand on good authority, that a number of American manufacturers are about to produce small and low-priced musical instruments of this kind, which may be used as an adjunct to any radio set; I am certain that we shall witness an avalanche of such instruments in the very near future.[119]

He was struck by the ability of electric musical instruments "to imitate anything from a piccolo down to a bass drum, all in the self-same instrument" by merely turning a knob.[120] Echoing the desires of French composer Edgard Varèse, he felt that "entirely new musical effects will be produced which we cannot now foresee . . . And as for the experimenters who are taking up this new electronic music, they will be in a paradise of their own, because there will be no limit to the number of new and novel sound effects which they will create in the very near future."[121] Contrary to Paul Théberge's claim that something new happened in the so-called digital age, namely the collapse of medium and musical instruments, it is clear that such conflation had been occurring back in the 1920s and '30s.[122]

The trautonium was becoming a legitimate musical instrument: it was featured on *Fox Tönende Wochenschau* (Fox Resonating Newsreel). And lest one thinks the trautonium was only useful for producing music, it made its debut in movies in 1930, when Sala was able to recreate the sound of a propeller airplane in Arnold Fanck's *Stürme über dem Mont Blanc* (*Storm over Mont Blanc*), one of the first German films with sound, starring Leni Riefenstahl.[123]

In July 1931, at the aforementioned Munich Radio Music Conference (which featured the trautonium, Oskar Vierling's electric organ and piano, the Neo-Bechstein grand piano, and other electric musical instruments), Schünemann once again stressed the necessity of collaboration between musicians and

FIGURE 5.11. "How to Build the 'Trautonium'—Musical Instrument." *Source: Radio-Craft*, vol. 4 (March 1933).

engineers, which was exemplified by the RVS. The curiosity of the Berlin press had been piqued, with reports stating that the audience was under the impression that they were entering an unknown world. Yet others delivered much harsher judgments. For example, referring specifically to the hellertion, author Arthur Koestler wrote that "the wildest Negro music" was "as soft as silk in comparison to the electric musical instrument. . . . [Arnold] Schoenberg and [Franz] Schreker sound like lullabies in comparison."[124]

Throughout 1931, Sala and Trautwein worked together to improve the trautonium. To gain a better understanding of the instrument's technical aspects, Sala enrolled in courses in mathematics and natural sciences, with a concentration in physics, at the Friedrich Wilhelm University of Berlin from 1931 to 1936.[125] He demonstrated the workings of the trautonium to students in the seminar of Walther Nernst, a Nobel laureate in chemistry and the co-inventor of the Neo-Bechstein grand piano. Sala took two semesters of experimental physics with Nernst; one semester of quantum theory with Erwin Schrödinger; analytical geometry, mathematics, and theory of functions with Ludwig Bieberbach; advanced mathematics and infinite calculus with Georg Feigl; general psychology with Max Dessoir; experimental chemistry with Wilhelm Schenk; two semesters of a physics practicum with Arthur Wehnelt; two semesters of differential equations with Adolf Hammerstein; two semesters of physical fields (*Feldphysik*) and the theory of matter with Gerhard Hettner; zodiacal light and cosmic radiation with Werner Kolhöster; and a course on the body and soul with psychologist Wolfgang Köhler.[126] His studies of the natural sciences enabled him to break away from Trautwein and work independently on the improvement of the instrument's design.

On May 6, 1932 a newer version of the trautonium was played in Munich's town hall auditorium to celebrate the opening of the Deutsches Museum library. This concert underscored the relationship between science, technology, and culture in a venue where the titans of science and engineering were celebrated. This trautonium was also very similar to the original one build two years earlier.[127] It had the formant switches that could be turned on and off, a formant filter, which controlled the blending of tones, and settings that could adjust the instrument's pitch and range.[128] The keynote speaker was Munich's mayor Karl Scharnagl. Oksar von Miller, the museum's founding director, wished to feature two electric instruments at the opening: the trautonium and the Neo-Bechstein grand piano. The evening featured a small chamber orchestra comprising four violins, two violas, two celli, a bass, and two horns in D, playing George Frideric Handel's organ concert in B-flat major and Leopold Mozart's trumpet concerto in D major.[129] With the aforementioned improvements in the design and parts of the trautonium, by late summer 1932 Sala was able to play more complex pieces, such as Bach's Gigue from the fourth suite for unaccompanied cello.[130] In October and November 1932, concerts featuring electric instruments were broadcast on Funk-Stunde Berlin.[131] The press felt that these electric instruments, including the trautonium, provided "a most valuable complement to the possibility of musical expressions."[132] It was, however, up to the composers to become familiar with these new instruments in order to realize their musical potential.

Throughout the 1920s and '30s, electric music was filling German airwaves and concert halls. The impetus that electrical engineering received through radio was critical to the development of electric music in general, and the trautonium in particular. The Prussian Ministry of Culture and the RRG financed the labor of musicians, engineers, and scientists. German electrical companies were well aware of the importance of the research being conducted at the RVS to their businesses. They also knew of the potential market for household electric musical instruments, even if such a market would not be fully realized until several decades later. Radio engineers, physicists, and physiologists working on the analysis, synthesis, and broadcasting of speech and music provided musicians with the long-coveted ability to generate new tones, thereby answering the challenges posed years earlier by Italian composer Ferruccio Busoni and French composer Edgard Varèse.

But, as we saw, the trautonium was also an instrument inextricably linked to physiological theories of the production of vowel sounds. As Trautwein stated, it was the electrical analogue to human speech organs, and he was convinced that it offered overwhelming support for Ludimar Hermann's theory of impulse excitation of formant formation for both vowels and musical instruments.

The early reviews of the trautonium vacillated between praise and condemnation. Trautwein and Sala fought hard to convince the public of the instrument's potential. As the world welcomed in the New Year in 1933, they seemed optimistic about the future of electric music in general, and the trautonium in particular.

6

The Nazis and the Trautonium

ON FEBRUARY 17, 1939, less than seven months before the Nazi invasion of Poland, Reich propaganda minister Joseph Goebbels opened the Berlin Auto Show with a speech underscoring the importance of technology to the Reich.

> National Socialism never rejected or struggled against technology. Rather, one of its main tasks was to consciously affirm it, to fill it inwardly with soul, to discipline it and to place it in the service of our people and their cultural level. National Socialist public statements used to refer to the steely romanticism of our century. Today this phrase has attained its full meaning. We live in an age that is both romantic and steel-like, that has not lost its depth of feeling. On the contrary, it has discovered a new romanticism in the results of modern inventions and technology. While bourgeois reaction was alien to and filled with incomprehension, if not outright hostility, to technology, and while modern skeptics believed the deepest roots of the collapse of European culture lay in it, National Socialism understood how to take the soulless framework of technology and fill it with the rhythm and hot impulses of our time.[1]

An ever-increasing number of Germans had seen their livelihood threatened by the rapidly declining economic situation during the Weimar Republic. The Nazis promised to recognize and reward individuals' desires to achieve. As Daniel Siemens has argued, liberalism

> was no longer the association of economically independent and free individuals that seemed to be able to further the country and its citizens, but, on the contrary, the imagined collective—that is, for the National Socialists, the "people's community"—was supposed to offer protection and the chance to develop to the individual.[2]

The German political landscape changed on January 30, 1933, when the National Socialist German Workers' Party assumed power. As a key tool of propaganda, the radio became a conduit for Joseph Goebbels's and his party's machinations. As is well known, music was not free from governmental interference in Nazi Germany. "Degenerate" art and music were banned in the Third Reich, and many musicians and composers fled. So how did the Nazi Party react to the new electric music genre? Early on, the future of electric music and the trautonium was far from certain. The key was to fuse technology with Nazi ideology.[3] The interests of the Nazi Party and German engineers, including Trautwein, overlapped in three critical ways. First, many engineers were keen to render technology consistent with the idealist culture of the German mandarins, a group that engineers both abhorred and wished to emulate. Second, many wanted to separate their intellectual work from that of their American counterparts. As Karl-Heinz Ludwig has argued, starting in the 1870s, German engineers were by and large anticapitalist. Such a sentiment was still prevalent during the early 1930s.[4] Engineers felt strongly that technology was part and parcel of German *Kultur*, dating all the way back to Goethe's *Faust*, rather than being a feature of disenchanted materialism of *Zivilisation*. In June 1934, the *Zeitschrift des Vereines Deutscher Ingenieure* (*Journal of the Association of German Engineers*) published an article by board member Heinrich Schult titled, "Organic Formation of the Economy: An Engineer's Assignment."[5] The crisis of Germany and, indeed, the world, in the 1920s was caused by greedy capitalists, not machines. The targets of those attacks against capitalism were predominantly Americans and Jews. As engineer Carl Weihe argued, it was not German engineers, but American ones, who transformed "the machine into a money-making machine, and industry into a stock exchange."[6] He warned his fellow Germans of the perils of "Americanism and all of its detrimental effects on the soul of the people."[7] Materialism and capitalism were anathema to the German spirit. The welfare of the German Reich needed to be in the hands of a powerful nation-state, not in the hands of private economic interests. Third, the Nazis directed their propaganda at enrolling engineers to their cause starting in the mid-1920s.[8] German engineers could provide the state critical expertise in the technology of warfare and propaganda.

The lack of social status accorded to engineers during the Weimar Republic, as discussed in chapter 2, and their lack of strong connections to the government, left them particularly vulnerable to Nazi propaganda. Since they sought to improve their standing by turning to culture and not restricting themselves to the machine or factory, engineers were susceptible to arguments appealing to the mystical spirit of the German *Volk*.[9] In late 1930 the Nazi Party promised the creation of jobs for over seventy thousand unemployed technicians (a figure provided by the party).[10] Engineers, as nearly everyone else, suffered greatly during the Depression. They were either unemployed or employed with ludicrously low wages: as Bertolt Brecht famously wrote in *The Threepenny Opera*, "Erst kommt das Fressen, dann die Moral" (First comes the gorging, then comes morality). In April 1933 the Reichsanstalt für Arbeitsvermittlung (the Reich's Institute for Job Placement) supplied public funds to assist young engineers.[11] Such a move was necessary since there had been a glut of engineers on the market during hyperinflation and the Depression: the lack of jobs forced many to turn to training at technical universities.[12] The Nazi *Gleichschaltung* was just as effective in the engineering disciplines as it was in other professional societies and political, social, and cultural institutions.[13] Many engineers spoke of the Nordic origins of technology while simultaneously espousing virulent anti-Semitic views. In addition, they saw fascism as freeing them from the yoke of bureaucracy and industry.[14] Two-thirds of the individuals elected to the Verein Deutscher Ingenieure's board in May 1933 were members of the Nazi Party.[15]

Though the trautonium was a product of Weimar Germany, it thrived during the Third Reich, thanks in part to Oskar Sala's political opportunism and Trautwein's commitment to fascism. While the first portion of this book considered the sociocultural, scientific, and engineering contexts that gave rise to the trautonium, the second half proffers a broader study of the trautonium in its various political and social contexts. The trautonium's ability to both reproduce the tone colors of more traditional orchestral instruments and to synthesize new tone colors and sounds proved to be critical to its future. While some Nazis prefered the imitative qualities of the instrument, many critics during the Third Reich encouraged Sala to go beyond imitation and to generate new sounds.

Nazi Radio

While the Nazis assumed power in January 1933, a number of Nazi officials had already gained control in Prussia, where the Nazi Party had received 36 percent of the vote in April 1932.[16] Erich Scholz became radio commissioner in August 1932. Scholz had been a member of the conservative Deutschnationale Volkspartei (DNVP), or the German National People's Party, but he subsequently and briefly joined the Nazi Party. Two days after Scholz's appointment as commissioner, Hans Bredow, who had been the commissioner, returned from his vacation and was informed that he had been fired. This move was the culmination of at least two years of increasing calls for nationalizing German radio. President Paul von Hindenburg spoke to his people via the *Reichssendung*, which was broadcast throughout the entire Reich via all regional radio stations. The German statesman exploited the radio as an instrument of political influence: in this case to stop Adolf Hitler. *Überparteilichkeit* was no longer in fashion. The Guidelines of the Regulation of Radio of 1926, which declared the nonpartisan aspects of radio, were replaced in 1932 with the Guidelines of Radio Reform. Its first guideline declared that "German radio serves the German people."[17] The fourth guideline stated that "radio serves all Germans inside and beyond the borders of the Reich" and that "maintaining the idea of the Reich is a duty of German radio." The fifth encouraged radio "to form the Germans into the people of the State and to shape and strengthen State thinking and the will of the listeners."[18] Those new reforms also resulted in the withdrawal of private capital from all aspects of radio broadcasting and placed the nation-state and regional governments in total financial control.[19] The radio's mandate was now to educate the listeners about the development of the German Reich and strengthen feelings of German honor.[20]

In addition to the guidelines of 1932, radio stations needed to grow accustomed to two important ideological changes. The first was the maintenance of the German language, its proper use and pronunciation. Little was mentioned concerning regional dialects, which were so predominantly featured in Weimar radio. A year later, an essay appeared in *Der Bericht der Reichs-Rundfunk-Gesellschaft* (*The Report of the Reich's Broadcasting Corporation* [RRG]), titled "Sprachpflege im deutschen Rundfunk" ("Concern for the Purity of Language in German Radio"). The essay echoed the new guidelines, stating that "the German language must be spoken astutely and purely, with dignity and clarity."[21] Radio announcers needed to be well acquainted with the *Hochsprache* as well as Theodor Siebs's *Deutsche Bühnenaussprache* (*Language of German Theater*) of 1898.[22] Whereas radio in the early Weimar Republic proudly broadcast works in regional dialects, radio of the late Weimar Republic and during the Third Reich stressed *Hochdeutsch*, proper (or high) German.[23] It was an attempt to standardize the German language and to culturally unify German citizens. It is interesting to note that Karl Würzburger of the Radio Experimental Laboratory (RVS) had rejected the aim of creating a national, standardized voice during the Weimar Republic.[24]

The second change dealt with the second guideline of 1932, namely the maintenance of "good music." What was considered good music? "The creations of the German masters form the core of the musical programs."[25] One no longer read about experimenting with radio broadcasting. The term "avant-garde music" was no longer a compliment. As a result of this shift, two important programming changes relevant to music occurred during the late Weimar Republic. First, the percentage of music played on the radio decreased by 10 percent to accommodate more discussions of election campaigns. Second, the percentage of patriotic marching music significantly increased.[26]

When Goebbels took control of German radio after the Nazis came to power, he did not need to change the structure and content of radio broadcasting all that much due to the conservative trends of the late Weimar Republic. This assisted him in maintaining continuity while exerting the influence of Nazi ideology. Despite the view that the Nazis ruptured all cultural links to the immediate past, they actually often hijacked preexisting bits of Weimar culture, including some musical trends, for their own nefarious intentions.[27] The audience was there for the taking. The radio was a unifying transmitter of culture from the early days of the Third Reich.[28] The selling of radio licenses continued to increase steadily. In 1932, around 3.9 million radios were sold. That number had risen to over 10.8 million by January 1939, as radios became much more affordable.[29]

In 1933 Goebbels began replacing radio employees deemed dangerous or unsympathetic to the Nazi cause. A number of the radio pioneers associated with the Funk-Stunde Berlin were sent to the Oranienburg concentration camp just north of Berlin, including Alfred Braun, Heinrich Giesecke, Kurt Magnus, and Hans Flesch. Goebbels introduced a new program in April 1933, called "Stunde der Nation" ("The Nation's Hour"). The nationally broadcast program could be heard in the primetime slots of 7 or 8 p.m. either daily or weekly on all radio stations throughout the Reich. It extolled the virtues of Nazi doctrine by providing the German people with the spiritual content of National Socialism.[30] As is all too well known, radio became the Nazis' most important mass medium of propaganda: the Führer's words into every factory and every home, as the party slogan proclaimed.

The broadcasting of music remained an important component of German radio. After falling 10 percent in 1932 and remaining at 57 percent in 1933, the percentage of broadcasting hours dedicated to music increased steadily thereafter, reaching 70 percent in 1938. Studies suggested that approximately two-thirds of that 70 percent comprised *Unterhaltungsmusik*.[31] Radio station managers and programmers felt that entertainment music was less challenging for the listener than more "serious" music, such as experimental, symphonic, chamber, or operatic works. As Nazi radio producer Eugen Hadamovsky explained, "Broadcasting must attract the listener by a program of light entertainment so as to make him receptive. Only then can it lead him toward higher aims."[32] Renowned radio journalist Gerhard Eckert concurred:

> The undisguised task of the National Socialist Movement is to make the listener receptive to the aims of particular spoken-word broadcasts by means of a sufficient number of [light] music broadcasts. The more music that is broadcast on the radio, the more open the listener becomes to the spoken word . . . [33]

Goebbels increased the average broadcasting day from just over fourteen hours before the Nazis assumed power to nearly twenty hours by 1938. He personally ordered the continued increase of Unterhaltungsmusik after 1939 to take German minds off the war. In 1942 he argued that "entertainment music of the German radio service, for relaxation and release on both the military and home fronts, is vital to the war effort. Special care must therefore be taken with this branch of the German radio broadcasting service."[34] While the importance of Unterhaltungsmusik increased throughout the Third Reich, serious music was by no means ignored. The works of Bach, Mozart, and Beethoven were often played. Richard Wagner's operatic works, of course, were a favorite, particularly of Adolf Hitler. Of the modern composers, Richard Strauss's works were frequently heard on the radio.

In general, modern music did not fare well under the Nazi regime, although it should be noted that the Nazis did not ban all of it. Some of the types of music supported during the Weimar Republic were also actively sponsored by the Nazis.[35] Of course, many Jewish musicians and composers either emigrated or suffered a horrific fate. Certainly by 1938, the musical enemies of the Reich were made known in the Exhibition of Degenerate Music, which specifically named Arnold Schoenberg, Anton Webern, Paul Hindemith, and Kurt Weill. "Modernism," an attribute proudly extolled by composers of the Weimar Republic, elicited a more ambivalent or skeptical response from Nazi officials.[36] The term was often associated with attributes that were anathema to the Nazis, such as "non-Aryan," "Jewish," "international," "progressive," and "bolshevik." These characteristics were antithetical to the romantic, spiritual, austere, and nationalist forms of music many Nazis coveted.[37] Famously, German music, according to Hitler, needed to be emotional, not intellectual: "It is not the intellectual mind that is the godfather of the musicians, but an overflowing musical soul."[38]

Trautwein and the Nazis

When the Nazis rose to power, Trautwein himself was well aware of what needed to done. "The future of electric music lies totally in the hands of the State. It comes down first and foremost to the ideal promotion of electric music through a positive attitude by the relevant party and state offices."[39] Despite the initial financial support of Telefunken and Siemens & Halske for the trautonium, in the autumn of 1934, Trautwein criticized German industry for their limited interest in electric musical instruments. That said, he did not want commercial interests to control research. The state needed to play the leading role, otherwise, "inferior products" from North America would prevail. This, according to Trautwein, was typical of the "liberal-capitalistic era."[40] In the same month the Nazis took control of the government, Sala wrote that "electric music stands and falls with the trautonium."[41] Trautwein and Sala's stance was reminiscent of the views of Thomas Mann's fictional character Adrian Leverkühn in *Doktor Faustus*. Mann reminds us that natural scientists were just as susceptible to the seduction of irrational fascism as those well versed in the arts, such as music. Trautwein joined the Nazi Party on April 1, 1933, the same day the Nazis began to boycott Jewish shops. He was one of around seven thousand engineers who joined that year.[42] Often using party connections to further his research and career, he had been the leader of an amateur radio group in Berlin-Neukölln since the late 1920s.[43] For example, the RVS received a license in August 1930 to erect a radio broadcasting and receiving station for experimental research.[44] After three warnings from June 1932 to January 1933—the final stages of the Weimar Republic—not to broadcast beyond what the license specified, the Nazis subsequently decided not to renew the license shortly after their rise to power.[45] On April 4 Trautwein responded that he had a license to broadcast from both his private apartment and the RVS. He protested:

> as a result of shutting down the short-wave radio in my private apartment as well as in the Academy of Music, my name has become associated with Jewish or Marxist-led organizations. I wish to inform you that I am a member of the local chapter [Alfred von] Schlieffen of the National Socialist German Workers' Party, and as a result of this position I applied to the regional post office administration (*Oberpostdirektion*) for a broadcasting license for my apartment. I am requesting to the Rundfunkversuchssstelle [RVS] that a broadcasting license for the Berlin Academy proceed in the same fashion . . . Although I have made little use of the broadcasting license in recent years, I can now declare that I refrained from working in the radio technical association and in the Academy of Music because I had concerns about the political views of the employees. I have just created a radio-training program in the local chapter Schlieffen of the National Socialist German Workers' Party and want to take up energetically my old plans of military exercises. I hereby declare emphatically to the Academy of Music and the National Socialist German Workers' Party that my radio-technological experiences are fully at the disposal of the establishment of the work of the Fatherland. I also note that I am a reserve officer in the Tel. Btl. [Telegraph Battalion] IV and that as a result of my well-received technical advice gleaned at the front [in World War I], I was in charge of the technical department for the development of technical novelties in radio in July 1918 and that I also have worked since then on problems, which I had originally taken up at that time.[46]

In addition, Trautwein drew the attention of Fritz Stern, the new director of the academy who replaced Georg Schünemann, to two politically problematic compositions of electric music pioneer Jörg Mager, the inventor of the spherophone. Trautwein called Mager a "psychopath: egocentric fantasist."[47] During this time, it was well known that the RVS would be shut down, and

Trautwein wanted to ensure that Mager would not be given a position in his group at the Berlin Academy of Music. Mager was allegedly a socialist, and he had composed a version of "Warszawianka" ("Whirlwinds of Danger"), a beloved Polish communist revolutionary anthem, for an all-male choir. He also composed an arrangement of the Russian Red Guards March, titled in German "Brüder zur Sonne zur Freiheit" ("Brothers to the Sun, to Freedom"), which became the anthem of the German Social Democratic Party after World War II, for a male choir.[48] Mager allegedly told Trautwein in November 1932 that he switched from the Communist Party to the Nazi Party because he hoped he could obtain more funds from the Nazis for his work.[49]

The RVS and the Early Years of the Third Reich

A concert featuring the theremin, trautonium, and Neo-Bechstein grand piano that originally aired on January 25 was rebroadcast on February 6, 1933: this was the first broadcast of the trautonium during the Third Reich.[50] Reviews in the radio newspapers praised the concert, commenting on how scientists, engineers, and musicians worked together to enrich both radio music and house music.[51] The *Bayerische Funk-Echo* published a manual for building a trautonium aimed at the technically gifted.[52] The *Vossische Zeitung*, however, was a tad more skeptical. It insisted that the initial sensationalism associated with electric music was long gone and that the key was now the practical application of the inventions. The reviewer felt that electric musical instruments would not succeed until they became easier to perform and when their range of tone colors would offer the possibility of a new form of house music based on musical tinkering (*musikalisches Probieren*) and invention.[53] The majority of the early reviews underscored the trautonium's ability to replicate the sounds of acoustical instruments.

Both the Berlin Academy of Music and the Heinrich Hertz Institute for Oscillations Research (HHI) were "cleansed" of Jews and Marxists. Since Hertz himself was a Jew, the institute's name was shortened to the Institute for Oscillations Research. Gustav Engelbert Leithäuser, a radio frequency engineer at the HHI, was dismissed because his wife was Jewish. HHI director Karl Willy Wagner was also forced to resign. In 1938 engineer and electric instrument inventor Oskar Vierling left for a faculty position in Hanover. As seen in chapter 3, both engineers worked on problems of military interest to the Third Reich.

Georg Schünemann was forced to step down after he was accused of being a Marxist by Bruno Kittel, who at the time was director of the academy's choir. Kittel had joined the Nazi Party back in May 1933.[54] In 1936 he became the director of the Konservatorium der Reichshauptstadt Berlin (Conservatory of the Capital of Berlin), earlier known as the Stern'sches Konservatorium (the

Stern Conservatory), whose origins dated back to 1850 as a private conservatory founded by Julius Stern. Schünemann was replaced by musician and theologian Fritz Stein, who was a member of the Nazi Party. Back in July 1933, Stein had been appointed *Reichsleiter der Fachgruppe Musik* (Reich leader of the music group) of the Kampfbund für deutsche Kultur (Militant League for German Culture), a nationalist and anti-Semitic organization founded by vehement Nazi Alfred Rosenberg. Sala later reported that there was a poster in the academy condemning electric music. A number of the professors were signatories.[55]

Goebbels shut down the RVS on March 31, 1935, although he had planned its demise as early as April 1933.[56] He initially wished to save money by slashing the RVS's budget for teaching, sound films, and electric music, even though a year before its closing, it was still being praised for its work on "perfect transmissions" and the use of technology in teaching music.[57] In August 1934, knowing full well that the RVS was to close, a new group, *Musik und Technik*, was proposed at the academy. Trautwein directed the group and was given a full professorship at the academy.[58] Trautwein's commitment to the fascists paid off. *Musik und Technik* was divided into three sections: radio and records, sound films, and electric music. Commencing operations a day after the RVS was closed, Trautwein's group continued the RVS tradition.[59] Technical problems that had been studied and solved at the RVS were now the focus of the department of music und engineering headed by Trautwein.[60]

The relationship between Paul Hindemith and the Nazis was complicated. He was, after all, a cultural icon of the Weimar Republic, and he played a critical role in the New Objectivity (Neue Sachlichkeit), a movement loathed by Nazis. His wife was Jewish, and his music embraced trends that the Nazis considered dangerous, like jazz, expressionism, parody, and constructivism.[61] By the late 1920s and early '30s, however, Hindemith had embraced German folk music, which was to become popular with the Nazis and which greatly influenced his compositions. Despite that, the Ministry of Education banned his music in 1930 in the state of Thuringia. His brief collaboration with Bertolt Brecht resulted in him being labeled a "bolshevik." Critics often attacked him in the *Zeitschrift für Musik*; however, in June 1933, critic Walter Berten applauded Hindemith in that same journal for a fundamental shift in his work, which now possessed clarity and reflected an allegiance to classicism.[62] Premiering in 1934, Hindemith's *Mathis der Maler* was lauded by critics: the symphony drew upon the cultural resources of German folk songs and chorales.[63] Many Nazi Party members, however, were not impressed. Attacks were leveled against Hindemith for his association with the International Society for Contemporary Music (ISCM) that was "dominated by Jews and encouraged musical bolshevism, dilettantism, and atonality."[64] He felt compelled to withdraw his string trio from the ISCM Festival in 1934 to avoid criticism. Later that year, when a number of radio stations debated whether or not to play Hindemith's works, Goebbels expressed his opinion:

> Although the fundamental spiritual attitude which appears in most of Hindemith's previous work cannot be sharply enough condemned, Hindemith should nevertheless be recognized unquestionably as one of the most important talents in the younger generations of composers.[65]

Goebbels was desperately trying to fashion Hindemith and Richard Strauss as the new standard-bearers of German music.[66]

Friedrich Herzog, the editor of *Die Musik*, took a more cynical stance against Hindemith than Goebbels initially did. He successfully conducted a boycott in the autumn of 1934 of the performances of Hindemith's works, particularly his *Mathis der Maler*. In an attempt to defend his colleague, renowned conductor Wilhelm Furtwängler penned an essay published in the *Deutsche Allgemeine Zeitung* titled "Der Fall Hindemith" ("The Hindemith Case"). It backfired. Furtwängler was forced to resign from his positions at the Berlin Philharmonic, Berlin State Opera, and the Reichsmusikkammer (the official musical organization of the Nazi Party whose goal was to promote Aryan music). Hindemith became the target of orchestrated attacks from various Nazi officials and musicians supporting the regime. In December 1934, Goebbels changed his opinion of the composer and launched a diatribe against him and *Mathis der Maler* at a party meeting in the Berlin Sportpalast:

> Technical craftsmanship never excuses, but is rather an obligation. To misuse it in writing purely motoric empty *Bewegungsmusik* [mechanical music] is a mockery of the genius that stands above every true art . . . when the occasion is ripe, not just thieves but atonal musicians arrive on the scene who in order to attain a particular sensation or remain close to the spirit of the time allow naked women to appear on the stage in obscene scenes in a bathtub, making a mockery of the female sex . . . and in general surrounding themselves with the biting dissonances of musical bankruptcy.[67]

Goebbels labeled the composer an "atonal noise maker."

Hindemith tried to remedy the situation. In 1935 he accepted a position offered by Nazi officials to establish the Ankara Conservatory in Turkey with the view of spreading German culture abroad. In March 1935 he permitted his publisher to send a letter to violinist Gustav Havemann, a prominent member of the Reichsmusikkammer, that attempted to distance himself from the "decadent intellectual efforts of a Schoenberg."[68] And in January 1936 he went so far as to sign an oath of loyalty to Hitler.[69] While some of his works were once again performed, this would prove to be ephemeral. With Hitler calling for an intensified cleansing of the arts, in October of that year Goebbels placed an official national ban on future performances of Hindemith's oeuvre. Henceforth, his work was featured one last time in the Third Reich, at a Nazi-sponsored exhibition in Düsseldorf of degenerate music (*Entartete Musik*) in 1938. At the exhibition, Hindemith was labeled "a theoretician of atonality" and "a rootless charlatan."[70] Shortly thereafter, the composer left Nazi Germany for Switzerland, settling in the United States in 1941 after receiving a professorship at Yale. He would never again work on electric music. After the war, when he visited Berlin and met with Sala, he said of electronic music, "It served its purpose. Today it is obsolete."[71] Sala apparently did not approve of Hindemith's new musical orientations: he referred to a concert in Berlin in 1954 featuring Hindemith's new work based on the singing of folk songs as "American-made hype" ("amerikanisch aufgezogener Rummel"). "So horrible that one actually must say that he is for us quasi-dead."[72]

Volkstrautonium

In August 1933, the next generation of trautoniums, the Volkstrautonium, debuted at a press review at the Singakademie in Berlin. Sala performed a movement of the flute concerto of Frederick the Great, an appropriate selection of music under the new regime.[73] It was also featured in the Radio Exhibition in Berlin sixteen days later along with other electric devices, including musical instruments (such as the Neo-Bechstein grand piano, the hellertion, the theremin) and the notorious Volksempfänger, or radio receiver, commissioned by Goebbels and built by radio engineer Otto Griessing.[74] One enthusiastic reviewer called the Volkstrautonium "das vollkommene Musikinstrument" (the perfect musical instrument) since musicians now had an instrument that could imitate (*nachgebildet*) a range of tone colors, from the piccolo to the lowest pipes on the organ to the drum.[75]

Similar to the May 1932 version of the trautonium, Trautwein recommended that a battery be used for the first two tubes, as a slight fluctuation in voltage would produce a large change in pitch and sound quality.[76] As was the case with the August 1932 version of the trautonium, the glow discharge (neon) tube was replaced with a thyraton, a gas-filled tube used as a radio-signal detector, specifically an RK 1 tube from Telefunken, which produced sawtooth waves. Telefunken's REN 904 tube was used for amplification. An additional circuit was added to produce the harmonics that more closely resembled those of a violin. A setting of the pitch could be accomplished by turning the dial (Figure 6.2, number 6, dial C), which regulated a capacitor.[77] The musical intervals could be adjusted by altering the resistance via dials A and B (4 and 5 in Figure 6.2). A scale running the length of the wire enabled the musician to find the requisite

FIGURE 6.1. The Volkstrautonium (people's trautonium), around 1933. Reproduced with the permission of the Deutsches Museum (Munich) Archives, BN_24831.

pitch.[78] In order to locate pitches quickly, particularly relevant when playing rapid passages in pieces, "auxiliary keys" (*Hilfstasten*) (Figure 6.2, number 13) were attached above the first and fifth pitches of each octave. They could be moved individually or all at once by pulling out the setting lever (Figure 6.2, number 7). Unlike the original trautonium, which possessed a range of two octaves, the Volkstrautonium enjoyed a range of three-and-a-half octaves, typical of most single-voice (monophonic) instruments of the period. In addition, its pitches could be transposed to the lowest end of the range of a cello or to the upper range of the violin and flute.[79] The Volkstrautonium could also imitate a clarinet in B and a horn in F.[80]

Control knob F (8) and control buttons I, II, 1, 2, and 3 (9) (Figure 6.2) changed the values of the capacitance of the electric circuit, thereby altering the formant frequencies either in a stepwise fashion or by sliding from one formant to another, creating a glissando effect. If the formant was abruptly changed, differing timbres appeared in rapid succession.[81] Dial KL (10 in Figure 6.2) could mix formants in the desired ratio of intensity.[82] In short, control knob F (8), control buttons I, II,1, 2, and 3 (9), and dial KL (10) changed the tone color to reproduce the desired musical instrument.[83]

Imitation was seen as critical to the instrument's early success: the Volkstrautonium's ability to reproduce familiar instruments to create Hausmusik

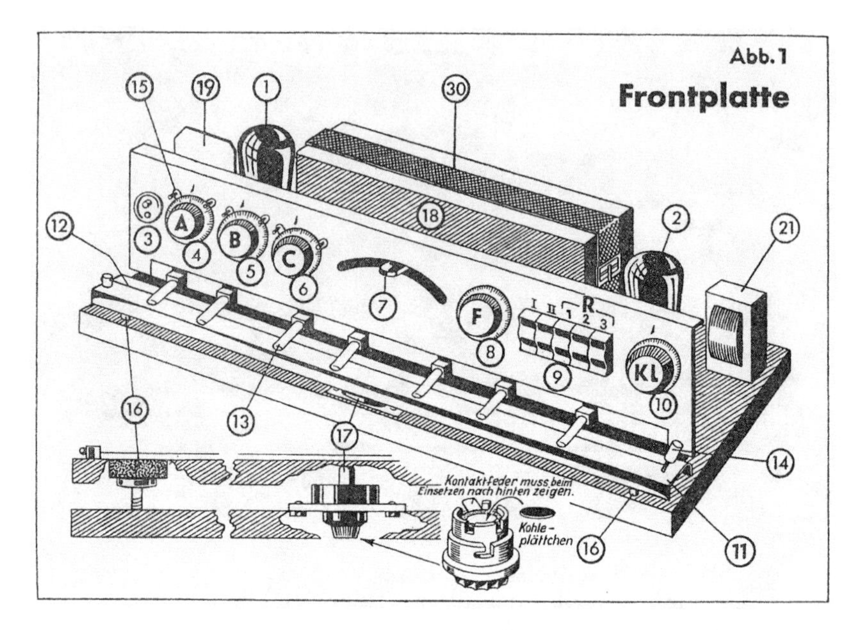

FIGURE 6.2. Diagram of the Volkstrautonium (people's trautonium).
Source: Trautwein, *Trautonium-Schule*, 6.

(or music for the household) certainly served as a key marketing strategy for Telefunken. In addition, the instrument could be used in professional orchestras, bands, and chamber ensembles. It could replace the trumpet in an orchestra by playing rapid passages much more easily than a trumpeter could. It could also be a solo instrument that could play the vast repertoires of the cello, violin, flute, trumpet, or oboe while imitating the tone colors of all of those instruments. Telefunken, which sold models to the public in 1933, hoped that it would find a home in bourgeois households and music academies and physics institutes around the globe. Finally, the company extolled the instrument's ability to produce sound effects for film and theater.[84] In 1934 Sala composed and played a trautonium melody for *Hanneles Himmelfahrt* (the *Assumption of Hannele*), directed by Thea von Harbou, and the trautonium could be heard in numerous UFA films during the Third Reich.[85]

One did not need to play the limited number of pieces that were specifically composed for the original trautonium: the repertoire was increased to play more challenging pieces. Its cost, however, thwarted any hopes that it would ever become such an instrument. It sold for 431.25 RM, including battery and tubes, or just over $3,400 in today's money. During a time of economic uncertainty, many German families could not afford a Volkstrautonium. We do not have an exact production total of the Volkstrautonium, although archives suggest that approximately two hundred were made by Telefunken, which clearly lost money in producing the instrument.[86] Trautwein wrote privately to Schünemann expressing his concerns over the economic problems affecting its construction. Germany had experienced its most recent currency crisis eighteen months earlier.[87] The RVS needed to intervene, since Telefunken might decide to forgo any further work on electric music altogether. He feared that Telefunken would make his invention "kitschy and would sell 9,999 of every 10,000 for scrap," as he claimed RCA had done with the theremin.[88] Archives from the firm dated to December 1937 indicate that the company spent nearly 302,000 RM ($2.3 million in 2023) on production. Even if all the copies were sold, which is highly doubtful, their loss would have totaled over 215,000 RM (over $1.6 million in 2023).[89] It certainly never came close to ten thousand despite the publication of *Trautonium-Schule* (Trautonium School), a 1933 guide to playing the instrument written by Sala and edited by Trautwein—although he is generally credited as the author—and featuring musical pieces composed by Hindemith.[90] All three had hoped that the Academy of Music would host a school of trautonium players. That hope was never realized despite an enthusiastic review of the instrument in *Allgemeine Musikzeitung* in 1934.[91] In addition to the lack of financial incentive to continue building the Volkstrautonium, there were other reasons why Telefunken lost interest in the trautonium. First, as we shall see, Telefunken wished to have other versions of the trautonium built. Second, in late 1934 and early 1935, Telefunken shifted its priorities. The company toyed with the idea of ending its contract with Trautwein and substantially cutting back support after 1935. In February 1938 Telefunken officially ended the relationship with Trautwein, allowing him to keep all intellectual property rights.[92] In addition, the instrument was not all that well suited for Hausmusik. If one wanted to produce a tone color of a cello or violin, surely it would be better to purchase those particular instruments. In addition, as one critic remarked, it was simply too difficult to produce the desired tone color on the trautonium.[93]

Many complained that too few composers were writing music specifically intended for electric instruments. As a result, performers were forced to play classical pieces despite all the early talk about how the trautonium could generate new sounds.[94] Some of the more conservative critics slammed Sala for experimenting with "our classical composers."[95] Even those who were less devoted to the classics felt that mere imitation was not the way forward. As Mexican composer Carlos Chávez argued in 1937, "These people undiscerningly want the new electric instruments to imitate the instruments now in use as faithfully as possible and to serve the music we already have."[96] He added, "It is a great mistake to wish the new electric instruments to equal the traditional ones, since the latter already exist and are at hand."[97]

Contrary to Chávez's commitment to this new music, most contemporary composers were partially to blame for not writing pieces for the instrument. Schünemann had observed back in January 1932 that "there is certainly no shortage of technical solutions. . . . But the musicians, both composers and performers, follow too slowly. There are only a few who can help tackle the technical challenges, but the engineers can only make progress by working hand in hand with musicians."[98] As Thomas Patteson astutely points out, the situation had reversed in less than three decades: Ferruccio Busoni had challenged engineers and scientists to produce new types of sounds back in 1907, but by the early '30s the musicians, rather than engineers, were the limiting factor in musical creativity.[99] Alas, the Volkstrautonium had not received the attention that both Trautwein and Telefunken had initially hoped back in the summer of 1933. After 1934, electrical musical instruments were no longer featured at the regular Berlin radio exhibitions.[100]

Improved Versions of the Trautonium

Sala, with waning assistance from Trautwein, began to tinker with and expand upon the capabilities of the Volkstrautonium in 1933.[101] For three years he experimented with increasing its range of sounds and tone colors. Of particular importance was his ability to generate subharmonics, a sequence of notes resulting from the inversion of the intervals of the overtone series.[102] Trautwein had discovered this musical phenomenon back in 1934 and obtained a patent for the circuits, generating and synchronizing them.[103] He had suggested to Sala that the best way to produce tone colors with subharmonics was to create an electric circuit with a specific capacitance and self-induction.[104] Unlike overtones, subharmonics must be artificially created. The first instrument featuring these subharmonics was called the "concert instrument" (*Konzertinstrument*), an improved trautonium not to be confused with the "concert trautonium," which he built several years later.[105] The concert instrument resembled the earlier Volkstrautonium; however, the instrument had two manuals, meaning a resistance wire over a metal rail, that rendered it "multivoiced" (*mehrstimmig*). With the two manuals, Sala could produce more tones, particularly subharmonics.[106] The instrument enjoyed a range of three octaves. During the next phase of improvements, he was able to form double tones, and subsequently triple and quadruple tones, through the addition of various subharmonics of the fundamental frequency. He would later call these *Mixturen*, or "mixtures."

To draw the attention of Nazi officials to the trautonium, Walther Funk, the press chief of the Reich, arranged a meeting with Joseph Goebbels on 11 April 1935, eleven days after Goebbels had shut down the RVS. Sala performed

FIGURE 6.3. The Konzertinstrument (concert instrument) performed for Dr. Joseph Goebbels, April 1935. Reproduced with the permission of the Deutsches Museum (Munich) Archives, CD_85669.

a Bach sonata, a Beethoven trio, an intermezzo by Harald Genzmer, and a movement of a sonata by Max Reger on the Konzertinstrument with piano accompaniment by Rudolf Schmidt and cello accompaniment by Herbert Lehmann.[107] In a recording of the recital, Sala demonstrated to Goebbels many unique sounds and timbres that the instrument could generate. Goebbels tried to play it, but with limited success.[108] The minister of propaganda seemed to be more interested in how the instrument could play well-known pieces than discussing the unique sounds it could generate. While we know few details about Goebbel's response, Sala informs us that the Goebbels apparently said, "We haven't ever heard anything like this before. It's fabulous: do carry on with it!"[109] He was interested to hear if the instrument could be used for mass rallies. Trautwein was quick to point out how his trautonium fit into the larger musical agenda of the Führer and Dr. Goebbels.[110] Sala publicly expressed his appreciation for Goebbels's support of the program for his concert on the trautonium at the Berlin Academy of Music on June 11, 1936: "The instrument was played for Dr. Goebbels, and thanks to his initiative, new types of problems could be tackled."[111]

On November 16, 1935, at the Berlin Academy of Music, Sala played a composition of Hindemith's titled "Langsames Stück und Rondo für Trautonium" ("A Slow Piece and Rondo for the Trautonium"), written the previous August, in honor of the composer's fortieth birthday.[112] It was one of Hindemith's last compositions before Goebbels's official ban less than a year later.[113] Hindemith commented on how it was a challenge to compose for the instrument; however, the "curious possibilities" intrigued him.[114] With two manuals, each containing a wire with a resistance value of 1800 ohms, one could play two different parts, and with the coupling devices, one could even generate four parts. These enable more virtuosic pieces to be played, such as Paganini's caprices for violin. Two more concerts later that month featured the instrument as bass reinforcement in Claude Debussy's "La cathédrale engloutie" ("The Sunken Cathedral") and as the solo instrument performing works by Bach, Marius-François Gaillard, Harald Genzmer, Handel, and Reger.[115] A third manual was added in 1936 to the concert instrument. It also possessed a "tone color pedal," whereby a large number of different timbres was combined into a single one. An increase in the intensity of the expression, however, needed to come at the expense of the number of different timbres that could be generated.[116] Sala called this final version of the concert instrument "the new trautonium" (Neues Trautonium). In subsequent versions, the third manual was discontinued.

In February 1936 Sala convinced the Ministry of Public Enlightenment and Propaganda to put on a concert at the Funkschule (Radio School) in Berlin. In June 1936 the trautonium was featured at the 67th Annual Musicians Conference of the General German Music Association in Weimar.[117] Accusations that the association was advocating for the atonal, *entartete Musik* of Jewish musicians and composers swirled around Weimar, the city that could boast the likes of Goethe, Schiller, Liszt, and Nietzsche. Peter Raabe, the president of the Reichsmusikkammer and a fan of electric music, felt that since the inventors of electric instruments were Germans, they should not suffer the consequences of those responsible for composing and playing degenerate music. He argued that electric music and mechanical music should not be conflated.[118] Sala had invited Raabe and his adviser and coauthor of *Von deutscher Tonkunst* (*Of German Musical Art*), Alfred Morgenroth, to inspect the concert instrument. Apparently, they were impressed with the labor required to render it an acceptable solo instrument for concerts.[119] On the morning of June 17, 1936, Sala introduced his instrument to the conference attendees in Weimar. The ensuing day, Harald Genzmer's piece for the trautonium and orchestra ("Konzert für Trautonium und Orchester") in three movements, debuted in Weimar, and Sala gave a repeat performance in Duisburg on October 26 of that year.[120] Critics were impressed by how the trautonium increased the palette of sounds upon which a composer could draw in order to produce different effects. In short, it was a promising instrument.[121]

At that time, Trautwein stressed the importance of technology to the Third Reich. He also revealed his fascist and anticapitalist tendencies.

> One needs to keep in mind that the reckless exploitation of the technical forms of musical dissemination [i.e., mechanical music such as the gramophone] was carried out in a time of individualistic capitalism. Today the forms of musical dissemination, as is the case with every technology and with every art, are placed in the service of the people as a whole (*Volksganzen*). The role technology will play with this new purpose can be seen from the example of March 28, [1936], when all technological means relevant to the radio and loudspeaker were put to the service of a spiritual uplifting of the German nation, where the entire fatherland was united by these technological means in a large church service, more powerful than the largest cathedral that could ever be seen or the largest organ that could ever be designed. Art has the obligation today to deepen and preserve the spiritual uplifting of the people. To that end the artist relies on the technological means of these recent times, and he violates his purpose if he rejects those means either totally or partially without good reason. Technology is not a demon: it is carried out by responsible-minded compatriots (*Volksgenossen*), with whom the artist can and should cooperate with the best form of camaraderie for the new Germany.[122]

This new view was in line with other engineers who binded themselves to the ideology of the Third Reich.

Trautwein also worked on the physics of loudspeakers and sound amplification as part of Telefunken's assignments for a number of large gatherings of the Nazi Party in Berlin, Vienna, Nuremberg, and Stettin, putting to work the experience he had gleaned at the RVS testing loudspeakers.[123] He often used sound checks to demonstrate his trautonium on the loudspeakers. In the autumn of 1935, a trautonium with amplifiers was installed in the Dietrich-Eckart Open-Air Theater (now called the Waldbühne, or Forest Theater) adjacent to the Olympic Stadium.[124] The trautonium was featured in a number of the sound checks leading up to the Berlin Olympic Games in the summer of 1936, including the bass accompaniment of two thousand choristers and two hundred orchestral musicians in George Frideric Handel's opera *Hercules*.[125] Apparently, twenty thousand spectators were listening in the open-air theater. A musical high point of the sound checks included a performance of one of Harald Genzmer's compositions on the trautonium.[126] The trautonium could be heard in the official radio broadcast of the Olympic Games with Genzmer's composition for trautonium and piano.[127]

FIGURE 6.4. Loudspeaker tower based on Friedrich Trautwein's recommendation. The architectural draft was made by Paul Walcker. *Source:* Trautwein, "Dynamische Probleme," 32.

Rundfunktrautonium (Radio Trautonium)

Sometime in 1935 or '36, the Reichs-Rundfunk-Gesellschaft (now under complete Nazi control) placed an order with Telefunken for a new version of the trautonium. This trautonium would later be called the Rundfunktrautonium to be housed in the Berlin's Haus des Rundfunks (Broadcasting House) on the Masurenallee in the Charlottenburg district.[128] Telefunken gave the order to Trautwein, who passed it on to Sala, who took the lead in constructing the instrument. He completed construction in late 1937.[129] While Sala claimed that Trautwein had nothing to do with the improvements to the instrument, it turns out that Trautwein was responsible for building a circuit responsible for the synchronization of the subharmonic generator with the main generator, for which he received a patent in 1937.

FIGURE 6.5. Oskar Sala playing the Rundfunktrautonium (radio trautonium), circa 1938. Reproduced with the permission of the Deutsches Museum (Munich) Archives, CD_85668.

Sala and Trautwein had completely stopped working on the trautonium together well before 1937 as a result of priority disputes and accusations of patent infringements.[130] After the outbreak of war, in addition to his duties at the Berlin Academy of Music, Trautwein became the director of the Kurt Eisfeld Company for Electromechanic Aircraft Construction in Berlin-Heiligensee.[131] In mid-1944 he received a teaching assignment for applied physics at the University of Freiburg; however, due to illness and the destruction of the university's Institute for Applied Physics by the Allied forces, Trautwein could not accept the offer.[132]

Two conductors in particular were interested in Sala creating this newer version of the trautonium: Otto Dobrindt, director of the radio orchestra of Germany's Deutschlandsender (Germany Channel), and Herbert Jäger, renowned for his daily radio show "Allerlei von 2 bis 3" ("A Number of Things from 2 to 3").[133] Much like the concert instrument, this new instrument contained two playing manuals (metal strips), two mixers, and two pedals, which were independent of each other.[134] Sala reconfigured the pedaling.[135] Two qualities were controlled by two separate movements of the pedals: moving them back and forth altered the volume—similar to the concert instrument, and a sideways movement changed the timbre. The hands could stay on the manuals rather than execute other tasks: more difficult, virtuosic pieces could now be played.[136] Sala immodestly boasted that the improvements enabled him to perform like a virtuoso. The radio trautonium vastly increased the number of tone colors upon which the performer could draw. Various timbres could be blended to produce an overall tone color (*Gesamtfarbe*). The right pedal unleashed "dark, soft, vowel-like, full tone colors," while the left pedal governed "higher, silvery, nasally, light tone colors."[137] The performer could now easily change the mixtures of tone colors while playing the piece; one could even alter the tone color while playing a single note. The new instrument could also generate two different subharmonics per manual by coupling two thyratrons in a circuit with a large resistor and small capacitor.[138] For example, with the fundamental tone at C6, its subharmonic series—C5, F4, C4, A flat 3, F3, and so on—could sound simultaneously. It possessed a frequency divider that enabled the instrument to generate accordion-like tones—a device that Sala would later feature in his Mixturtrautonium, discussed in the next chapter.

According to Sala, the various broadcasts of the radio trautonium before the war resulted in the trautonium's continued developments. More pieces, many of which had been beyond the abilities of the original trautonium, could now be included in the repertoire.[139] From January 1938 to September 1939, the radio trautonium was featured more than fifty times on the radio throughout Germany and its occupied territories in Deutschlandsender, which featured a series called "Musik auf dem Trautonium" ("Music on the Trautonium"), which lasted approximately fifteen to twenty-five minutes.[140] In addition, it was featured twenty-seven times on "Musikalische Kurzweil" ("Musical Pastime"), "Musik am Nachmittag" ("Music in the Afternoon"), and "Unterhaltungskonzert" ("Entertainment Concert").[141] The trautonium was without a doubt the most popular electric musical instrument during the Third Reich. Critics were impressed by "this instrument of wonder," the tone colors of which appeared to be "inexhaustible."[142]

In April 1939, in the large Berlin broadcasting hall in the Haus des Rundfunks, Deutschlandsender broadcast the premiere of Genzmer's "Konzert für Trautonium und Orchester" ("Concerto for Trautonium and Orchestra"), an expansion of the original composition with the same title (discussed above). It comprised three movements that featured the trautonium. Karl List conducted.[143] Musicologist Hugo Heurich noted:

> Some radio listeners who simply turned on their radio without previously looking at the program listened with admiration to an extraordinary virtuoso [Sala] beautifully playing a violin solo and were then astonished when the alleged violin suddenly descended from the highest pitches of the instrument to below the lowest possible tones such that it suddenly sounded like a cello, and then a clarinet, then a horn, and after that a flute. And then one learned that the solo was played on the trautonium.[144]

Another critic claimed that the new instrument was "a legitimate tool of art" and "the most perfect of the electroacoustic instruments."[145] Another still described the sounds as "indescribably clean" and "almost celestial."[146] A significant percentage of the reviews extolled the technological advances that enabled the "unfolding of limitless tone colors," "a welcomed enrichment of the palette of expressions" at the composer's fingertips.[147] With this technology, virtuosity could now be demanded of all keyboard and string-instrument players. "Until now, the unimaginable—the long awaited—becomes a reality through technology. A product of electrical engineering lends wings to the fantasy of the composer."[148] Rather than being antithetical to creativity, technology enabled the composer to be creative once again. Similarly, in a review titled "Technically and Creatively Transformed," Heinz Joachim—a Berlin music student who was infamously murdered in Plötzensee Prison in 1942 for his resistance to the Nazis—praised the role that electroacoustic instruments had played in the renewal and broadening of Western musical instruments.[149]

In addition to playing the trautonium on the radio, Sala traveled throughout Germany with his concert instrument. After a live performance by Sala in Düsseldorf in May 1939, critics spoke of the "nearly unlimited opportunities" the two-manual instrument provided.[150] Due to the trautonium's ability to alter its tone color, the only criticism of the performance was that at times it was difficult to differentiate between the solo instrument and the orchestral accompaniment. Other reviewers commented on the instrument's ability to change tone colors throughout a piece, praising "the seemingly sensational, pioneering work of the engineer and musician in the field of instrument building. The nearly limitless possibilities of the trautonium rendered the performances of Paganini and contemporary composers G. A. Schlemm and Harald Genzmer a resounding success."[151]

Sala performed live with the concert instrument in Cottbus on the Führer's fiftieth birthday, April 20, 1939.[152] A Cottbus critic explained how the invention of the trautonium was predicated on the invention of radio. He gave a brief history of the instrument dating back to Trautwein and thought that its external appearance resembled an interesting conglomeration of a piano, organ, and radio. He also detailed the procedure by which the trautonium could change tone color and produce sounds that resembled a violin, flute, clarinet, or horn: by adding frequencies together that had been filtered out and that constituted the overtones of those instruments. He felt that the most impressive aspect of the instrument was its "seemingly limitless tone-color palette."[153] Another music critic who listened to the performance in Cottbus was also impressed by the range of tone color featured; however, he did predict that this technical innovation would be unfamiliar to the majority of listeners, who might have preferred the specific sounds of each of the individual instruments.[154] A musical instrument possessing the tone colors of other instruments lacked one of its own. A reviewer in the *Dresdener Neueste Nachrichten* saw the concert instrument as a true competitor to the more traditional instruments.[155]

Konzerttrautonium (The Concert Trautonium)

The broadcasts and tours were so popular that in 1939 the Reich's Broadcasting Corporation requested that a new version—to be called the Konzerttrautonium (concert trautonium)—be built.[156] It was partially financed by the Nazi Reichsmusikkammer, whose explicit goal was to promote Aryan music, and completed a year later.[157] Peter Raabe and Alfred Morgenroth, whom Sala invited to listen to the radio trautonium, were so impressed that they agreed to its financial backing.[158] The concert trautonium also possessed two manuals and two pedals. The major difference was that it was portable and purposely designed to be transported throughout the Reich, unlike the radio trautonium. Each of the manuals possessed an ability to create additional subharmonic tones.[159]

It contained numerous tuning knobs and two manuals, each of which comprised a metal rail on which a resistance wire was attached. Auxiliary keys

FIGURE 6.6. Oskar Sala playing the Konzerttrautonium (concert trautonium) with Harald Genzmer accompanying on the piano at a performance in the Mozartsaal of Vienna's Konzerthaus in 1942.

indicated where on the manuals musical intervals such as thirds, fourths, and fifths were located. As with the radio trautonium, the tones were generated electroacoustically when the finger pressed against the string, thereby making contact with the metal rail. Sala insisted that the concert trautonium could create ten thousand shadings of timbre resembling the sounds produced by the clarinet, horn, oboe, flute, tuba, cello, and violin by increasing the number of circuits that comprised two capacitors and two self-inductors.[160] Similar to the radio trautonium, the pedals controlled both the volume and tone color. The range of the instrument was increased over the years from three to five octaves.[161] As was the case with the earlier trautoniums, the principal circuit consisted of various capacitors, resistors, and thyratrons, and one could reproduce the vibrato and glissando effects typical of string instruments.[162] This instrument, however, possessed a powerful amplifier and loudspeaker.[163] Two thyratrons in conjunction with a large resistor and small capacitor produced subharmonics. By studying the oscillograms, Sala was convinced that he could demonstrate experimentally that the timbres of vowels and musical instruments—including the trautonium—were determined by the number of natural frequencies

(*Eigenfrequenzen*), their position in the frequency domain, their dampening, their phases, their amplitudes (and variations thereof), and the manner of their impulse (or shock) excitation.[164] He also provided the mathematical analysis of the oscillation modes as depicted by those oscillograms.[165]

Sala's Tour of the Reich

Sala toured with the concert trautonium throughout Nazi Germany and allied and occupied countries from 1940 to 1943, often accompanied by Genzmer on the piano. In late August 1940, at the Leipzig Trade Fair in the State Conservatory, Sala presented the concert trautonium.[166] On October 28, 1940, or nearly five months after the Royal Air Force dropped its first bombs on Berlin, Sala debuted his concert trautonium with the Berlin Philharmonic (conducted by Carl Schuricht). He played Genzmer's "Concerto for the Trautonium and Orchestra," the first time the piece was performed before a live audience.[167] Walter Steinhauser congratulated Schuricht for his daring, labeling the Berlin premiere "an artistic experiment."[168] Alfred Burgartz was not as convinced as his colleagues, arguing that the instrument was "an incomprehensible Proteus, a shrill chameleon," "a beast."[169] On November 10, Sala performed in Berlin's Schumann Auditorium accompanied by Genzmer on the piano playing Niccolò Paganini's "La Campanella," "Witches' Dance," and "Carnival of Venice," the last one arranged by Genzmer; Feruccio Busoni's "Kultaselle"; and Genmzer's "Fantasie-Sonata für Trautonium und Klavier" ("Fantasy Sonata for Trautonium and Piano").[170] Paganini's "Adagio" and "Rondo" from "La Campanella" were arranged by Sala, who added various tone-color changes and more virtuosic passages. He admitted that he did not wish for his trautonium to imitate the violin. Rather, he strove to create "a new tonal representation from the essence of the trautonium."[171] He felt the variations enabled the performer (himself, of course) to reach a new level of expression and virtuosity. Sala also arranged Busoni's piece, particularly the bass register, such that the piece could only be performed by the trautonium. Genzmer's "Fantasy Sonata" arose from his earlier work for the trautonium and orchestra. Sala was convinced that the new possibilities of sound afforded by the trautonium enabled composers and performers alike to develop their creative imaginations. The trautonium perfectly illustrated how new compositional possibilities could be generated with electrical musical instruments. The variations of the themes contained many tonal surprises—even to Sala himself—which became evident while he was performing. Genzmer's arrangement of some of Paganini's movements in "Carnival of Venice"—adding countermelodies and reinforcing echoing effects—gave the work a different character and rendered Paganini's piece a "small compositional work of art."[172] Finally, Paganini's "Witches' Dance" was

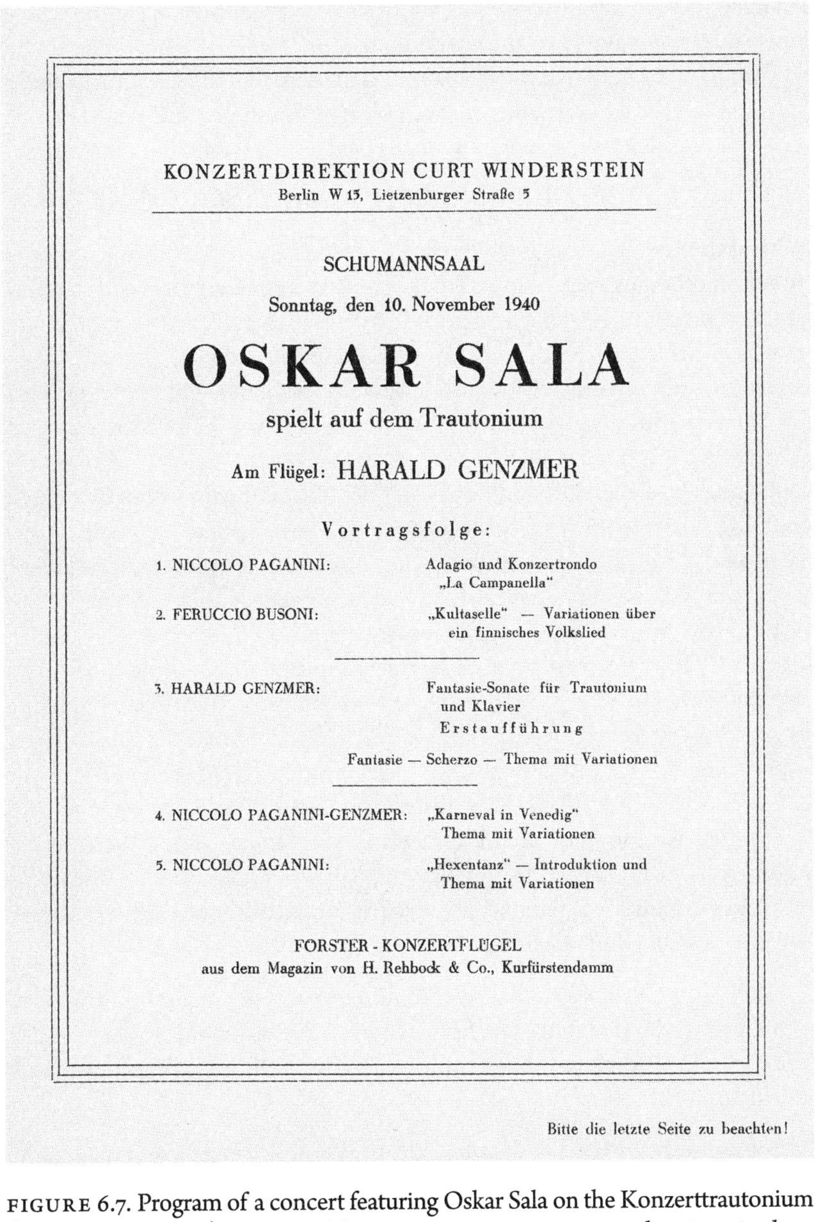

KONZERTDIREKTION CURT WINDERSTEIN
Berlin W 15, Lietzenburger Straße 5

SCHUMANNSAAL

Sonntag, den 10. November 1940

OSKAR SALA

spielt auf dem Trautonium

Am Flügel: HARALD GENZMER

Vortragsfolge:

1. NICCOLO PAGANINI: Adagio und Konzertrondo „La Campanella"
2. FERUCCIO BUSONI: „Kultaselle" — Variationen über ein finnisches Volkslied

3. HARALD GENZMER: Fantasie-Sonate für Trautonium und Klavier
Erstaufführung
Fantasie — Scherzo — Thema mit Variationen

4. NICCOLO PAGANINI-GENZMER: „Karneval in Venedig" Thema mit Variationen
5. NICCOLO PAGANINI: „Hexentanz" — Introduktion und Thema mit Variationen

FÖRSTER - KONZERTFLÜGEL
aus dem Magazin von H. Rehbock & Co., Kurfürstendamm

Bitte die letzte Seite zu beachten!

FIGURE 6.7. Program of a concert featuring Oskar Sala on the Konzerttrautonium (concert trautonium) with Harald Genzmer accompanying on the piano in the Schumannsaal in Berlin on November 10, 1940. Reproduced with the permission of the Deutsches Museum (Munich) Archives, CD_77473.

arranged in such a way as to translate the Italian's virtuosic antics into new tonal realities generated by the trautonium.

The concert critics reiterated Sala's original claim that the instrument could summon ten thousand shades of tone color.[173] The soprano Gertrud Runge was impressed by the instrument's range of tone colors: initially sounding like violins, it metamorphosed into a choir of horns and then a flute. She did, however, feel that the trautonium, which she described as a "*Wunder*," would make its breakthrough if it developed its own character as opposed to imitating the sounds of other instruments.[174] A review underscored that "the trautonium is neither a substitute nor an imitation, but rather much more an electric melody instrument of its own kind."[175] Music critic Ludwig Lade insisted that it could not and should not replace more traditional instruments, but rather increase the number and range of the mixtures of tone colors. That was the crucial point: what was needed was "a new composer for the new instrument."[176] Wilhelm Zentner agreed, cautioning that the trautonium must not be seen as imitating other instruments, as it possessed its own unique "character of expression."[177] Another critic spoke of the "romantic warmth" the instrument generated.[178] Walter Steinhauer of the *Berliner Zeitung* praised Sala's virtuosity on the instrument and described Genzmer's piece as "fresh" and "lively inspired."[179] One often read about how the instrument could bring the fantasy of the musician to life.[180] Other reviewers were impressed with the frequency range of the instrument, from contrabass to piccolo.[181] Many remarked on the improvement of the instrument since its invention a decade earlier.[182]

One important attribute was underscored in a number of reviews: the trautonium was not a mechanical musical instrument—such as a musical automaton—a characteristic that Nazi officials would not have condoned.[183] Heinz Kirschninck insisted that, just like the more conventional instruments, this instrument could reach "the depths of the soul." He added:

> the trautonium is neither an automaton nor a radio apparatus. It may be the prototype of a new type of instrument development; the very same mystical phenomenon that lies in the singing violin, also rings out of this instrument.[184]

Lothar Band concurred, insisting that electrical musical instruments had nothing in common with penny automata, or *Groschen-Automaten*.[185]

On November 18, 1940, in agreement with the Reich Ministry of Public Enlightenment and Propaganda, Sala performed in the Hercules Auditorium in Munich.[186] By that time, Sala had already performed 150 concerts on the radio.[187] From late 1939 to late 1940, he performed thirty new compositions on his concert instrument and concert trautonium.[188] One reviewer recommended the instrument for lyrical, humorous, grotesque, and fantasy-filled passages.[189] Rudolf Hofmüller felt that the instrument would not threaten more traditional instruments because it was too difficult to play. He also admitted that only time would tell how the trautonium would be used, postulating that it could potentially achieve success in theater or film.[190]

In January 1941, Sala continued his tour, often accompanied by Genzmer, this time in Stuttgart. The concerts were promoted by the Nazis' leading leisure organization—Kraft durch Freude (or KdF), Kulturgemeinde (Strength Through Joy—Cultural Community)—which by the late 1930s had become the world's largest tourism operator. With a view to render bourgeois leisure activities accessible to the lower classes, the KdF provided working-class families with day trips, holidays, and concerts. On January 10, Sala performed Genzmer's "Concerto for Trautonium and Orchestra" in the Liederhalle (Choral Hall) with the State Orchestra of Gau Württemberg-Hohenzollern, directed by Gerhard Maasz. The Stuttgarters were impressed with the instrument's range of tone color, variability of playing possibilities, and its dynamic shading. One reviewer echoed the Berlin critics' view that this instrument was not meant to replace others.[191] Several days later, Sala played Paganini at the State Conservatoire. Alexander Eisenmann of the *Württembergische Zeitung* said that the violin possessed more soul than the trautonium, but that the demonstration of the "technical work of wonder aroused the curiosity of the visitors."[192] An article in the periodical *Funkschau* claimed that the trautonium was a fully legitimate and concert-worthy musical instrument that potentially could become popular.[193]

The trautonium became a powerful tool of cultural propaganda in occupied and allied territories.[194] For example, in February 1941, Sala played before an audience in Utrecht.[195] Dutch papers praised the concerts, calling them "a historic event" and the instrument "a technical work of wonder"[196] In May 1941, Sala and Genzmer performed in Teatro Verdi Florence. The repertoire was their standard one: Pagnini, Busoni, and Genzmer.[197] The concert was attended by Dr. Rosen of the Ministry of Public Enlightenment and Propaganda.[198]

The ensuing October and November included performances in Chemnitz and Beuthen in Upper Silesia (now Bytom, Poland).[199] On October 8, Sala performed the premiere of Genzmer's "Musik für Trautonium und Blasorchester" ("Music for Trautonium and Wind Ensemble"), directed by Rudolf Schulz-Dornburg, in the large Berlin broadcasting hall.[200] By the end of November, the pair played in Hanover, Lower Saxony, also in conjunction with the KdF.[201] In early December, Sala performed in his hometown of Greiz, Thuringen.[202] The program featured Paganini; the concert trautonium was described as a "wonderful instrument . . . so perfect that it cannot be compared with any other instrument."[203] It was then on to Wilhelmshaven, where the concert was sponsored by the city's Nazi Party in conjunction with the

KdF.[204] The final performance of 1941 took place on December 9 in Strasbourg, where a reviewer praised the trautonium's ability to change tone color; however, he too stressed that the instrument possessed a truly new character and should therefore not be used to imitate other instruments.[205]

The jubilation felt by many within the Reich was at its zenith during that performance. Nazi Germany declared war on the United States, after the United States declared war on Japan for attacking Pearl Harbor. By the time the spring thaw arrived in 1942, the general mood in the Reich had soured. Operation Barbarossa, launched in June 1941, had failed to yield total victory against the Soviet Union in a single campaign. Aerial assaults on Germany by the US Air Force and the Royal Air Force increased in frequency.

Harkening back to the instrument's ability to produce sound effects for talkies, the concert trautonium was also featured in various musical works. In 1942 it produced the sounds of gongs for Richard Strauss's "Japanische Festmusik" ("Japanese Celebratory Music") at a performance of the Berlin Philharmonic, which Strauss himself conducted. A review in the *Deutsche Allgemeine Zeitung* in January 1942 lauded the instrument:

> When Richard Strauss conducted his "Japanische Festmusik" in a recent recording session in Berlin, the problem of the gongs [which could not be obtained from Japan during the war] was solved by the trautonium. As is well known, this electroacoustic prodigy can do everything: it fiddles, flutes, drums, trumpets, and now it even imitates gongs! Strauss was so pleased with this substitute gong that the instrument with its many mysterious knobs will now take the place of a whole battery of giant gongs in the performance by the Philharmonic Orchestra.[206]

The concert was repeated in Dresden in February 1942 with Karl Böhm conducting.[207] Similarly, in December 1942 the instrument was used as a substitute for the ondes Martenot in Arthur Honegger's "Jeanne d'Arc au bûcher" in the Teatro Adriano in Rome.[208] In this instance, the trautonium was meant to produce "the unearthy, demonic—so to speak—associations of new sound material."[209] It was also entrusted with generating the howling of a dog.

The concert trautonium was featured in performances in Dresden, Budapest, Vienna, and Oldenburg in March and April 1942. In June it was back with the Berlin Philharmonic, and we are told that audiences—particularly in Budapest—warmly welcomed Sala and Genzmer and energetically applauded their performances.[210] On November 7, the instrument performed in Erfurt. Heinrich Funk of the *Thüringer Allgemeine Zeitung* stressed—as so many did—that the concert trautonium should not replace or imitate a musical instrument; its ability to create new timbres and sound qualities was crucial.[211]

A November 19 article in *Fuldaer Zeitung* announced a forthcoming concert with the trautonium. It stressed the uniqueness of Sala's instrument. It was neither a mechanical instrument, such as a pianola or orchestrola, nor a Neo-Bechstein grand piano or theremin. On the contrary, the concert trautonium possessed a "soulful tone."[212] On 21 November 1942, the KfD sponsored a concert by Sala in Fulda. Sala played works by Paganini, Liszt, Handel, and F. Ries, as well as compositions written specifically for the trautonium by Georg Häntzschel, Josef Ingenbrand, Gustav Adolf Schlemm, and Genzmer.[213] One critic exclaimed that the significantly improved trautonium was "yet another example of German spirit [or intelligence, *Geist*] that was not only conquering German music halls, but European music halls as well, and doing so in the middle of a war."[214] The next stop was Brunswick (Braunschweig): the concert took place at the Technical University, and the audience included students from the SS music academy.[215] One critic stressed that Sala wished to dispel the myth that the instrument suffered from "soullessness." He felt that the numerous tone colors generated by the instrument were "tiring." Perhaps, he argued, that was why it was seen as "soulless." He also feared that the instrument did not possess its own unique "sound character."[216] In early December Sala and Genzmer were in Danzig (Gedansk, Poland) and finished the year with a concert in Rome.[217]

The continued Allied bombardment of Germany rendered tours more challenging. Sala performed in Essen on January 17. Two days later, Franz Clemens Gieseking's review of Sala's concert in Münster appeared, claiming that trautonium was "a German invention that was absolutely unrivaled in the world."[218] The KdF organized concerts in Dessau and Meißen in mid-February 1943.[219] The final live performance in Germany of the concert trautonium during World War II occurred on May 8, 1943, in the industrial city of Gelsenkirchen in the Ruhr region.[220] The following November, the Foreign Organization of the Nazi Party (Auslands-Organisation der NSDAP) organized a performance of the trautonium in Paris's Palais d'Orsay titled "Sphärenmusik—Was ist ein Trautonium?" ("Music of the Spheres—What is a Trautonium?"). The *Pariser Zeitung* spoke of the beauty of its tone.[221]

These numerous concerts with the trautonium featured classical works, such as Beethoven's Romance in F major (opus 50); Handel's Flute Sonata in F major (HWV 369); various movements from Paganini's Violin caprices; works by Giuseppe Tartini, Henryk Wieniawski, Pablo de Sarasate, Carl Maria von Weber, Ferdinand Ries, and Ludwig Spohr; pieces by contemporary composers, such as Busoni's "Kultasella," Georg Haentzschel's "Stück für Trautonium und Klavierbegleitung," Julius Weismann's "Thema mit Variationen und Fuge über ein eigenes Thema, Opus 143 für Trautonium und Orchester,"

Gustav Adolf Schlemm's "Capriccio über Sechs Okatven," Christian A. Pfeiffer's "Suite für Trautonium und Klavier," Hermann Ambrosius's "Concertino," Wolfgang Friebe's "Capricio für Trautonium und Orchester," Helmut Riethmüller's "Drei Stücke für Trautonium und Klavier," Fried Walter's "Elegie für Trautonium und kleines Orchester," pieces by Josef Ingenbrand, and, of course, the aforementioned works of Genzmer.[222] By the final performance in 1943, Sala had played (often with Genzmer) in Altenburg, Beuthen, Berlin, Chemnitz, Danzig, Dessau, Dresden, Duisburg, Erfurt, Essen, Gelsenkirchen, Greifswald, Halle, Hanover, Kassel, Marburg, Meißen, Munich, Oldenburg, Plauen, Strassburg, Stuttgart, Wilhelmshaven, Zittau, Brussels, Budapest, Florence, Hilversum, Paris, Rome, Utrecht, and Vienna.[223]

As we saw, critics generally stressed two points: first, the trautonium should not be used to imitate other instruments, and second, it was not a mechanical instrument, but one that possessed a soul. Perhaps the most interesting reviews of the electric musical instruments in general, and the various versions of the trautonium in particular, were the ones that saw the instrument as a return to romanticism, albeit with a new twist. They singled out a particular aesthetic that resonated with fascist ideology. On January 16, 1942 the *Mitteldeutsche National-Zeitung* reported that electric musical instruments "touch ever so lightly the sounding regions of the romantic imagination." They "seek the ideal of literary romanticism in incorporeal, indeed supernatural sounds."[224] When referring specifically to the trautonium, one often reads the phrase rendered famous by Goebbels, that it was an "instrument of steely romanticism." Many greeted the instrument "with passionate hearts."[225] One thinks of the philosopher Ernst Cassirer's phrase "romantics of technology," meaning the praise of technologies that people perhaps did not quite understand. Cassirer famously warned us that these technologies could be used by those harboring nefarious, antimodernist views.[226] One critic said that it sounded similar to the glass armonica, the preferred instrument of late-eighteenth- and early-nineteenth-century Romantics. Electric musical instruments regenerated the romantic aesthetic, or at least this is what so many critics echoing Goebbels's phrase argued. Austrian composer and music critic Fritz Skorzeny held a similar view of the trautonium, labeling it an "elektrisches Wunderinstrument der 'stählernen Romantik,'" or "an electric instrument of wonder of 'steely romanticism.'"[227] The instrument was billed as a "Grosstat Deutschen Erfindergeistes" ("feat of German spirit of inventors").[228] Another critic commented on how the trautonium's success was an honor for German music, "because Germany has defeated all foreign competition in this field [electric musical instrument building], thanks to farsighted support from interested parties."[229] The author was most likely referring to Goebbels. A reviewer of the trautonium stressed Goebbels's foresight in financially backing the instrument.

> By the way, it was Reich Minister Dr. Goebbels himself, who at the decisive moment recognized the future significance of the trautonium, especially for large mass rallies and celebrations. And in contrast to some hostility and skepticism on the part of music experts (what new invention would have been spared the fate of being misunderstood?), he emphatically supported and promoted the endeavors of Trautwein and Sala.[230]

In October 1944 Sala was drafted by the Wehrmacht and was sent to the eastern front to fight the Soviet Red Army as a grenadier. He broke five ribs after falling into a ditch in East Prussia. In addition, shrapnel struck him over his left eye. He was transported from Königsberg to Danzig and then to a hospital in Münster. He concluded the war as a member of a radio squadron.[231]

While one might have thought that the Nazis would dismiss all aspects of the new genre of electric music as "degenerate," the contrary proved to be case. The trautonium and its various instantiations were promoted by the Nazis. The instrument was rarely heard playing microtonal pieces, an attribute extolled by its creators, as microtonal music was frowned upon in Nazi Germany. Rather, works composed for more traditional instruments, such as the piano, flute, or violin, were featured, often arranged by Sala to reflect the instrument's unique characteristics. In addition, pieces specifically written by contemporary composers for the instrument were also featured.

While the trautonium was often featured in performances under the Third Reich, it did at times struggle with its identity. Reproduction of the original sounds of the acoustical instruments had its supporters, particularly among the Nazis. Hausmusik was an important cultural characteristic of the Volk. In addition, Telefunken's advertising acumen ensured that Germans identified the instrument as one that could take the place of a violin, cello, flute, or Hammond organ. Yet the trautonium enjoyed a massive tone-color palette and could therefore create new sounds, thereby enriching the musical experience. Recall that Sala's performances of variations of classical pieces, such as those composed by Paganini, were intended to exhibit the trautonium's ability to produce new timbres. The Nazis did not disapprove of such performances; on the contrary, they actively sponsored them.

This chapter also illustrates the Nazis' labyrinthine relations with technology. While Nazis constantly espoused the superiority of "the natural," they also used technology to enhance musical experiences. The Nazis promised engineers a future in which they would be respected and gainfully employed.

Many engineers intensified their critiques of capitalism and stressed the importance of technology to German culture, its spirit and soul.

German inventors created a new aesthetic for the trautonium that uniquely resonated with Nazi ideology: one of steely Romanticism. While the Nazis drew upon preexisting conservative musical trends that were present during the late Weimar Republic, steely Romanticism was a new musical aesthetic created by radio equipment like resistors, capacitors, LC-circuits, thyratrons, amplifiers, and loudspeakers. Trautwein actively worked to support fascist tools of propaganda by providing the sound systems required for broadcasting at mass gatherings. In short, the trautonium became an instrument of propaganda throughout the Third Reich as it embodied the virtues of Germany's musical, scientific, and technological prowess.

7

The Trautonium after the War

JUST AS the future of the trautonium had been uncertain after the dissolution of the Weimar Republic and the rise of National Socialism, no one knew what the future of the instrument would be in a completely devastated land. Shattered lives and relationships abruptly severed by the war needed to be mended. Seeking employment in ravaged postwar Germany as a member of the vanquished Nazi Party was not an enviable position to find oneself. The war had taken its toll on Friedrich Trautwein. He lost one son, Albert, in Ukraine in 1943, and another lost an arm.[1] Trautwein found work in a small factory for electrophysics and electromusic run by his son Werner in Uiffingen, Baden, making electronic bells whose sounds were sent through loudspeakers for the Lutheran church in Walldorf some 12 kilometers south of Heidelberg.[2] In 1948 he went to Paris to work for Compagnie Sans Fils (CSF), where he continued research along the lines of what he had done at the Kurt Eisfeld Company during the war. He returned to Germany in 1949 and taught sound engineering at the Bild-und Klangakademie (Picture and Sound Academy) in Düsseldorf, which ceased operations shortly after he arrived. In addition, he attempted to revive interest in the trautonium.[3] Commenting on the economic disaster after the war, Trautwein made a pitch for his instrument as a potentially lucrative export. He added that trautoniums could be used by impoverished orchestras, as they could save on the labor of musicians.[4] Imitation might just be a profitable trait after the war. Drawing upon the success of his estranged colleague, Sala, Trautwein recommended the trautonium for movie soundtracks. He also pointed out that the trautonium could create numerous strange noises that other instruments could not.[5]

The city of Düsseldorf provided Trautwein a cellar free of charge in exchange for teaching music students techniques for generating sounds and images.[6] In 1949 the state of North Rhine-Westphalia gave him 2,000 Deutsche marks (DM) to establish a program of study in sound engineering (*Tonmeisterausbildung*) at the Düsseldorf Conservatoire (now called the Robert Schumann Musikhochschule Düsseldorf). A year later, the program accepted its first students, and Trautwein was the founding director. By the mid-1950s, the field of sound engineering was in far greater demand than it had been back in the 1930s, as radio, talking pictures, and sound-recording industries were booming. Successful applicants needed to possess musical talent and a basic knowledge of piano playing as well as a good command of physics and mathematics at the level taught in high schools. The length of study was three years and was divided into two general parts: music and technology. Training in music included: harmony, reading musical scores, music history, musical form, ear training, instruments, and choral singing. Technological expertise could be gleaned from course work in the mathematical and physical principles of electroacoustics and radio-freqeuncy technology, musical acoustics, electronic music, and the fundamentals of room and building acoustics. Instruction was based on lectures and practical exercises, which included a physico-technical practicum that involved studying the techniques of sound recording, sound films, sound studies, radio dramas, and the acoustical design of rooms. We know that while the students who finished their study with Trautwein were successful in obtaining gainful employment, the numbers were minuscule, and he began to worry about his financial well-being.[7]

With a hope to improve his financial situation, he accepted the occasional invitation to collaborate with colleagues. For example, Berlin conductor Hermann Scherchen, with whom Trautwein had collaborated on projects back in the early 1930s at the Rundfunkversuchsstelle, invited him to conduct research at the newly created Electroacoustic Experimental Studio in Gravesano, Switzerland. In 1952 he worked briefly with Herbert Eimert in the Studio for Electronic Music of Nordwestdeutscher Rundfunk (Northwest German Radio, NWDR) in Cologne, where he built an electronic monochord that was similar to his trautonium.[8] As tensions between the two heightened, Trautwein returned to teaching students in Düsseldorf.[9]

Trautwein felt bitter over the lack of recognition of his earlier accomplishments with vowel production on his trautonium. Oskar Sala wanted to reassure him that his early research, and not Karl Willy Wagner's "speech model" (*Sprachmodell*), was the actual precursor to the Vocoder, a human voice synthesizer invented by Homer Dudley at Bell Telephone Laboratories. In a letter addressed to Trautwein, Sala wished to provide him with some solace:

> You are the one who translated [Ludimar] Hermann's theory . . . with a truly brilliant intuition into electrical form, and the vowel formants are doubtlessly excited simple natural frequencies (*Eigenschwingungen*). Don't allow your great idea to slip away. I think in this respect that science has outrageously underestimated your merits. Everywhere K. W. Wagner and his "speech model" are called the precursors to the Vocoder and similar

> apparatus. I have tried hard to fight this fight everywhere, where I had something to publish, and I shall continue to do so . . . [by] calmly pointing out the amazing results, namely that your first trautonium was able to say a, o, and u to the astonishment of all listeners.[10]

Sala attempted to enter into more collaborations with Trautwein shortly before the war's conclusion: alas, they proved to be unfruitful. The two had quarreled over patent rights for at least sixteen years, and it would not be until February 1953 before the two men cosigned an agreement that finally settled their arguments concerning priority claims and patent infringements.[11] Sala visited him the following September, and by December they had reestablished a professional, working relationship.[12]

Sala informed Trautwein that he could present the aging engineer as "the father of all things" relevant to the trautonium. "I would honor your work, which would be highlighted by my concert series."[13] Similarly, in January 1954, Sala wrote to German director and music editor Siegfried Goslich that Trautwein should be invited as president by seniority (*Alterspräsident*) to open the electronic music series that Werner Meyer-Eppler—physicist, phonetician, and cofounder of the Studio for Electronic Music in Cologne—was planning at the Technical University of Berlin.[14] As discussed in the next chapter, most of Sala and Trautwein's working relationship was dedicated to defining and preserving their legacy in postwar Europe. By presenting a united front, they were far more likely to succeed.

Trautwein and Musical Aesthetics

Late in his life, a pensive Trautwein commented and wrote on the influence of electronic music on musical aesthetics and engineering. He argued that if a science (*Wissenschaft*) of art wants to provide an objective foundation, it can only be through aesthetics in the purest sense of the Greek definition, meaning, the study of perceptions. He also firmly believed that the fundamental aesthetic principle in music is that "the ear is the measure of all things."[15] For Trautwein, human perception needed to be based on experimentally determined properties.

> Should physico-mathematical assumptions contradict the properties of sensory perception, then priority must be given to the experimental findings until theory and experiment can be successfully reconciled. On the basis of natural science, conclusions of largely objective importance for the humanistic direction of aesthetics can also be added. On this basis, the question of the effects of electronic sound creation on the development of music must be answered fully in the positive; future musicians will from

> now on be freed from material restrictions, as every possible acoustical form is now available. This freedom, however, imposes on them an enormous responsibility in which the spirit predominates over the material.[16]

Trautwein sharply criticized those who wished to disallow any "objective, scientific approaches" to the subject of aesthetics.[17] He accused them of focusing on the inferiority of technology and of asserting that the natural sciences were merely products of "pure, rational, and soulless activity, in contrast to art, which by itself can create from the deep resources of the soul."[18] Aesthetics could only become the doctrine of the beautiful when it was based on the natural sciences of perception.[19] Science and technology required intuition just as much as a work of art needed a logical plan of implementation and manual skill. He blamed this false dichotomy on the new, postromantic worldview, which criticized exuberance while simultaneously assuming a supernatural imagination: "Art and technology draw upon the same sources of human spirit (*Genius*)."[20] Without discovery, art and technology would be unthinkable. In 1953 he drew a critical link between engineer and artist during an interview with Radio Bremen.

> There is no difference between the artistic and technical creative tendency. . . . This demon, this divine epiphany, this little spark must be added, and that will come in a totally unexpected moment. I have troubled myself to unite these two things, creative activity in the fields of technology and art.[21]

While technology could certainly assist art by generating a creative impulse, it was in no way subordinate to art. According to Trautwein, history was replete with examples whereby technology had either given rise to new forms of art or improved upon previously existing ones.[22] As Germans tend to do, Trautwein summoned Goethe's spirit to defend his point. Not only did Goethe successfully link art and science, he employed technology against the machinations of Mephistopheles in order to improve the human condition.[23] His sentiment was reminiscent of a declaration made by the Nationalsozialistischer Bund Deutscher Technik (National Socialist Association for German Engineering) in 1939: "It is impermissible that artistic knowledge is socially accepted, but technical competence is placed on a lower level."[24]

Once again, Trautwein was drawing upon a trope that had been popular since the end of World War I. The role of *Geist,* or spirit, within technology was paramount to both Trautwein and engineer, physicist, and philosopher Friedrich Dessauer.[25] As German historian and philosopher Oswald Spengler argued in the last chapter of his 1918 *The Decline of the West,* modern technology possessed a Geist. Geist and technology were not antithetical, since it was a Faustian technology that demonstrated a "will to power over nature."[26] That sentiment was also expressed by Manfred Schroter, an associate of Spengler, who, in his *Deutscher Geist in der Technik* (*German Spirit in Technology*) of 1935, interpreted the Faust legend as a "demonic technical destiny" that called for the domination over nature by German engineers.[27]

Trautwein began to lecture widely on the relationship between technology and music; he wished to carve out a space for engineers, once again, in defining *Kultur*, a space that hitherto had been dominated by humanists, social scientists, and natural scientists as *Kulturträger*.[28] Much like the engineers some thirty years earlier, he decided to write on the relationship between technology, philosophy, and society.[29]

In a lecture delivered at the Volkshochschule (Adult Education Center) Düsseldorf less than a month before his death, Trautwein bemoaned the infamous distinction between Kultur, which was organic and superior, and *Zivilisation*, which was mechanical and inferior.[30] He felt that "the *totality* of the human creative expressions of life (*Lebensäußerungen*), as Goethe and Rudolf Steiner underscored," illustrated that there was no true schism between Kultur and Zivilisation.[31] Such a view had been rather typical of what Jeffrey Herf has called reactionary modernism during the Third Reich: engineers wished to convert technology from being a component of mechanical Zivilisation into being a part of German organic culture.[32] Trautwein's rhetoric eerily mirrored that of a number of other engineers during the Third Reich.

The renowned Spanish philosopher José Ortega y Gasset maintained that the role of the engineer, however magnificent and venerable, was inevitably second-rate.[33] The perfection of scientific technics inevitably resulted, he argued, in the diminution of imaginative abilities. Engineers lacked creativity: technics is "an empty form—like the most formalistic logic; it is unable to determine the content of life."[34] Trautwein felt that if Ortega y Gasset meant that the engineer was seen by society as a person of inferior dignity with inferior rights, then he would unfortunately be correct. If, however, he was asserting that this was just and accurate, then he needed to be challenged. "The saying of Struwwelpeter forces itself upon us: 'What can the Moor do, since he is not white like all of you?' What then can the engineer do if he is neither philosopher, artist, nor literary figure?"[35] In essence, Trautwein was echoing Julius Schenk's *Die Begriffe "Wirtschaft und Technik" und ihre Bedeutung für die Ingenierausbildung* (*The Concepts "Economy and Technology" and Their Importance for Engineering Education*) of 1912. Schenk, a professor at the Technical University of Munich, had declared that engineers and visual artists both dealt with creative forms and images; therefore, engineers needed to shed their inferiority complex and society needed to change their view of engineers.[36]

It turns out that a number of engineers from 1950 onward turned to apologists such as Friedrich Dessauer to raise their self-image and gain respect so they could once again defend the importance of technology to culture.[37] Evidence suggests that during the mid-1950s, engineers once again felt unappreciated. While West German engineers faced backlash from antitechnology cultural critics, such criticism was not as extreme as it had been after World War I.[38] In 1956–57, the Verein Deutscher Ingenieure surveyed their members on their opinions about status and respect. The response rate was quite high, with 71 percent (or nearly 24,000) answering the survey. The first question was: "Do you think that engineering as a profession enjoys the social recognition and respect in public life that you believe it deserves?" Only 13.2 percent said yes, 60.3 percent said partially, and 24.7 percent said no. The remaining 1.8 percent did not answer that particular question. Another relevant survey question was: "Do you think that engineering services are valued and remunerated according to their fair share overall in the economy and its future development?" Only 13.4 percent responded that such services were correctly valued ("richtig bewertet"), while 83.1 percent felt that they were underestimated ("zu wenig bewertet").[39] Over a quarter of respondents felt that their profession was "underappreciated."[40] Interestingly, some sociologists, including René König, felt that the relationship between culture and civilization was again (or perhaps still) in play.[41]

After World War II, engineers in the western zones of what would become the Federal Republic of Germany began to ask themselves questions about the role of the human vis-à-vis technology. In 1947, the Technology University of Darmstadt sponsored a conference titled "Technik als ethnische und kulturelle Aufgabe" ("Technology as an Ethical and Cultural Duty") supported by funds provided by the US military government. Throughout the 1950s, the Verein Deutscher Ingenieure organized special conferences on similar topics, such as "On the Responsibility of Engineers," "The Human and Labor in the Technological Age," "Human Changes through Technology," "Humans as the Force Field of Technology," "Technology in the Service of the World Order," and "The Engineer and his Tasks in New Economic Areas." In addition, the first lecture in 1956, celebrating the 100th anniversary of the Verein Deutscher Ingenieure, was titled "Mensch und die Technik" ("Humans and Technology").[42] That same year, a working group was created with the same name, chaired by Paul Koeßler, professor of vehicle technology at the Technical University of Brunswick. The main missions of this group were:

> The internal effect: to illuminate the essence of technology and engineering work and to lead the creative engineer out of technical narrowness and one-sidedness. The external effect: for a fair view of our engineering work, for a fair distribution of responsibility for technology, and for a humane application of technical means.[43]

Interestingly, in 1954 Trautwein revisited an issue that both he and Sala had discussed from the outset in 1930, namely, the thorny issue of imitation. By the early 1950s, Sala's emphasis had been placed squarely on the unique sounds the instrument could generate. For example, in the autumn of 1951, Sala asked physicist Werner Meyer-Eppler, a pioneer of electronic music, "But who today seriously thinks to imitate sounds already well known?"[44] Indeed, Sala referred to imitation as "an ominous concept in electronic music."[45] Trautwein conceded that imitation should have a role in electronic music. Commenting on a conference on electronic music in Cologne's Conservatoire in July 1954 at which he spoke, Trautwein admitted that imitation was important:

> The conference convinced me that we should not deal too lightly with the imitation of familiar sounds. Not that you should make it the final goal in and of itself. But if you don't manage to create well-known sound qualities, you don't have the right to talk about new musical territory. . . . Always say: to be able to imitate, but not to do it, only to prove we can. A painter who cannot paint does not have the authority to scold other painters.[46]

Electronic musical instruments should certainly not be limited to imitating other musical instruments, although imitation had certain advantages.[47] Once again, instrument makers and listeners alike shuttled between imitation (fidelity to acoustical instruments) and creativity (the invention of new musical sounds). The two epistemological approaches went hand in hand: one needed the other.[48]

During the last two years of his life, Trautwein had clearly been thinking about his legacy. As Sala began to dominate the trautonium scene, Trautwein reflected on aesthetic and philosophical issues, such as the role of technology and engineers in culture and society, the relationship between imitation and creativity, and the roles of the engineer and electronic devices in forging a musical aesthetic. These were common themes for engineers during the 1950s. On December 20, 1956, Trautwein passed away in Düsseldorf at the age of sixty-eight.

Oskar Sala

In contrast to Trautwein, Oskar Sala's career thrived in postwar Germany, albeit not initially. At the war's conclusion, Sala was imprisoned by US soldiers in Sinzig on the Rhine for eight weeks. He then traveled approximately 450 kilometers to Greiz, his hometown in Thuringia, where he was shocked to find his concert trautonium only slightly damaged and awaiting his arrival, thanks to a US Army sergeant who personally knew Paul Hindemith.[49] In August his wife joined him at Rudolf-Breitscheid-Strasse 52 in Greiz after experiencing numerous horrors in Berlin. She had been taken with a high fever to a Russian makeshift hospital in the unheated cellar of a Berlin building, the Shell-House. The Salas' apartment in Berlin in Lützowstrasse went up in flames the last day of the war.[50]

Sala remained in Greiz until 1947, continuing to play the concert trautonium.[51] He played numerous pieces for the Weimar Radio Station accompanied by Gerhard Schael on piano throughout 1946.[52] He registered at the Weimar Conservatoire in January 1946, and in 1947 he offered a course on "Musikalische Akustik und ihre allgemeinen Grundlagen" (Musical Acoustics and its General Principles) at the Weimar Academy of Music. The course included instruction in elementary wave theory and the fundamental principles of acoustics, electroacoustics, and physiological acoustics.[53,54] He hoped that the trautonium would be used by the Weimar Academy of Music to attract students. While he wanted more trautoniums (particularly quartet trautoniums) built, he realized that that was not a high priority in the aftermath of the war.[55]

Sala occasionally traveled to Berlin to play the radio trautonium. It was still housed in the Haus des Rundfunks in the Masurenallee in Berlin-Westend, miraculously undamaged by the war. He played it for a radio broadcast on January 1, 1947, and several other times later that year.[56] Accepting an invitation to perform on his concert trautonium from Robert Heger, the music director of the Berlin City Opera (now called the Deutsche Oper, Berlin), then located at the Theater des Westens, Sala returned to live in Berlin in June 1947.[57] Sala played his trautonium in Swiss composer Arthur Honegger's hybrid opera and oratorio, *Jeanne d'Arc au bûcher*, directed by Heger, which premiered on December 9.[58] The piece originally included an an ondes Martenot, not a trautonium, in its orchestration. But an ondes Martenot was not available in Berlin at the time. Much like the 1942 performance, the trautonium provided the unearthly sounds and the howling of the dog.[59] Indeed, in the postwar period, more so than during the Third Reich, the trautonium was called on to summon "extraordinary sound effects" rather than imitate acoustical instruments.[60] The trautonium once again replaced the ondes Martenot in Honegger's *Jeanne d'Arc au bûcher*, this time for its Hamburg production in 1950.[61] One reviewer thought that the trautonium was a bit over-the-top and that perhaps in the future it should be used more sparingly, particularly for the creation of unique sounds and timbres.

> The garish tones of an electroacoustic instrument (trautonium) [. . .] enable dynamic, uncanny, and even hellish effects, but it already reached the limits of the artistic and should only be used with extreme sensitivity.[62]

Another had a similar reaction. "The over-dynamic 'trautonium'" drowned out the choir and orchestra and made it appear as if there were an angry omnipotence present.[63]

Sala performed numerous pieces, including those composed by Genzmer and Julius Weismann for the trautonium in 1948 and 1949 and played for Rundfunk im Amerikanischen Sektor, Berlin (Radio in the American Sector, RIAS-Berlin); Berliner Rundfunk (Berlin Radio); the Berlin division of the Nordwestdeutscher Rundfunk (Northwest German Radio, NWDR); and Mitteldeutscher Rundfunk, Leipzig Station.[64]

He indefatigably readvertised his concert instrument during those early postwar years. In an article about the trautonium published in *Der Rundfunk* (*The Radio*), Sala once again stressed that the aim of his instrument was not to replace or imitate other instruments.[65] Contrary to earlier reviews, however, one no longer heard about the aesthetic of "steely romanticism." Sala wrote that the trautonium lacked "fundamental aesthetic principles." He called on his audience for assistance.

> Any interested listener can help us here. We need material, a lot of material, in the form of detailed listening reports, spontaneous listening experiences, astute observations, and critical appraisals.[66]

The decision to be democratic and not to rely on Goebbels's aesthetic was a wise and opportunistic one after the war.

In early 1948 Trautwein contacted Sala with a view to reach a financial and legal agreement on resurrecting the trautonium: he hoped to iron out any of the ill will that had festered for over a decade. On May 3, 1948, Sala wrote Trautwein about the future of their intellectual property rights after the devastating loss at the hands of the Allies. Trautwein attempted to assuage his former assistant's fears, insisting that his intellectual property lawyer, Dr. Walter in Augsburg, had assured him that their patents would be respected.[67] Trautwein also reminded Sala of their correspondence in 1938 that outlined their intellectual property agreement. He was skeptical about the possibility of reviving interest in the trautonium given the lack of funds in postwar Germany for purely cultural activities.[68]

Later that year, Sala sketched out a course of study for the trautonium and sent it to composer Werner Egk.[69] Shortly thereafter, he gave a performance of his trautonium to a number of composers, including Grete von Zieritz, Ludwig Richard Müller (also the music adviser to the Potsdam Broadcasting Station), and Günter Kochan, as well as to students from the Berlin Academy of Music. It was an attempt to propagandize for his instrument. Sala once again stressed that the instrument should be seen as a unique one with an impressive range of tone colors. The key to success was in part predicated on composers creating works specifically for the new instrument, as Genzmer and Weismann had.[70] He met with another group at the Berlin Radio Station (Haus des Rundfunks), including composer and musicologist Siegfried Borris;

composer Max Butting; Günter Griep of the Deutsche Zentralverwaltung für Volksbildung (German Central Committee for the Education of the People); poet and lyricist Peter Huchel; translator and storyteller Else von Hollander-Lossow; journalist and political activist Cläre Jung; and radio and TV broadcaster Eva Baier-Post. Those present at the gathering discussed how the claim that electroacoustic music was "soulless" was simply fallacious. Stressing that new music needed to be composed for the instrument, including works for films and theater, Butting argued that the special qualities of the trautonium had neither been fully realized nor appreciated. While classical pieces that were composed for other instruments could be used for "propaganda purposes for the instrument"—as Butting called them—Siegfried Borris felt that the works of Bach and Paganini were far less important to the future of the trautonium.[71] Rather, pieces needed to embrace the diversity of sounds that could be produced. They could either be radically new works, or they could be music played in factories and at mass gatherings. He recommended that the compositions be short, around one or two minutes. He did warn, however, to not hint at the revolutionary aesthetic implications of the instrument.[72]

The Trautonium, *Faust I*, *Abraxas*, and Opera

The year 1949 was an important one for the trautonium, as it marked its recapturing of the German imagination. The performance of *Faust I* broadcast by the NWDR Studio in Berlin[73] in April 1949 was the first performance in which Sala experimented with recording on reel-to-reel tape. He had been working on the tape music for the production for over a year and had been tinkering with tape music since 1947. That broadcast caught the attention of a German composer of film scores, Hans-Martin Majewski, with whom Sala would later collaborate. Majewski was delighted with the overall reception of the trautonium in the play.[74] One reviewer claimed "the modern instrument" was "on the way to become the music of the future."[75]

As impressive as that radio performance was, it was not nearly as important as the production of *Faust I* on August 28, 1949: the bicentennial of the birth of the titan of German culture, Johann Wolfgang von Goethe. After the war the Germans now needed to think hard about how to celebrate their cultural icon in a fashion that did not resurrect the memories of the fascist celebrations during the Nazi regime. A modern look at the past was necessary, and the renowned Jewish and leftist composer Paul Dessau turned to the trautonium for just that. Dessau had immigrated to France in 1933 and then to the United States in 1939. He returned to East Berlin to live permanently in 1948. That particular staging of *Faust I* was based on the renowned production by Max Reinhardt of 1909. Reinhardt, who was Jewish, left his native Austria after the

Nazi Anschluss in 1938. Rather fittingly, the trautonium was not meant to perform the types of music it had from 1933 to 1945. It was to offer new, futuristic sounds, as the Germans wanted to look forward to a new future and to rupture links with its recent fascist past.

On the evening of the bicentennial, Dessau's stage music to *Faust I* premiered in the Deutsches Theater in Berlin.[76] The trautonium provided "celestial sounds" and "the music of the spheres" when the archangel Raphael spoke in the "Prolog in Heaven" and "the grotesque banjo pizzicato" when Mephistopheles shouted in Auerbach's Cellar. When the black poodle brushed up against Faust's feet, the trautonium unleashed the sound of a barking dog, and it generated sounds that were reminiscent of an orgy for the *Walpurgisnacht*.[77] Sala himself spoke of the "rustling and rumbling ghostly sounds of the devil" that caused "a cold fear to rise up among the audience."[78] German critic and columnist Friedrich Luft felt that the trautonium "was becoming more and more legitimate as the most important musical instrument of the stage."[79] Reviews stressed that the perfection of electroacoustic music had led to a new type of polyphony in which one could play a mixture of chords that could be executed on no other instrument.[80] Such reviews much reflected Dessau's desires, as he wished to solve the problems of sound in *Faust I* by means of a novel electroacoustic method.[81] The trautonium for Dessau was meant to generate unique sounds and timbres. Musical instruments could not produce the sounds that the trautonium could. That was now the trautonium's critical attribute.[82]

In 1949 Sala also performed in a new production of Werner Egk's controversial ballet, *Abraxas*, in Berlin. A year earlier, the Catholic Church in Bavaria vehemently protested the ballet's broadcast for its alleged obscenity. A lengthy debate erupted in the Bavarian Ministry for Education and Culture about freedom of expression. The trautonium was again called to generate its modern and unique sounds. The ballet was composed of five scenes and was based on Heinrich Heine's *Der Doktor Faust: Ein Tanzpoem*, which in turn drew upon Goethe's *Faust I*.[83]

The instrument appeared in numerous operas after the war. First, it provided the sound of the bells of Monsalvat (also known as the Grail bells or Parsifal bells) in Richard Wagner's *Parsifal* at the Teatro di San Carlo in Naples in 1950 at the request of the director, Karl Böhm.[84] A year later it again provided those sounds at the Berlin City Opera under the direction of Leo Blech.[85] Richard Wagner had originally commissioned Bayreuth piano builder Eduard Steingraeber to build the Gralsglockenklavier to generate those particular sounds. In September 1951, Sala played his trautonium for French composer Marcel Delannoy's opera *Puck* at Berlin's City Opera in Theater des Westens.[86]

FIGURE 7.1. Oskar Sala's newer version of the Konzerttrautonium (concert trautonium), rebuilt after World War II, circa 1950. Reproduced with the permission of the Deutsches Museum (Munich) Archives, CD_65730.

The trautonium was featured in the March 1951 premiere of Bertolt Brecht and Paul Dessau's opera, *Die Verurteilung des Lukullus* (*The Condemnation of Lukullus*), directed by Hermann Scherchen, at the Berlin State Opera on Unter den Linden. The original scoring by Dessau called for the trautonium. The twelve-scene opera was based on Brecht's radio drama titled "Das Verhör des

Lukullus" ("The Trial of Lucullus"), written back in the autumn of 1939 when he was in exile in Sweden. It was modified several times, including when it became a radio opera with music provided by Dessau in 1949 for NWDR-Hamburg.[87] Brecht's pacificism back in 1939 was reflected in the story, which is about the Roman General Lucullus. After his death, the general appears in a court in the underworld to determine his fate in the afterlife. In the original radio drama of 1939, the court withdraws to consider the verdict, and the play ends without the audience knowing the verdict. In the aftermath of the Nuremberg trials, Dessau urged Brecht to alter the conclusion for the radio opera. The new concluding scene featured the judge and jury members condemning Lucullus.[88]

Party functionaries of the SED (the Socialist Unity Party of the German Democratic Republic) attended the closed premiere at the Berlin State Opera in March 1951. They insisted that Brecht rewrite various bits, including scene four with the children ("In the School Books") to include a commentary that the grave of the conqueror is a useful classroom tool, to differentiate between wars of aggression and of defense, and to remove the self-criticism of jury members.[89] Brecht acquiesced. The more serious criticisms, however, were leveled against Dessau's music. He was accused of composing in the style of formalism, meaning that the composition was determined by its form and that its musical meaning is inherently intellectual. Joseph Stalin loathed formalism and considered it elitist, thereby mandating socialist realism. Music was not meant to reflect the current reality but a future utopian reality that socialism would inevitably create.[90] It turns out that Dessau made very few stylistic changes to the music. The slightly altered opera was performed ten times for the 1951–52 season, beginning in October 1951 under the new title, *Die Verurteilung des Lukullus* (*The Condemnation of Lucullus*).[91]

The music for *Die Verurteilung des Lukullus* was meant to provide a critical commentary of social gestures. The witnesses at the trial represent different social classes, which require different musical expressions. In the case of the defense witnesses, the music of the trautonium attempts to prevent the viewer from experiencing uncritical empathy for Lucullus. In a typically Brechtian manner, with the assistance of the trautonium, the audience distances itself—what is referred to as *Verfremdungseffekt*, or the distancing effect—from the witnesses in order to be more judgmental. The music is not meant to support or illustrate the words of the play; rather, it is used to undermine the oral message, enabling the audience to distance itself from the actors and to engage in criticism.[92]

The trautonium is first heard in scene six, "Wahl des Fürsprechers" ("Choice of the Advocate"). It plays chords as dotted half notes, which support the *Sprechgesang* of the court of the afterlife's spokesperson (*Sprecher des Totengerichts*), who narrates the appearances of Lucullus, the judge, and five members of the jury in the courtroom.[93] The trautonium generates otherworldly and foreboding sounds. In scene seven, "Herbeischaffen des Frieses" ("Procurement of the Frieze"), the trautonium accompanies, and is dissonant with, the spoken voice of Lucullus as he describes the frieze.[94]

The trautonium also appears in scene eight, "Das Verhör" ("The Interrogation"), opening the scene with the cellos and double basses while the court of the afterlife's spokesperson tells Lucullus to bow down before his witnesses. After Lucullus complains that his witnesses are his enemies, the trautonium reenters ominously as the spokesperson repeats that these are indeed the witnesses.[95] The trautonium reappears and accompanies the spokesperson, who recognizes the courtesan's question to the queen, and continues when the courtesan asks the queen where she is from.[96] The chords and notes played are dissonant. Once again, it elicits a sense of foreboding. The instrument is next heard when the spokesperson instructs the jury to contemplate the queen's testimony.[97]

One hears it again in the beginning of scene eleven, "Das Verhör wird fortgesetzt" ("The Interrogation Continues"), generating an eerie, uneasy, supernatural effect as the audience refocuses on Lucullus's interrogation.[98] During the trautonium solo, the female commentator exclaims, "Look, the jury has returned!"[99] It accompanies the *Sprechgesang* of the spokesperson when he instructs the jury to listen to one of its members, the teacher, thereby evoking a sense of suspense.[100] The spokesperson then readdresses the jury, instructing them to consider the cook's testimony.[101] He continues to speak, "Here the judge, once a farmer, has a question." The trautonium again accompanies the voice of the spokesperson to create a sense of trepidation.[102]

The trautonium is also heard in the final scene, "Das Urteil" ("The Judgment"), accompanying the jury as they sing: "There are still hungry mouths, of which you have so many up there."[103] It continues by supporting the next line: "What dust we can heap on eighty thousand slaughtered there!"[104] It provides consonant and dissonant chords to set up the final scene of the opera when everyone sings the verdict.[105] It resumes when the jury repeats their verdict: "Oh, yes, to nothingness with him!" It generates the sounds of condemnation.[106]

A second opera by Brecht and Dessau, *Deutsches Miserere,* features the trautonium.[107] Several times between March and May 1943, Dessau met with Brecht in New York City (both were in exile from the Third Reich) to discuss *Deutsches Miserere.*

I precisely remember my visit to Brecht on 52nd Street, where he was living. I said, "You know, Brecht, I'd like to put a bee in your bonnet. There's

> something I'd so very much like to write, a sort of German Requiem, but not like Brahms's—just the opposite, in fact. A big *miserere*, a German work depicting the horrible tragedy of our country." That seemed to whet his interest enormously, and he began to look for material.[108]

They completed the opera in April 1947. It was divided into three parts: part one deals with the early history of Nazism until the outbreak of war, part two covers the war, and part three discusses life for Germans after the war. It was a piece that was staunchly antiwar and was written for the German public after its liberation from fascism. It did not premiere until 1966, a decade after Brecht's death, in the city of Leipzig in the former German Democratic Republic. The trautonium was meant to appear in the score in part one, number four, "Auf der Mauer stand mit Kreide: Sie wollen den Krieg" ("On the War Was Written with Chalk: They Want War"); part two, numbers one, two, three, twenty, and twenty-one: "Wie einer, der ihn schon im Schlafe ritt weiß ich den Weg" ("Like One Who Rode in His Sleep, I Know the Way"); "'Was macht ihr, Brüder?' 'Einen Eisenwagen'" ("What Are You Doing Brothers?—An Iron Wagon"); "Und Feuer flammen auf im hohen Norden" ("And Fires Blaze up in the Far North"); "Ihr Brüder, hier im fernen Kaukasus" ("You Brothers, Here in the Distant Caucasus"); and "An jenem Junitag nah bei Cherbourg" ("That June Day Near Cherbourg").[109] Critics spoke of the piece's "dark colors" written for the trautonium.[110] They were meant to convey a sense of foreboding and shake up the audience. The instrumentation masterfully disrupts what the audience expects and what they actually hear, whether it is a result of the anticipated (but missing) tone colors or the totally unique timbres.[111] One reviewer spoke of Dessau's musical style as one that "alienates" and "encourages people to think along with it."[112]

While its predominant use after the war was in the creation of supernatural and futuristic sounds, the trautonium did not altogether forgo its ability to imitate the tone colors of acoustical instruments. It was also featured as a substitute for various instruments. In October 1948, Sala was asked to play his trautonium as the substitute for the F-trumpet in Bach's Brandenburg Concerto No. 2 for RIAS-Berlin in a performance.[113] Sala performed his trautonium at the Berlin Music Festival (Musiktage) at the City Opera in 1949. According to writer and director Hans Borgelt, the sound both fascinated and repelled the audience. The energy the instrument "unleashed seemed inexhaustible, capable of achieving great effects."[114] The critic was therefore bewildered by Sala's desire to imitate acoustical instruments: "Was there a need for imitation as long as there were still sufficient numbers of these instruments?"[115] Apparently, though, there were numerous "enthusiastic responses from proponents of the new material."[116] In April 1950 it once again replaced the trumpeter, this time in J. S. Bach's Brandenburg Concerto No. 2. Renowned music critic Hans Heinz Stuckenschmidt, director of "new music" at RIAS-Berlin, referred to the trautonium performance as "a miracle, an incomparable trumpet virtuoso."[117] He insisted that engineers must transgress disciplinary boundaries that had limited human virtuosity. In January 1950, Mitteldeutscher Rundfunk (MDR) incorporated the horn from Bach's Brandenburg Concerto No. 1, played by Sala on the trautonium, in the pauses between works.[118] Later that year, the instrument was used to create the sound of a plane crashing in the desert in Günter Rutenborn's radio drama, "Durst," broadcast on Radio Bremen in May 1950.[119]

In short, from the late 1940s to 1950, the trautonium was once again being featured on German radio and in theater pieces, radio dramas, operas, and even a ballet. Sala continued to perform and record classical and virtuosic works, such as J. S. Bach's Cantata No. 147, Paganini's "*Carnival of Venice*", Franz Schubert's "The Bee," Antonio Corelli's "La Follia," and Giuseppe Tartini's Sonata in A minor and "Il Trillo del Diavolo," as well as more contemporary works, such as Josef Ingenbrand's "Humoreske" ("Humoresque") and "Tanzweise" ("Dance Melody"). But now, more so than before, the trautonium was being featured for its unique sounds and tone colors. Sala labored tirelessly to return the instrument to where it had been in the public's eye before he was drafted in the Wehrmacht. However, he was not resting on his technological laurels. Instead, he was working on a new trautonium.

Mixturtrautonium (Mixtur-Trautonium)

Sala had been refining his trautonium after the war. It was metamorphosing from the concert trautonium into the mixture trautonium. For example, he increased the number of rotary controls to six in 1949 and thirteen in 1951.[120] During that period he often referred to the instrument in his publications as the mixture trautonium, since he could generate many tonal mixtures.[121] It was to be Sala's crowning achievement, his most influential instrument. One can see the similarities between the postwar version of the concert trautonium and the final version of the mixture trautonium by comparing figures 7.1 and 7.2.

In addition to more rotary controls, as with the radio trautonium and later versions of the concert trautonium, there was also a liquid under the rails that acted as a resistor and controlled the volume: the more forcefully the string was depressed, the louder the tone. The manner and pressure in which the finger pressed the rail affected the initial sounding of the tone, thereby altering the sound's envelope. Two pedals could subsequently blend together various timbres and complete the sound envelope of the tone.

FIGURE 7.2. Oskar Sala's Mixturtrautonium (mixture trautonium), 1952. Notice the similarity between it and the later version of the Konzerttrautonium in figure 7.1. Reproduced with the permission of the Deutsches Museum (Munich) Archives, BN_22679.

He continued to tinker in a room in the Haus des Rundfunks in Berlin.[122] By far the most important improvement was his invention of a circuit that would enable him to use mixtures of subharmonics. He patented the circuit in Germany and the United States in 1952 and in France in 1954.[123] The new circuit was critical to the final version of the mixture trautonium. He finished its creation in 1952 with funds provided by the Notgemeinschaft der Deutschen Wissenschaft (Emergency Association of German Science) in 1949.[124] Weighing approximately 100 kilograms, the mixture trautonium cost approximately DM 25,000 to build ($6,000 in 1952, or just under $62,000 in today's money).[125] When traveling with the instrument to concert performances, it was disassembled, its parts placed in ten boxes. Its impressive audible range was from 16 Hz to 16 kHz.[126] It generated sounds and tones based on both subtractive and additive synthesis.[127]

Oscillations were generated by a thyratron, tuned to the pitches between G2 and G5, that produced periodic discharges of a capacitor and gave rise to the sawtooth-like, discontinuous oscillations, also known as the excitation frequency. Much like a number of the other earlier versions, the instrument possessed two manuals, each comprising a resistance wire that was stretched over a metal rail.

By changing the resistance and capacitance of an RC circuit, one could regulate the natural frequencies (*Eigenfrequenzen*, eigenfrequencies) and timbres. If an RC circuit was excited by a sudden surge of voltage, it decayed to its natural frequency. By damping the RC circuit, the natural frequency decayed exponentially: a succession of decaying vibrations gave rise to numerous timbres, including those resembling musical instruments, or vowel sounds, or totally unique ones never before heard.[128] As the natural frequency increased, the tone color became lighter. As it decreased, the tone color became darker. If more than one natural frequency was used, the tone color became richer, fuller.[129] The right pedal blended together "the darker" tone-color groupings, while the left blended together "the lighter" ones. These tone-color groupings were then sent to the two amplifiers and finally to the loudspeaker.

Subharmonic tones could be easily generated. For example, if the thyratron played the C4 pitch, a secondary generator could be set to play a frequency of the subharmonic series of that pitch. If one changed the frequency of the thyratron, the frequency of the secondary generator ran at the equivalent subharmonic interval. These parallel-running intervals and chords generated from harmonic tones were defined as the mixtures.[130] Whereas earlier trautoniums could produce two different subharmonics per manual, the mixture trautonium could produce four with four secondary generators.

While constructing his mixture trautonium, Sala busied himself with understanding the importance of the psycho-physical aspects of sound synthesis by means of electronic musical instruments.[131] It was clear to him that Helmholtz's theory of hearing (based on the ear performing a Fourier analysis) needed to be tweaked or at least fleshed out by other, more recent experimental work.[132] He felt more research was needed on the mechanics of hearing—something that Hungarian biophysicist and Nobel laureate Georg von Békésy was undertaking—and on the role of the mind (*Psyche*) in assimilating and processing sounds.[133]

By means of coupling and taping, the mixture trautonium was capable of playing multivoiced parts. Sala would record a part of a piece on a Telefunken M5 reel-to-reel tape recorder and play it while he played another part. He would then record that instantiation of the work, and so on. Genzmer composed a piece for orchestra and the new instrument, "Konzert für Mixturtrautonium und großes Orchester" ("Concerto for Mixture Trautonium and Large Orchestra"), which debuted in 1952 on Südwestrundfunk (Southwest Radio), directed by Hans Rosbaud.[134] To the instrument he added an electronic metronome that created staccato and glissando effects. This produced sounds that resembled those of a xylophone. He also included a noise generator and a decay device that converted continuous tones into decaying tones with a metallic timbre that sounded like drums.[135]

Hans Heinz Stuckenschmidt, who had earlier praised the concert trautonium for its imitation of a trumpeter, described the mixture trautonium as

> a brass band, super violin, and universal organ all in one, simultaneously an ichthyosaurus and homunculus of the world of sound, an instrument that can whisper divinely and roar infernally.[136]

According to another reviewer, "the trautonium, much ridiculed in its early phases, has now become an actual musical instrument."[137] With coupling (similar to an organ), eight chord tones could be played simultaneously. The mixture trautonium "had conquered its own world of sound."[138]

Another critic was less enthusiastic. While he thought that "something promising for the future had been created here," he complained that "a fatal aftertaste of the sound of a Hammond organ and often even the barrel organ could not be ignored."[139] A reviewer of a concert in Tübingen in May 1953 felt that one was attracted to the electronic tones, groups of sounds, and mixtures while simultaneously feeling alienated by "a strange, impersonal, dematerialized (*entstofftlicht*), baseless oscillating music, whose vibrations seem to have renounced everything sensual."[140]

In 1954 renowned musicologist and freelance music critic of Südwestrundfunk Fred K. Prieberg proclaimed that the trautonium had been essential for film, theater, and radio drama music.[141] Its popularity in these genres was increasing. Even renowned composer Carl Orff was interested in the mixture trautonium.[142] In that same year, Sala performed on the instrument in Munich's Hercules Hall in a version of the composer's "Entrata," replacing the organ.[143] Several days after the performance, Sala spoke about his mixture trautonium and played it for an audience in the Hercules Hall. One audience member was not at all impressed:

> The audio recordings played afterward, to which the lecturer added rhythms, fully demonstrated not only the intrusion of technology into the manual art of music reproduction, but also demonstrated the icy coldness of those sounds, which will now creep more and more into the old symphonic sound, just at a time when one seeks to rediscover people as individuals with all of their values of personal diversity.[144]

After listening to Sala performing Genzmer's piece with the Berlin Philharmonic in 1954, one critic expressed that the electronic part of the performance—referred to as the "soulless part" of Genzmer's composition—did not mesh well with the orchestra. He compared his experience in the audience to watching an endless composition of an imaginary film. He did, however, admit that the youths in the audience were enthralled by the "technological experiment" and gave their thunderous applause.[145] Another review of that same concert declared that Sala "earned unlimited recognition" for his technical inventions and musical musicianship.[146] Most reviewers praised Sala's musicianship. However, many were not as enthusiastic about Genzmer's composition. Yet another reviewer noted the uniqueness of the instrument's sounds. That said, he felt that the mixture trautonium should not be used as a solo instrument, but—contrary to the opinion of the previous critic—should only be part of an ensemble for music and sound in film and radio dramas.[147] According to a contemporary reviewer, Sala "alone commanded the field."[148] Another felt that while the combination of electronic music in the form of a tape and the interpretative instrumental music had a "certain charm," they were not sure if it possessed "a convincing style."[149]

Sala performed Genzmer's "Concerto for Mixture Trautonium and Large Orchestra" at several venues in the summer and early fall of 1954. Music critic Josef Eidens insisted that "there is no doubt that the composer made imaginative use of the tonal possibilities of the solo instrument. It was extremely clever how the trautonium was built into the orchestra's sound effects."[150] Echoing the mixed opinions discussed above, another reviewer doubted whether the mixture trautonium would always work with a conventional orchestra. Harsh tone colors were a problem, although at times it could "open the door to a thoroughly 'human' expression that penetrated the soul."[151] Another reviewer called the instrument "a lifeless, artificial work of art," but admitted that "one was nevertheless spellbound" by its "miraculous sound" (*Klangwunder*).[152] Some were underwhelmed with the instrument, claiming that "the apparent sensation" was "more of a product of technical experimentation" than a musical achievement.[153] Still another held the antithetical view.

> In addition to the sparingly applied imitation of familiar instrument sounds, there is a wealth of new effects, from gently gliding sounds to brutal noises; artistically arranged with taste and still comprehensible as music.[154]

Similarly, *kapellmeister* and music critic Friedrich Herzfeld described the mixture trautonium as "currently the most perfect instrument for electronic music." He also stressed that it was not a substitute for other instruments, but rather should be employed for music that sought to evoke "cosmic forces."[155]

A less sympathetic critic was convinced that the experiment to combine the trautonium with a traditional orchestra was a complete failure. It could synthesize impressive sounds, but it was not to be coupled with an orchestra. "The solo instrument does not harmonize with the spirit of an orchestra."[156]

The Various Genres of the Mixture Trautonium

Similar to the concert trautonium, the mixture trautonium was also used in operatic performances. For several years beginning in 1955, Sala once again provided the sounds of the Parsifal bells in Wagner's *Parsifal*. This time the opera was not performed in liberal-minded Berlin, but in Bayreuth under the direction of Wieland Wagner.[157] The mixture trautonium also produced the hammering in Wieland Wagner's production of *Das Rheingold* starting in 1955.[158] Apparently, Carl Orff recommended its use to Wieland for the anvil sounds in the Nibelheim scene.[159]

Wieland was renowned for his minimalist staging of his grandfather's operas. This was in stark contrast to the productions at Bayreuth during the Third Reich. His task was a daunting one: he was the first opera director at Bayreuth after the war. According to the opinions of many artists in postwar Germany, a tabula rasa was necessary. His grandfather's works were infamously favorites of Adolf Hitler, and Richard himself was an anti-Semite. Bayreuth was a bastion of right-wing ideology.

Wieland's productions of the 1950s were influenced by Bertolt Brecht's earlier literary pieces. Wieland saw his own productions as radical experiments breaking with earlier productions, much as Brecht insisted that his own work was both experimental and a conscious break with contemporary literary forms.[160] Austrian critic Franz Willnauer has argued that Wieland's operatic productions of the 1950s owed much to the Brechtian *Verfremdungseffekt*.[161] Given the influence of Brecht, perhaps it should not come as a surprise that Wieland used a trautonium to shock the audience and force them to think about the moral of the story rather than identify with one of the Wagnerian characters.

One purist critic called the sounds generated to replace the Grail bells of *Parsifal* and the hammering in *Rheingold* "coarse." The critic asked: "Is it not indicative of the dubious nature of such efforts that our sound engineers, with their science of frequencies and overtones, can basically only achieve an 'unrefined' [*Vergröberung*] effect?"[162] Apparently all of the music critics insisted that Wieland return to the more traditional productions of his grandfather, since "Richard Wagner had conquered the calculations of the technical age."[163]

Another critic called Sala's electronic playing of the Parsifal bells "insulting to the ear."[164] Yet another agreed, claiming that both the Parsifal bells and the hammering in *Rheingold* were "disjunct" (*elementfremd*) and simply too loud.[165] A third reviewer complained that the mixture trautonium was out of tune for the Parsifal bells during the first performance of the season, but that it had improved by the season's fourth and final performance such that the audience was reminded of the "blissful, mystical sound . . . as the old magician Richard Wagner once devised it."[166] Peter Otto Schneider commented on the "jarring electronic grail bells and the Nibelungen hammer," which, in his eyes, were completely unsuccessful. This

> suggests that the acoustics of the *Festspielhaus* were being tampered with by using microphones and electronics. The rejection of these attempts is general, and Wieland Wagner, with his own liberality and nonchalance, admitted without further ado that it was a mistake and that these things need to be dispensed with.[167]

In short, the reviews of the mixture trautonium were mixed—pun intended. Many of the reviews were critical. One referred to the mixture trautonium as "howling,"[168] while another labeled the mixture of sounds "from grotesque to eerie."[169] Others were more sympathetic, insisting that the suggestive tone colors "point to new and undoubtedly very productive kinds of possibilities."[170]

Sala also found a niche for his mixture trautonium in the film industry. He embarked on a series of projects to compose and play music for numerous types of short films, such as industry films, cultural films, and documentaries. Back in 1950 Sala had received a letter from Hubert Schonger of Schongerfilm Produktion, a leading German film company. Composer Werner Egk had convinced Schonger that Sala's trautonium was perfect for providing musical background to modern films.[171] Sala agreed and felt that the types of music employed in older films should no longer be used.[172] It was time for a complete change, a time to be modern, and his mixture trautonium could offer just that.

In 1953 the fruits of his first film project appeared. He composed the music and played his mixture trautonium for *Verzauberter Niederrhein* (*Enchanted Lower Rhine*). In 1955 he did the same for the fairytale film, *Schneeweißchen und Rosenrot* (*Snow White and the Red Rose*) and for *Dein Horoskop—Dein*

Schicksal? (*Your Horoscope—Your Destiny?*). His music and instrument were featured in *Schöpfung ohne Ende* (*Creation with no End*), a color industry film of 1956 produced by the Gesellschaft für bildende Filme (Society for Picture Films, or GbF) of Munich for Bayer Leverkusen to extol the virtues of chemical research. It featured the process by which a chemist's research in the laboratory led to myriad consumer goods. For that film, Sala received a *Bundesfilmpreis* (Federal Film Prize) for best film music at the Seventh International Film Festival in Berlin in 1957.[173]

That same year, a Veit Harlan film, *Anders als du und ich: Das dritte Geschlecht* (*Different than You and I: The Third Gender*) was released. It was the first film in post–World War II Germany that discussed the theme of homosexuality. The parents of a seventeen-year-old adolescent were concerned that their son might be gay. His friends were all males who listened to electronic music and were fans of abstract art.[174] Harlan wished to condemn the view that homosexuality was strange, perverse, and unnatural.

In 1958, *Aluminium: Porträt eines Metalls,* also produced by the GbF in collaboration with the General Association of the Aluminum Industry of Düsseldorf, was released. Similar to the theme of *Schöpfung ohne Ende,* this scientific film for a lay audience told the story of aluminum from the moment it was mined to the moment it was incorporated into final products critical to our daily lives.[175] He also performed the mixture trautonium for the 1958 film *Gefahr Nord-West* (*Danger Northwest*), which depicted a rescue of a ship off the North Sea coast.

In that same year, Sala established his own sound studio in the Mars-Film headquarters at Charlottenburger Chaussee 51–55 in Spandau, West Berlin, where he continued his work on music and sounds for film and theater. The studio enabled him to be much more efficient at cutting and splicing tapes, composing new timbres and sounds, and synchronizing sounds to film.

By this time, Sala was lecturing on how electronic sounds could be incorporated with great effect into films. He performed the mixture trautonium for Hans Martin Majewski's music in *Labyrinth* (1959). When discussing his work on the music for *Labyrinth* in his Spandau studio, Sala stressed how, using the studio-taping technique, he altered natural sounds and well-known musical tone colors.

> With the electric production of sound and with the assistance of tape recordings in the studio, one can change natural noises and well-known musical sounds. However, among the possible variants there are seldom those that can compete with the original in terms of aesthetic effect and expressiveness, and even more rare are the "acoustic miracle flowers" (*akustische Wunderblumen*) that surpass the original. We were looking for them for the

FIGURE 7.3. Oskar Sala's Berlin Studio recreated in the Deutsches Museum, Munich. Reproduced with the permission of the Deutsches Museum (Munich) Archives, CD_65725.

> film *Labyrinth*. They cannot be conjured up by calculation and circuits alone. Extensive experience, continually trying things out (*Probieren*), and a studio technique that has been highly developed for such possibilities of variations must, in our case, luckily, be combined with the composer's intuition and the art of electronic interpretation.[176]

His performance technique played a defining role in contrasting his style of electronic music to the styles of Werner Meyer-Eppler and Hebert Eimert, which were forged in the Cologne Studio for Electronic Music, as well as Pierre Schaeffer's techniques of *musique concrète* (discussed in the next chapter).

Sala produced the 1959 soundtracks for Rolf Thiele's *Rosemary* and Fritz Lang's *The Indian Tomb*. One of his most famous contributions to the genre of industry films was the sounds he set to the Mannesmann GmbH film about steel, *Stahl—Thema mit Variationen* (*Steel: Variations on a Theme*) of 1960, directed by Hugo Niebling, which won the Grand Prize in Rouen that year.[177] In 1962 he won Cannes's Palme d'or for his music in the Swiss nature film, *A fleur d'eau*.[178] Perhaps this should come as no surprise. After all, when Sala was working at the RVS during the early 1930s with Trautwein, one of his projects was the improvement of sound movies. As Georg Schünemann had written back in the summer of 1931, "One of our areas of responsibility is sound films. In particular, our students should get the impression of exotic music through sound movie presentations."[179] During the early 1960s Sala turned more and more to Gebrauchsmusik.[180]

In 1962 Sala composed and played his mixture trautonium for the successful film, *Alvorado—Aufbruch in Brasilien* (*Alvorado—Departure in Brazil*). That same year, the chemical company Badische Anilin- und Sodafabrik produced the color film, *Der Fächer* (*The Subjects*), which received second prize for music in the industry film category. Once again it involved propaganda for the medical and natural sciences industries, this time illustrating how plastic helped make life easier.[181] He also played the mixture trautonium for a 1964 film for another chemical industry giant A. G. Hoechst, *Mit Farben begann es* (*It Began with Colors*).[182] Sala provided the soundtrack for Hesse Broadcasting's 1976 production *Eine Reise zum Mond* (*A Journey to the Moon*), directed by Manfred Durniok. The film included NASA photographs and film clips from various Apollo missions and Skylab.[183] By the end of his career, Sala composed sounds for over three hundred films, including numerous industrial movies—a major genre during the 1960s and '70s in West Germany—and one hundred commercials, including a 1950s Coca-Cola TV advertisement.[184]

In 1960 Sala and his mixture trautonium moved into another musical genre: electronic ballets. On May 29, 1960, the world witnessed the premiere of *Päan*

(*Paean*) in the Städtische Oper Berlin, choreographed by Tatiana Gsovsky, with Yvonne Chauviré as the ballerina.[185] No musician was present, only a reel-to-reel tape of the mixture trautonium providing all of the music composed by Remi Gassmann, a student of Hindemith.[186] On March 22, 1961, the ballet *Electronics* enjoyed its world premiere with the New York City Ballet. The music was once again composed by Gassmann and performed by Sala, while George Balanchine directed.[187] And once again, the mixture trautonium recorded on tape was the only music played during the ballet.[188]

Balanchine's work with electronic music in general, and the trautonium in particular, should come as no surprise. As Whitney Laemmli has so astutely shown, the Russian-born American choreographer and dancer was seduced by twentieth-century technology, just as he was devoted to the technical qualities of dancing, particularly the construction of the pointe shoe.[189] Stressing the importance of artifice, and furthermore often referring to his own work as engineering, Balanchine characterized his classic *Agon* as "a machine."[190] He often viewed his dancers as technological tools, and he fetishized the efficacy and standardization of the performer's body that engineering could bring. Marian Horosko, one of his dancers at the New York City Ballet, complained that she and her colleagues were "used like machines or skilled robots and have become interchangeable, almost like atoms, endlessly changing places, making no mark of our own."[191] Apparently, when Balanchine asked how he felt about electronic ballet music, he sardonically responded: "It has one advantage over normal ballet scores—no orchestra, no musicians to pay."[192]

WNYC music critic Edward Tatnall Canby wrote a lengthy review of the ballet for *Audio* centering on the themes of aesthetics and the technological. He first discussed the aesthetic aspects of the piece. He was struck by the fact that despite the music being generated electronically by the mixture trautonium, he thought he was listening to a live orchestra. He added, "I was really amazed to hear—or seem to hear—a sort of supernatural 'super' orchestra."[193] Struck by how "unelectronic" and "real" the music sounded, he was convinced that the mixture trautonium could "serve as a glorious imitation of many an aspect of literal, acoustic music" and was not at all surprised when the audience yelled, stamped, and shouted bravo.[194] He was shocked that the piece, with such a compelling sound, had been created in Sala's West Berlin studio, in a room that was considerably smaller than the New York City Ballet. He did admit that the musical score was "oddly conventional, even old-fashioned," calling *Electronics* "a work of virtuoso conservatism . . . a forward-looking conservatism."[195] He was quite correct: Gassmann's goal was not to reject absolutely traditional sounds, but rather—as Gassmann himself said—have the music be "a logical extension of orchestral means."[196] Balanchine combined the innovations proffered by electronic music in the dancing "with standard, human-like movement, in proportions just right for the music."[197] As for the technical aspects of the piece, Canby insisted that *Electronics* was the most advanced taped music he had ever heard.[198]

In the same issue of *Audio* as Canby's review was another by Harold Lawrence, a distinguished classical music producer and music director of Mercury Records.[199] Lawrence was impressed by the instrument's ability to replicate the timbres of the trumpet, trombone, piccolo, bass clarinet, violin, snare drum, cymbal, gong, bells, bassoon, oboe, harpsichord, and viola as well as generate new electronic sounds. Unlike Canby, however, Lawrence thought that something was missing, namely a "quality of 'liveliness' associated with traditional instruments."

> We found ourselves yearning for an honest-to-goodness open-window sound, say, of a trombone glissando, a plucked string, or the snap of a twig—anything to relieve the instrument's built-in "color."[200]

Specifically referring to Balanchine's *Electronics, New York Times* dance critic John Martin wrote that "the advantage of the electronic medium" lies "in its limitless range and variety of sounds, volumes, intensities, speeds, harmonics, manipulations of new areas. We would do well to open our minds to its possibilities . . ." He continued,

> It is, as it were, the eighth day of Creation, and it cannot be ignored out of existence, by the artist any more than the scientist. The artist must inevitably learn how to be used to forces thus opened up to him; how to be inspired, perhaps even frightened by them.[201]

The trautonium had been resurrected, and a new world awaited its music.

The Music of the Future

A number of American reviews of *Electronics* alluded to the mixture trautonium as the music of the future. Given the mixed reviews, what did the future have in store for the mixture trautonium? While many critics saw the instrument as "modern," some were unsure if it would go on to represent the music of the future. Modernism was not always seen as a positive musical trait.[202] One keen observer praised the instrument as "a triumph of technology and the spirit of invention" after a Sala performance; however, he was not sure about where "electron music" was going. Its future was up to the musician, not the physicist. "It is hard to imagine that the soulful sound of a Stradivarius would be relegated to the junk pile of the past. Music of the future?"[203] Yet another reviewer queried whether tomorrow belonged to Sala's instrument after his performance of Genzmer's piece in Elberfeld's City Hall. The answer: "We dare to doubt it."[204]

A review of the same concert claimed that the mixture trautonium's music ranged from "music of the spheres" to "discordant sounds" (*Katzenjammer*—literally, wailing cats: it can also refer to a hangover) to "steam whistle screeches." He added: "Anyone who loved bizarre and effective musical thrills got their money's worth."[205] A spectator at the International Musicology Congress in Cologne in October 1954 concluded: "Electronic music is still awaiting its future, but there should no longer be any doubt that it was not created to supplant conventional music, but to conquer its own sonic realm."[206]

A similar view was expressed by a reviewer of the Basel concert in May 1955. They asked whether or not this was, "the end of music." The critic spoke of "robots" and queried if humans were disappearing, capitulating to vacuum tubes and electrons.[207] He went on to say that the layperson, when viewing the mixture trautonium's cables and wires, is constantly warned: "Careful! High Voltage Line! Danger of Death!"[208] Still another reviewer proffered a harsh critique referring to the mixture trautonium as "a squeaking, howling . . . and singing monstrosity."[209]

A subsequent concert on Radio Stuttgart by Sala on his mixture trautonium elicited similar responses. That concert in November 1955 featured the world premiere of Sala's "Elektronische Tanzsuite" ("Electronic Dance Suite"), which comprised his mixture trautonium and previously recorded tapes of the instrument.[210] One reads about "the eerie sounds and noises swelling from the loudspeakers." Labeling the music "from another world," Friedrich Siebert wondered whether one could call it music at all.[211] He conceded, however, that Sala's performance had made a bigger impression on him than any of the other performances he had heard during the radio station's series, "Woche der leichten Musik" ("Week of Light Music"). Fritz Hammes was not impressed either. He queried whether Sala's "Elektronische Tanzsuite" was anything more than "experimental music." "These macabre tones and puffs can no longer be called music . . ."[212] Another critic ridiculed the radio's program title, "Woche der leichten Musik," because the works featured were anything but light. Sala's work was seen as "an experiment" that produced completely new sound combinations, which could be successfully employed for films, radio dramas, and ballets. But was it "the music of the future? We don't think so."[213] "The peculiar ghostly electronic music, driven by spirits from Hell, amazed more than convinced."[214] One review was particularly devastating:

> Oskar Sala on the mixture trautonium was a bad derailment. A pioneer virtuoso played a miserable piece of his own ("Elektronische Tanzsuite"), which greatly affected the enjoyment of an interesting experiment—it was a duo with a tape previously played by Oskar Sala. Even in the uncharted territory of electrons you can't do without music . . .[215]

A review of a studio concert featuring Sala playing Genzmer's "Concerto for Mixture Trautonium and Large Orchestra," on the other hand, spoke of the "impressive perfection" of the sounds and the wide range of tone colors.[216]

In one of his last public appearances, on July 6, 1956, Trautwein traveled with Sala to Cologne to present the mixture trautonium to the Cologne Conservatory.[217] A review of the events discussed the various sounds that were produced: the trills of canary birds, wailing sounds, and the hissing of a train and a defective bicycle tire. It concluded that the possibility of sounds was no longer limited to those relevant to film or radio drama. Such a review was not meant as a compliment: the critic felt that the sounds were lost in "chaotic boundlessness." That was "the stigma of Sala's compositions."[218] The instrument could produce so many possible sounds that it was now disorderly and possessed no single signature sound.

Perhaps the most fascinating query about the future of music that the mixture trautonium raised was whether or not the composer and performer were now redundant. After hearing a Sala performance in Basel in 1955, Prieberg was convinced that the electronic music emanating out of the loudspeakers provided the final step in rendering the conductor and performer superfluous.[219] The music of the future would only need reel-to-reel tapes—a sentiment that Sala vehemently denied, but that other electronic musicians, as we shall see in the next chapter, strongly supported.

To sum up, similar to engineers during the Weimar Republic, Trautwein wished to reforge a path for engineers to philosophize. His particular interest was in the relationship between technology, culture, and philosophy, and the role and status of engineers during the early years of the Federal Republic of Germany. Trautwein's affiliation with the Nazi Party and his obsession with and bitterness over his intellectual property severely hampered his success after the war. His last two years were largely dedicated to contemplating aesthetics, the relationship between humans and machines, the role of technology in culture, and the future of his music.

By the early 1960s, Sala, on the other hand, had successfully transformed the trautonium into an instrument with myriad applications. The trautonium served as a source for new sounds and timbres for radio dramas, theater pieces, films, operas, and ballets. Other than a few instances after the war when it replaced missing musicians and their instruments, the trautonium was generally not called upon to play classic works, as it had often been during the 1930s until World War II. It became an instrument of modernism, of strange and futuristic sounds. It is also not surprising that it was used for artistic expressions that subsequently provoked the wrath of the Catholic Church, as was the case with Egk's ballet, *Abraxas*, and the state, as happened with Brecht and Dessau's opera, *Die Verurteilung des Lukullus*. Dessau and Brecht used the trautonium to shock the audience and to enforce a type of alienation or distancing between certain characters. Dessau used the instrument in *Faust I* in 1949 to break with the fascist depictions of Germany's cultural icon. And Wieland's implementation of the mixture trautonium in the bastion of Wagnerism was a clear and deliberate signal that, to survive, Bayreuth needed to be completely severed from its fascist past. In short, the instrument was cleverly employed by composers who wished to rethink the past while simultaneously critiquing fascism and embracing a new, modernist future.

The instrument's aesthetic was no longer "steely romanticism"; indeed, it seemed to lack an aesthetic. Or at least Sala himself looked to others to supply one. Some saw its aesthetic as futuristic and otherworldly—the music of the spheres, the future, modernity. Others labeled it an ichthyosaurus, a homunculus of the world of sound—a monstrous and incomprehensible Proteus, a shrill chameleon, a beast. Its musical creations ranged from the grotesque to the eerie, similar to howling cats, steam-whistle screeches, and a leaky bicycle tire, to name just a few. For Sala, flexibility was the key after the war. He was opportunistic and entrepreneurial and convinced individuals in various genres that the trautonium was the instrument of the future. He and his trautonium seemed to be the perfect fit for the milieu created in the land of the capitalistic *Wirtschaftswunder*.

8

Sala & Trautwein vs. the Cologne Studio for Electronic Music

MANY MUSICIANS and music fans alike, when discussing the origins of electronic music, point to postwar Cologne. In 1951 the Cologne Studio of Electronic Music of Nordwestdeutscher Rundfunk (NWDR, now called Westdeutscher Rundfunk) was founded by musician Herbert Eimert, physicist Werner Meyer-Eppler, and sound engineer Robert Beyer. While Eimert, Karlheinz Stockhausen, and John Cage are well known to electronic music aficionados, I often receive a blank stare or a quizzical look when I mention the names Friedrich Trautwein and Oskar Sala. This chapter addresses what constituted electronic music and who were (and were not) its major contributors. These were hotly contested topics among the leading electronic musicians of the 1950s. The roles of composer and performer were at the crux of these debates. Sala's use of tapes from 1947 onward, as described in the previous chapter, was fundamental to his understanding of the relationship between composing and performing, perhaps since he often did both simultaneously. An analysis of his view sheds light on the various groups of the new music after the war.

Another critical musical movement in postwar Europe that Sala and Trautwein needed to confront was the so-called *musique concrète*, with its adherents firmly ensconced in Paris. Its founder, Pierre Schaeffer, drew upon his training as a radio sound engineer and investigated the characteristics of natural sounds, particularly their attack and decay, and the processes by which their timbres evolved over time (what is now called the acoustical envelope). He was unaware of Trautwein's work from the early 1930s on the subject.[1] He and his cohort famously manipulated reel-to-reel tapes, much in the same way Sala did, but with a view to modify familiar sounds and decontextualize them, thereby creating completing new ones. While Schaeffer preferred to think of his work as electroacoustic music that was totally distinct from the music produced in Cologne, musique concrète and the Cologne School are now often lumped together under the rubric of electronic music.

This is the untold story of the jousting between Sala and Trautwein on the one hand and the Cologne Studio for Electronic Music and, to a much lesser extent, musique concrète, on the other. The trautonium was at the center of those debates.

The Cologne Studio for Electronic Music

In July 1951 Herbert Eimert, Robert Beyer, and Werner Meyer-Eppler organized an International Summer Course of New Music in Darmstadt under the rubric of "Music and Technology." It was attended by such luminaries as Trautwein, Schaeffer, and Theodor Adorno.[2] Eimert spoke on "Music in the Borderline Situation," while Beyer addressed the importance of the production of electronic sounds for the future development of music.[3] Beyer commented on how "the equipment, with its arrangement of different electroacoustic procedures, outwardly resembles more a research laboratory" than a traditional music studio.[4] Meyer-Eppler introduced an early model of synthetic sounds by using Harald Bode's melochord and an AEG magnetophone.

A review of the summer course discussed how improvements to technology rendered interpretation redundant. Electronic instruments and the tape recorder enabled the composer to do away with any intermediary steps—such as the performer—that would spoil his authentic composition. The review went on to criticize Trautwein, who seemed to be stuck in the past with his invention of 1930. Trautwein, the reviewer insisted, called on electronic musical instruments to play alongside the acoustical ones with a desire to play the traditional forms of music. The reviewer thought that electronic music necessitated a complete break with acoustical instruments and the music composed for them.[5]

The sounds created at that Darmstadt International Summer Course were subsequently reproduced and broadcast with lectures by Eimert, Beyer, and Trautwein in an evening program of the NWDR station in Cologne on October 18, 1951, titled "The World of Sound of Electronic Music."[6] The afternoon before the broadcast, Eimert persuaded the studio's general manger, Hanns Hartmann, to host a number of musicians, radio engineers, and a physicist to discuss the role of radio in the creation of electronic music. Attending that meeting were Meyer-Eppler, Eimert, Beyer, NWDR technical engineer Fritz Enkel, the general director of NWDR Werner Nestel, the NWDR's chief engineer Karl Schulz, and NWDR film director Wilhelm Semmelroth. Meyer-Eppler had convinced the group that magnetic tape opened up new perspectives for radio: they were "to follow the process suggested by Dr. Meyer-Eppler to compose directly onto magnetic tape thereby opening up new perspectives for radio."[7] And so the Cologne Studio for Electronic Music was born.[8]

The early equipment in the Cologne Studio reflected their interest in tone-generating devices and filters. They wished to work directly with the physics and electronics of sound production.[9] Composer and music critic Hans Heinz Stuckenschmidt concurred: "We are not concerned with works for the trautonium or ondes Martenot concert instruments, but with music conceived purely for the electronic sound generator and which for its realization does not require, indeed excludes, human interpreters."[10] The composer's intentions would not be foiled by narcissistic musicians. Eimert proclaimed that "today, modern, differentiated methods of acousticians, engineers, and musicians have led to collaboration on the 'electronic' turn of music—not to technologize the music, but to serve it with new sources."[11]

The young Karlheinz Stockhausen, who would become one of the leading composers of electronic music, was well aware of the new music, having attended the renowned Darmstadt International Summer Courses for New Music in 1951 and 1952. He recalled:

> I began working at Cologne Radio Station in 1953. Among the sound-sources in the Cologne studio at this time were electronic music instruments—a melochord and a trautonium, which served in some experiments as sound-sources, but which were soon discarded when the idea of sound-synthesis asserted itself.[12]

The more "traditional" electric musical instruments (the melochord, monochord, and trautonium) were in his view slowly becoming less relevant. His emphasis was on the production and manipulation of sinusoidal waves, not experimentation with musical instruments. In short, he was able to compose a piece in which the electronic sources of sound, producing only sinusoidal oscillations as simple harmonic tones, could combine to form pitch, volume, and duration.[13] He organized those parameters using mathematical methods borrowed from compositional methods, physics, acoustics, psycho-acoustics, and phonetics.[14] For him the challenge was to generate sound spectra from sine tones, requiring numerous synchronization procedures whereby those tones were played simultaneously on two reel-to-reel tapes and recorded onto a third. And a challenge it was, requiring the intensive collaboration with the studio's sound engineers Erhard Hafner and Heinz Schütz. These sine tones served as the basis of the studio's early works, such as Eimert's "Klangstudie I" (1952), "Klangstudie II" (1952) and "Glockenspiel" (1953), and Stockhausen's early works, such as "Studie I" (1953) and "Studie II" (1954).

During the early days of the Cologne Studio, Stockhausen was committed to exploiting the purity of the sine tone.

> We return to *the element*, on which all sonic variation is based: the pure wave, which one can electronically generate, and which one calls the *sine-tone*. Every existing sound, every noise, is a mixture of such sine-tones—we call it a spectrum. Number-, interval-, and volume-relationships between such

> sine-tones account for the individual character of each spectrum. They determine timbre. And so for the first time, it was possible to *compose the timbre* of a [sic] music in the true sense of the word, that is to say from combining together elements, and so that the universal structural principle of the music will come into operation in the sonic proportions.[15]

Stockhausen felt that inharmonic sounds could be musical. He and other composers associated with the Cologne Studio focused their attention on the creation of inharmonic mixtures. Rather than creating those mixtures by filtering out the unwanted frequencies (as was the case with subtractive synthesis used by Trautwein and Sala), Stockhausen and his cohort needed to tune the sine generators to the desired frequencies and combine them, or what is called additive synthesis.

Werner Meyer-Eppler

A key member of the Cologne Studio, with whom Sala had a detailed correspondence lasting several years, was physicist and phonetician Werner Meyer-Eppler. The first exchange between Sala and Meyer-Eppler dated back to early 1949, the year in which Meyer-Eppler published his influential work, *Elektrische Klangerzeugung: Elektronische Musik und synthetische Sprache* (*Electrical Production of Sound: Electronic Music and Synthetic Language*).[16] Meyer-Eppler was a Belgian-born German who received his doctoral degree in physics from the University of Bonn in 1939, with a dissertation titled "An Arrangement for Direct Photoelectric Measurement of Radio Spectra."[17] From 1942, when he finished his *Habilitationsschrift* in sound analysis, titled "Distortions That Are Caused by the Finite Bandwidth of Physical Instruments with an Application to Periodic Research," until the end of the war, he worked as an assistant at the university's physics institute researching acoustical problems relevant to the war effort.[18] Like Trautwein, he was a member of the Nazi Party. As Iverson has shown, many of the physicists and engineers involved in postwar electronic music had been members of the Nazi Party and assisted in the war effort in one way or another.[19] After the war, he turned his attention to phonetics and speech analysis, joining the university's Institute for Phonetics in 1947 and working under the supervision of Paul Menzerath on the link between phonetics and acoustics, particularly in relation to telecommunications. In 1949, he became a research assistant at the institute, and in February 1953 he finished his second *Habilitationsschrift*, "Investigations on the Structure of Voiced and Voiceless Noise," this time in phonetics and communications research. He was particularly interested in information theory.[20] In 1956 he became a *Privatdozent* and a year later a professor and the director of the Institute for Phonetics and Communications Research.[21]

Elektrische Klangerzeugung summarized the numerous electrical devices that generated musical sounds and the various techniques employed to modulate amplitude and frequency. Meyer-Eppler also discussed the physics and engineering behind various electronic musical instruments, including the trautonium, electroacoustical piano and organ, the Neo-Bechstein grand piano, spherophone, melodium, Hammond organ, and theremin. He concluded the work with a brief section on synthetic speech. His expertise would prove crucial to a generation of young musicians wishing to experiment with new modalities of sound provided by physicists and engineers.

Starting in 1949 Meyer-Eppler began to create a synthetic language by means of a melochord and vocoder (a speech coder for telecommunications). He presented his research to a physics conference in Bonn in late September and then traveled to the small city of Detmold in the state of North Rhine-Westphalia to attend a conference titled "Developmental Possibilities of Sound" for sound engineers (*Tonmeister*).[22] At the gathering Meyer-Eppler made the acquaintance of the sound engineer, kapellmeister, and sound film expert Robert Beyer of NWDR in Cologne. Meyer-Eppler lectured on the developmental possibilities of synthesizing speech and singing based on the vocoder, which he had seen a year earlier while hosting Homer W. Dudley of the Bell Telephone Laboratories and after reading Claude Shannon and Warren Weaver's classic essay, "The Mathematical Theory of Communication."[23] Both Meyer-Eppler and Beyer were convinced that Germany could contribute to the new genre of electronic music, so they decided to participate in the International Summer Course for New Music in Darmstadt, which had started in 1946. Up to this point, German electronic music was only a subject of interest to engineers and physicists. As composer Herbert Eimert quipped, electronic music was unknown to most composers and musicians: "People who have had nothing to with music, physicists and engineers, suddenly speak about 'music' and 'composition' . . ."[24] That was about to change, albeit slowly at first. At the conference, Meyer-Eppler was made aware of Edgard Varèse's essay, "Music on New Paths." After obtaining Varèse's address from the music critic Stuckenschmidt, Meyer-Eppler wrote how he was interested in learning about the Frenchman's experiments with electronic musical instruments. Varèse replied that "these electric instruments are the most meaningful first step to the liberation of music . . . No distorting prism will exist between the composer and the listener."[25] The performer was apparently that distorting prism.

Meyer-Eppler read Sala's 1949 article in *Frequenz* and wanted to know how many trautoniums there were, as some three years earlier on a trip to visit Telefunken in Berlin, he had been informed that the instrument no longer existed.[26] In a subsequent letter, Meyer-Eppler explained that his interest in the trautonium was sparked by its ability to create artificial vowel sounds, as his laboratory was working on the synthesis of speech sounds with a view to create a classificatory scheme for vowels.[27] His curiosity intensified when he read with great interest about the trautonium in *Melos* and *Physikalische Blätter*. In November 1950 he requested that Sala send him a tape recording of the instrument.[28] Sala obliged.[29]

They discussed experiments on tone color, which Sala performed some twenty years earlier at the Rundfunkversuchsstelle. He told Meyer-Eppler that a physics laboratory alone could not achieve the desired results: a musically informed performance technique was critical.[30] Sala stressed that the early tapes did not do justice to the trautonium's true sound. The ranges of volume and tone colors were lost on the medium. For example, the "dogs-of-Hell" sounds in "Joan of Arc" sounded more like "a cat whose tail had been stepped on."[31]

"Authentic Manner of Composition"

Perhaps the most important theme with which the two engaged was Meyer-Eppler's notion of authenticity. In 1952 Meyer-Eppler wrote several paragraphs on what he called "authentische Kompositionsweise" (authentic manner of composition), which was a hallmark of electronic music.[32] Each composition, he insisted, was based on a twofold structure: a chronological subdivision into successive sections and a simultaneous superimposition of different layers. Say, for example, the composer was working on a piece lasting two minutes. They divided up the piece into ten-second sections, each one comprising three layers. Each layer could be created, for example, by an instrument, a type of distortion (linear or nonlinear), or the addition of reverberation. The first layer was recorded on reel-to-reel magnetic tape and then immediately checked to see if it corresponded to the composer's desired effects. Once the composer was happy with the layer, the tape was played back on a playback machine, and a copy was simultaneously made on a second machine. The composer played the second layer into this copy until they were happy with the two layers. The third layer was recorded in a similar fashion.[33] Note how powerful the composer was, and how absent the performer was, in Meyer-Eppler's account. There was no room for interpretation. Meyer-Eppler's sentiments echoed Stuckenschmidt's mechanical view of music, in which human interpretation was to be eliminated, some thirty years earlier.[34]

A year later Meyer-Eppler published a rather provocative essay, "Elektronische Kompositionstechnik" ("Electronic Technique of Composition").[35] The essay made three points that ruffled Sala's feathers. First, Meyer-Eppler wished to define electronic music as exclusively as possible: instruments such as the electric piano, electric cello, and electric guitar, as well as the musique concrète, were not included since their sounds were not generated electronically.[36] Second, he spoke of how the early electronic musical instruments were used as "substitutes" for traditional acoustical instruments rather than creating new sounds, which we know was not quite true. A picture of Sala's radio trautonium was included in that section, as was a picture of the Volkstrautonium, with Trautwein seated next to it.[37] Third, and perhaps most disturbing to Sala, Meyer-Eppler held a view of composer and performer as interpreter that was antithetical to Sala's and contradicted everything the trautonium virtuoso had achieved musically. Meyer-Eppler wrote:

> While until now a musical composition could only be translated into aural reality by going through the *performance* process, sound storage technology [i.e., a tape] allows the musical work to be prepared in detail on a sound carrier in such a way that it can be played back concretely and at any time (i.e., by the playback apparatus). The work that can be transferred into the acoustic field can be created by the composer himself. Such a work no longer needs any subsequent interpretation; it is presented by the composer in an *authentic* form. The comparison with a painter, who by also bypassing the interpreter brings his picture into a form that is sensual to the viewer, is obvious.[38]

This was the musical aesthetic of perfect reproduction by machines, one that had been epitomized by the Industrial Revolution.

Sala could not disagree more with view that the performer's interpretation was no longer a part of "authentic" electronic music. He thought "it was a joke" that "they [the Cologne Studio] have to torment themselves because of a 'lack of qualified instruments, composers, and performers' with 'fully synthetic engineering music.'"[39] Authenticity was, in Sala's view, the studio's excuse for their lack of musicality.

While playing on the mixture trautonium, Sala began to rethink what it meant to be a composer: "With electronic music one should not speak of 'composing': one would better speak of 'improvising.'"[40] Similarly he later wrote to Meyer-Eppler that those who came with a prewritten score would be disappointed. The key was to have a general conception of the sound (*Klangvorstellung*) that one wanted to generate and to let that serve as a guide. Then, allow "the sometimes crazy sound images" to emerge.[41] Those sound images were often under the control of the performer, not the composer, who should come with suggestions in the form of sketches or notes and general ideas and instructions about the directions in which they wished the sounds to go.[42] Now more so than ever—for better or for worse—the performer had a far greater role in interpreting what the composer desired. Sala concluded:

> The sonic surprises that come to light enliven the mood a great deal. One should not underestimate that, because with such musical-technical

> challenges, there must be a good mood as well as a desire to experiment from all sides.[43]

His view of the composer, performer, and composition stood in direct contrast to Meyer-Eppler, whose "authentic composition" was unsullied by the performer.[44] In Meyer-Eppler's view, electronic music rendered interpretation superfluous.

Much along the same lines as Sala, Trautwein stressed that electronic music, unlike mechanical music, actually freed the performer from slavishly following the will of the composer.

> The tendency of composers to lay down the rules of performance as precise as possible, and thereby limit the freedom of the performer (*Interpreten*) as much as possible, to make them perform mechanically, is unthinkable in the new music. Stravinsky included records with his musical scores in order to show how he wanted his works to be performed. Schönberg writes in numerous places that the performer (*Interpret*) must follow the examples that the composer prescribes. The machine is the best at executing these instructions, and it is only necessary for the composer to give exhaustively his commands to the machine regarding how he wishes his work to sound. Technology does not mechanize art; rather, it carries it upward into spheres of higher creative achievements.[45]

Unlike mechanical devices such as the metronome and records, which were used by composers to impose their will on performers, Trautwein and Sala's version of the new music encouraged improvisation and freedom of the performer. According to Trautwein, technology was liberating, not enslaving.

Sala voiced his opinion in a fascinating letter to Meyer-Eppler detailing his composing and playing techniques. Composers of electronic music were generally unable to play electronic instruments well enough to compose superior works. According to Sala, less time was needed for composing and more time for successive playing and recording, whereby new ideas arose. He was taking direct aim at Meyer-Eppler's view of an "authentic manner of composition."[46] The requisite technical interventions for successive recordings resulted in a completely different representation than if one did not use tape, even if the composer was simultaneously the performer. His so-called sound illustration (*Klangillustration*) generated by his music for the NWDR radio broadcast of *Faust I* of 1949 was the perfect example.[47] Successive recordings enabled a type of improvisation between performing and composing that resulted in a work that could not have been composable, or even produced by previous technological means. This is what Sala referred to as "a certain inevitable 'relationship of inexactness'" (*eine gewisse unvermeidbare "Ungenauigkeitsbeziehung"*).[48] If one attempted to be rigid from the outset, the end result would ironically be much more imprecise.

> The more sharply one fixes the original image, the less exact the transfer will be, the more one only sketches it and allows improvisation to prevail during the recording, the more precisely one can feel one's way toward the envisaged effect.[49]

Sala insisted that "the composer comes with suggestions, partly as sheet music sketches, partly as sound direction instructions. . . . The sound-direction instructions are specified in such a way that the player becomes quasi author."[50]

The tape music he created with his mixture trautonium was not aimed at mimicking known sounds, but generating new ones. Recall from the previous chapter how he asked Meyer-Eppler who in the world would want to use electroacoustic sounds to mimic the sounds of traditional acoustical instruments.[51] Sala stressed the importance of new sounds to his work in *Joan of Arc, Parsifal, Faust I,* and *Lucullus.* Composer Harald Genzmer concurred, wondering why anyone would desire to imitate acoustical instruments with the new music.[52] Meyer-Eppler agreed; however, for Sala and Genzmer that would not preclude the mixture trautonium from performing with a traditional orchestra. In early 1954 Sala wrote to Genzmer that Meyer-Eppler was adamantly opposed to coupling the mixture trautonium with a conventional orchestra. It would, in a sense, be superfluous. He added, "To my knowledge he [Meyer-Eppler] hasn't heard anything original. Hence my current strategy with the 'authentic' music."[53]

In addition, Sala felt at that point in time that audiences would not wish to hear a piece played on a tape through a loudspeaker without seeing the actual performance, an issue for electronic musicians even today. The technology was not good enough, and the audience would want to see the performers play.[54] While tapes were helpful in certain instances for creating multivoiced compositions, they still could not be used to convey the totality of the piece in terms of range of dynamics and tone colors.[55] Sala felt that his work offered a small practical contribution to the question of an authentic reproduction (not authentic composition) in music with the aid of the current technology of tapes.[56] If Meyer-Eppler was claiming his priority of using magnetic tapes to record an "authentic composition," certainly Sala himself had been doing that since 1947 with his work for various films and the radio drama music of *Faust I* for NWRD. He added, "I do not think that Mr. [sic] Eimert or Dr. Meyer-Eppler can dispute our priority here. Even P. Scheffers [sic and musique concrète] first began with that in 1948."[57]

Similar to Sala's claim of using tape to capture authentic reproduction, Trautwein was convinced that what Meyer-Eppler, Eimert, and Schaeffer were doing he had already done. He had called it the "phonomontage," or the creation of musical sounds and pieces by taping, cutting, overlapping, and rearranging other musical bits.[58] Trautwein's notion of "man [human] and measure," discussed in the previous chapter, was seen as an antidote to the "technological mysticism of the Parisian school of noise technicians (*Geräuschemonteure*) orbiting around Pierre Schaeffer."[59] His liked to use the term "phonomontage," as "everyone sees something mechanical in it, i.e., without an artistic component."[60] They were in his eyes a tool (*Hilfsmittel*) that Sala and he used, as did Meyer-Eppler and Schaeffer. According to Trautwein, however, Sala and he went beyond the phonomontage, while Meyer-Eppler and Schaeffer did not. Trautwein added that he and Sala should use this term much more often,

> because it is factually good and above all characterizes the mechanical aspects that are involved. "Authentic composition" and "musique concrète" suggest too much of artistic merit than is actually the case.[61]

He often spoke of music of the Cologne school as being "too machinelike."[62]

Sala liked the term, particularly when it referred to Meyer-Eppler's *Klangkulissen* (soundscapes), as they lacked any creative, artistic merit: they were simply mechanical pieces. That was precisely Sala's point about interpretation. Interpretation, which was needed to give artistic merit to the phonomontages, was the antithesis of authentic composition.[63]

Trautwein, not surprisingly, weighed in on the differences between Sala's and Meyer-Eppler's views of music of the early 1950s, specifically with reference to an "authentic manner of composition." According to Trautwein, Meyer-Eppler's musical synthesis was based on "technical tricks," including playing the tape forward and backward on a magnetophone, temporal expansion, artificial echoes, the systematic application of nonlinear distortions, and the cutting and pasting of the tape. Echoing Sala's stance, Trautwein responded sardonically:

> Meyer-Eppler and his group (Dr. Eimert and Beyer) trivialize the importance of artistic play by means of the dogma that every musical process arises primarily in the composer's mind. It is then a mere technical question how the composer realizes the process.[64]

In addition, according to Trautwein, Meyer-Eppler's technique was too laborious and time-consuming to create the desired effects. While Trautwein felt that Sala possessed the superior artistic experience and qualifications, he warned Sala not to underestimate the resources that a radio station provided Eimert and Beyer, particularly in terms of equipment and personnel.[65]

In May 1953 the NWDR station in Cologne hosted the "New Music Festival." It was the first time "authentic compositions" by Eimert and Beyer were broadcast.[66] Trautwein complained:

> The people want to present their own music as champions of electronic music waiting thirty years after [our] unsuccessfully groping. And in this conception, neither you nor Genzmer fits in. An analogous rivalry exists in the technical sector between Meyer-Eppler and me.[67]

Sala was unimpressed by what he had heard over the radio. "Overall, it was truly depressing and an antiadvertisement for the new music. Very unskillfully made."[68]

Legacies

The correspondence between Sala and Trautwein during this period reflected their strategizing: How could they play a role in defining electronic music while comparing and contrasting it with their own? Should they simply distance themselves from the Cologne Studio, insisting that what they were doing was not electronic music, or should they argue that theirs is a different form of electronic music? Was it best to challenge Meyer-Eppler and Eimert head-on and publicly? Trautwein wrote:

> I recommend the greatest restraint in our behavior toward Meyer-Eppler. We should not use words such as "anti-Meyer-Eppler" even in our internal correspondence. Dr. Meyer-Eppler is a significant scientist, who is very easy to get along with personally. He can do much more than mix tapes, and he will definitely jump off this hobbyhorse as soon as it does not cross the finish line as a champion, and that will definitely be the case sooner or later.[69]

Nearly three months later, Trautwein informed Sala that

> Dr. Meyer Eppler, who understands you very well as an artist and who also has a great future ahead of him, told me during the Cologne conference that he was very sorry that you seemed to be withdrawing from him. He regrets that very much. There is no resentment or apathy on his part.[70]

In addition to the notion of "authentic composition," Sala also fundamentally disagreed with what he perceived to be Meyer-Eppler's restrictive definition of electronic music. Meyer-Eppler basically defined electronic music as only the works emanating out of the Cologne Studio. Radio Bremen, RIAS, and other radio stations were labeling Sala's works, as well as the compositions coming out of Cologne, as electronic music. Why wouldn't Meyer-Eppler expand his definition? There was no legitimate reason to call a music tape created by electronic means something else other than electronic music, just because it was created in Berlin, Paris, or anywhere else. Sala found it absurd to refer to a work as belonging to either a correct or incorrect electronic musical style.

Under the general category of electronic music, there are those who worked with sine tones (Cologne), tape (Paris, Cologne, and Sala), mixture trautonium (Sala), and a melochord (Trautwein and Cologne).[71] Sala wrote to Trautwein complaining that Meyer-Eppler spoke of "real" (*wirkliche*) electronic music as only originating from the Cologne Studio: "Just between us: the next available opportunity will be used to present a factual opinion to the other side."[72] Trautwein held a similar view: "Active music making is the main point in which we want either to set ourselves apart or differentiate ourselves from 'the' electronic music, i.e., the Cologne things (*Dingen*) and musique concrète."[73]

Claiming that the electronic music composed by the Cologne Studio was "pathetic," Trautwein was sure that the studio did not have "a single musical idea, only a mixture of well-known sine tones that had a primitive and boring effect when all clumped together."[74] Trautwein compared the studio's dictating the future of electronic music to George Orwell's *1984*. He bemoaned that Meyer-Eppler, "himself an initiator of composition that cannot be interpreted," had recommended a terminology that was allegedly applicable to all electronic music without discussing any of the terms with either him or Sala.[75] Sala concurred: an agreement between everyone on the terms was critical. He recommended that performances using electronic devices be called "electroacoustic music"—a term used earlier by Schaeffer to describe musique concrète—and that music recorded on reel-to-reel tape be called "electronic music."[76] Interestingly, Trautwein disagreed with Sala's definition. While Trautwein thought the two of them should have come to an agreement on the musical terms earlier, he felt that what he and Sala were doing was a different type of electronic music than what Meyer-Eppler was defining. He argued that the term "electronic music" was Anglo-Saxon in origin and was used universally as any type of music based on the passage of electrons through vacuum tubes. Trautwein accused Meyer-Eppler of manipulating the definition in order to discount Sala's and Trautwein's work. Furthermore, he stressed that there was nothing specifically electronic about magnetic tape and that he was opposed to the phrase "electroacoustic music," since all music is acoustical.[77]

In response, Sala accused Trautwein of not being consistent when using terms such as "electronic," "electric," and "electroacoustic" and felt that they should both use "electroacoustic" to describe their work: "With all due respect, we cannot lump our stuff together with everything under the motto of electronic."[78] Meyer-Eppler, on the other hand, felt that one should not spend too much time defining the various musical styles, but rather continue working undeterred by such distractions: "Time will decide what lasts and what doesn't."[79]

Whereas Trautwein and Sala entered into a dialogue with Meyer-Eppler, they accused Eimert of ignoring them. In May 1953 Sala wanted Eimert to include his latest compositions in the electronic presentations of the Cologne

Studio. He told Trautwein: "Don't smile: I know he will refuse."[80] Trautwein whined that

> Eimert obviously wants to usurp electronic music, like twelve-tone music, for himself. He has already ousted his dear friend [sound engineer] Beyer and prevented someone else from publishing anything at all on this subject. Of course, we are uncomfortable characters: hence, his behavior toward you when you were last in Cologne.[81]

Trautwein was incensed by Eimert's piece "Formanten I und II" since "not even a hint of them [formants]—as we understand them—could be heard."[82] In addition, Eimert withdrew his tapes from a series of lectures and performances titled "Technology and Music," organized by Meyer-Eppler at the Technical University of Berlin. Sala bemoaned that this was "a regrettable and disconcerting gaffe on all sides, which clearly showed where the intriguing clique stands and with what means they work" to marginalize other musicians.[83] Finally, Sala had experienced difficulties convincing Eimert to coordinate a visit to the Cologne Studio to have a look at its technical equipment.[84] Sala approached Meyer-Eppler for assistance in the matter, and Meyer-Eppler was able to arrange Sala's visit. Beyer and NWDR engineer F. Enkel showed him around.[85] Afterward, Sala told Meyer-Eppler in confidence that the studio's electric monochord—built by Trautwein—was "more than feeble" (*mehr als dürftig*) and, not surprisingly, that it needed a mixture trautonium.[86] Sala hoped that the visit would improve relations between the Cologne group and him.[87]

While Sala was rather fond of Beyer, he admitted to Trautwein—also in confidence—that he was shocked when touring the studio by the engineer's fundamental lack of understanding of tone color. According to Beyer, "Tone color is currently an unresearched area." Sala added that Helmholtz would have rolled over in his grave had he heard such nonsense.[88] Beyer was often in contact with Trautwein, Sala, and Genzmer: it turns out that he was Trautwein's neighbor in Düsseldorf. All of the interlocutors hoped that an "east-west tension" would not arise in electronic music.[89]

Their harsh views about each other's music notwithstanding, Sala and Meyer-Eppler continued correspondence and collaboration. Meyer-Eppler received from Siegfried Goslich of Radio Bremen a tape of Sala's new compositions with the mixture trautonium.[90] Meyer-Eppler was impressed with what the instrument could achieve, and he wanted to include a recording of it for his radio program on January 7, 1954, that dealt with the development of electronic musical instruments.[91]

In August 1954, in a session with Meyer-Eppler and Maurice Martenot—the inventor of the ondes Martenot—Sala demonstrated new experiments on electronic sound designs on his mixture trautonium in Gravesano,

Switzerland.[92] This would become a major nodal point in the network of early electronic music. Two months later, the small broadcasting room of the NWDR Cologne transmitted a rather interesting concert: it was the first one based solely on tape recordings of electronic music. It featured the works of Karlheinz Stockhausen, Herbert Eimert, Karel Goeyvaerts, Henri Pousseur, and Paul Gredinger. Trautwein called the concert "miserable" (*kümmerlich*) and said that the electronic music featured "lacked any musical idea: it was merely a mixture of the well-known sine tones with damping and reversed decays, which in my view sounded awful."[93] Sala had not heard the concert, and Trautwein anxiously awaited his opinion.

Sala played Genzmer's piece for orchestra and mixture trautonium at the Conference for Electronic Music and Musique Concrète in Basel, May 19–21, 1955. The conference was sponsored by the Swiss Broadcasting Corporation, and the leaders of electronic music were present, including Genzmer, Trautwein, Martenot, Schaeffer, Meyer-Eppler, Eimert, and Jacques Poullin.[94] Attempts were made to forge peace among the various parties who had been sparring over the definition of European electronic music.

During an interview with Canadian music critic Norma Beecroft around 1977, Sala spoke about his views of Schaeffer and musique concrète, reflecting back on the time when they appeared together at various conferences in Switzerland presenting their work. During the early and mid-1950s, Sala did not yet have a studio. Schaeffer, of course, did. Sala added:

> I found the tape methods of Schaeffer very interesting. I saw many possibilities with tape. . . . What was interesting for me was the possibilities to form simple sounds into a complicated mixture of sounds.[95]

Sala did, however, differentiate between their techniques:

> Schaeffer had very primitive sounds for [sic] the beginning, noises and so on, and he must have very complicated equipment to make tapes, but I have very complicated sounds from the beginning, and my tape methods cannot be so complicated, as Schaeffer's.[96]

In 1995 Sala was once again asked about the relationship between his music and Schaeffer's. He admired how Schaeffer and his crew managed to work with what Sala considered among the ugliest tones, pure sine tones, which for Sala should only be used for measuring technologies. Echoing his view from some four decades earlier, he insisted that "an electronic art of interpretation was unknown to them. I myself would have never come to electronic music under such conditions."[97]

While the various musicians discussed in this chapter shifted from the immaterial realm of radio to the material world of tape, the ways in which they reconceived their work was quite different. The disagreements between various electronic musicians had critical implications for how this new postwar music would be defined and who would be included in the genre.[98] Sala and his trautonium represented a different form of musical aesthetics—one of tinkering, playing, and recreating. Contrasting this form with the others' methods is informative, as it occurred precisely when Sala's and Trautwein's legacies were at stake. Clearly, Sala dedicated much time and thought to how the technology of tape recording affected musical composition. In the 1950s, because of technological advances, tape was emerging as an alternative to radio and the world of musical instruments. It was yet another venue where reproduction and the theme of tone color played out. Interestingly, the Cologne Studio's desire to go beyond acoustical instruments and to disregard the performer is reminiscent of how radio manufacturers neglected sopranos back in the 1920s, as discussed in chapter 2.

The jousting during the early 1950s between Trautwein and Sala and the Cologne Studio reveals forgotten debates that were critical in shaping the future of electronic music worldwide. What, if any, is the role of musical instruments in electronic music? What is the relationship between the composer, the performer, and the instrument? How much does or should a composer "control" the piece? How much of the interpretation should be in the hands of the performer? Or should it be out of the performer's hands entirely? Should the performer be considered the coauthor of the composition? The trautonium played a fundamental role in those debates. It was a perfect source for tinkering with and thinking about compositions. It exemplified the importance of the musical instrument as the mediator between composer and performer.

9

Epilogue

I REMEMBER watching Alfred Hitchcock's *The Birds* on television as a young child. The movie had premiered in 1963 in theaters around the globe. I distinctly recall watching it with my father and older brother, as my mother would not join us. She had been traumatized by her older brother chasing her around with a dead pheasant when they were children. Ever since then, birds terrified her. Being the mischievous and devilish child that I was, I wanted to watch those scenes that would potentially terrorize my mother. I remember thinking they were not all that horrendous. I never thought of the movie again until a number of years ago, when I began research on how science and technology shape musical aesthetics. That research would eventually evolve into this book.

Sala employed his mixture trautonium to create the sounds of screeching birds flapping their wings in Hitchcock's film.[1] During an interview with the journal *Sight and Sound*, Bernard Herrmann, the movie's sound consultant, was asked about the "musical innovation" of the movie. He insisted that it was not music, that he saw the mixture trautonium as imitating the sounds of the birds. "We developed the noise of the birds electronically because it wasn't possible to get a thousand birds to make that sound. I guess you could if you went to Africa and waited for the proper day."[2] Hitchcock had a somewhat different take on the matter. He felt the artificial sounds were far ghastlier than any potentional natural sounds: "I can hear that sort of thing from the gulls and the crows. I need something out of the ordinary that will frighten people."[3] He did not want the typical musical background to his film; he wanted something unique, a distinctive "style and nature of sound."[4]

> Conventional music usually serves either as a counterpoint or a comment of whatever scene is being played. I decided to use a more abstract approach. After all, when you put music to film, it's really sound, it isn't music per se.[5]

Fittingly, Herrmann's and Hitchcock's views of the bird sounds in the film encapsulated the two major characteristics of the trautonium: imitation of well-established sounds and the creation of new, frightening ones.

FIGURE 9.1. Oskar Sala and Alfred Hitchcock at the Mixturtrautonium (mixture trautonium) in Sala's studio in West Berlin in December 1962 working on the sound of the birds for Hitchcock's classic *The Birds* of 1963. Reproduced with the permission of Ullstein Bild.

Remi Gassmann, who had collaborated with Sala on the ballets *Electronics* and *Paean*, had contacted script supervisor Peggy Robertson, who was also Hitchcock's assistant, to recommend that the producer employ electronic music for his films: "For the first time, we have at our disposal, through electronic generation, what has aptly been called the totality of the acoustical."[6] Robertson forwarded Gassmann's letter to Waldon Watson, the director of Revue Studios, who then gave it to William Russell, his sound director. Russell listened to a demonstration tape that Gassmann had made. He told Hitchcock that they now had the sound they wanted for the film.

Hitchcock visited Sala's studio in West Berlin from December 16–20, 1962, to see how Sala was progressing.[7] As a teenager, Hitchchock had taken courses in engineering and became a technical clerk at the Henley Telegraph and Cable Company in London. Hitchcock and Sala tinkered together until the director was satisfied. On December 20, he sent a cable to his assistant: "Work in Berlin completed to my satisfaction, Hitch."[8] In an interview with *Screen* magazine the following February, Hitchcock conceded, "All I did is to listen and offer a few changes. It took me all of four days . . . upon which I retired to St. Moritz to contemplate."[9]

Hitchcock decided to use both artificial and natural bird sounds. For the scenes of birds attacking, electronic sounds were used. For example, during the birthday party scene, electronic sounds were used to distinguish screeching birds from screaming children. The sparrow attack was to represent "shrill anger," the crow attack to show "anger." Electronic sounds were further used during the attack scene at the restaurant The Tides, the attic sequence, the death of Annie, and the final sequence, where the sounds were meant to represent "brooding." Natural bird sounds were used in the opening scene on the streets of San Francisco, at the bird shop, Melanie's journey to Bodega Bay, the journey to and from the Brenner house, at Fawcett's farm, and the boarding-up sequence.[10] Hitchcock turned to the mixture trautonium because it enabled him to enrich the sound effects. It augmented the possibilities of sound used in films.[11]

While *The Birds* undoubtedly represented the trautonium's zenith, its story certainly does not end there. Up until the early 1990s, Sala continued to perform and record on his mixture trautonium, and he still composed music for films and commercials.[12] In the early 1980s, three scholars from the communications technology faculty of the Fachhochschule der Deutschen Bundespost in West Berlin—Hans Jörg Borowicz, Dietmar Rudolph, and Helmut Zahn—attended a lecture and performance by Sala in 1980 at the "Augen und Ohren" ("Eyes and Ears") exhibition at the Academy of Arts. They were so impressed with the uniqueness of the instrument and its sound spectrum that they expressed the desire to build a new mixture trautonium with the assistance of Sala and their engineering students.[13] They visited his studio to study the instrument. In 1982, they started working on their version of the instrument, which would be easily transportable. They finished the first version of it a year later. It debuted, rather appropriately, at the International Radio Exhibition in Berlin in 1983. They continued to tinker with the instrument, changing its structure and attempting to adapt their manuals to the original.[14] The final round of improvements was completed in 1988, and the improved version debuted at the Berlin Congress Hall on August 18 as part of a celebration of Berlin being named the cultural capital that year. The sawtooth waves were now produced by transistors rather than thyratrons. Sala played a new composition for the occasion, his "Fantasie-Demonstration für Mixturtrautonium."[15] In May 1991 he performed yet another new composition of his own creation at the Osnabrück City Hall.[16] On February 20, 2002, Oskar Sala passed away at the age of ninety-one in Berlin, the city where it all started.

The work of Trautwein and Sala lived on in subsequent electronic musical instruments that drew upon the design of the trautonium. The story of the trautonium does have a German Democratic Republic chapter. Back in August 1948, Sala received a contract from the general director of the radio station of the Soviet-occupied zone in Berlin for a quartet trautonium to be completed by 1950.[17] The instrument never truly functioned as intended, so, in 1957, the project was terminated.[18] In 1958 engineers in the former German Democratic Republic, under the direction of project engineer Ernst Schreiber at the Laboratory for Frontier Problems in Acoustics and Music in East Berlin, began working on an instrument similar to the mixture trautonium.[19] That instrument was called the subharchord. Established two years earlier, the laboratory was the first electronic music studio in the Soviet bloc. Explicitly based on the Rundfunkversuchsstelle of the 1920s and '30s, the GDR laboratory employed sixteen people, including technicians, sound engineers, and assistants, all of whom tinkered on devices to solve acoustical problems related to recording and radio transmission.[20] Laboratory director Gerhard Steinke had first thought of the subharchord after having met conductor Hermann Scherchen in 1948. His idea was shaped by a subsequent visit with Oskar Sala in West Berlin.

The subharchord could produce subharmonics and formants that imitated other traditional instruments, such as oboes and trumpets. While the sound was also based on subtractive synthesis like the trautonium and so many mid-century devices, a keyboard replaced the metal strips. Nikita Khrushchev had expressed disdain for what he referred to as the "cacophony" of modern music in a lecture titled "Declaration on Music in Soviet Society," delivered to party members and artists in March 1963. Yet East German engineers further developed the instrument.[21] Forty-three works for the subharchord were composed from 1962 to 1974.[22] The subharchord contributed to many of the same music genres that the trautonium had contributed to. Twelve works featured the subharchord as a solo instrument, seven for an ensemble with the subharchord, five for Rundfunkmusik, six for television, twelve for film music scores, and one for theater.[23]

Austrian composer Max Brand, whose most renowned work was the 1929 opera *Maschinist Hopkins,* had heard Sala play his mixture trautonium and was fascinated by it. As a result, in 1966 he asked Bob Moog, the inventor of the Moog synthesizer, to build him a similar instrument.[24] Completed two years later, it was called the Moogtonium. Herbert Deutsch, who worked on the Moog synthesizer, maintained that the trautonium played a role in Robert Moog's subharmonicon, a semi-modular synthesizer.[25] And in 1994, computer software engineer Jörg Spix created a computer program to generate a digital trautonium using an Atari ST computer and a MIDI synthesizer.[26]

As many modern music fans will tell you, Germany (and particularly Berlin) became the electronic music and techno music capital of the world during the 1980s and '90s. As a frequent visitor to Berlin to work in the archives in the summer, I made sure I hightailed it out of the capital for a weekend to avoid the (eventually) million-plus electronic dance fans who would descend on the capital to attend the Love Parade, which took place from 1989 to 2003, and again in 2006. From 2007 to 2010, the venue was changed to the Ruhr area. In 2010 twenty-one people were tragically crushed to death in Duisburg. As a result, the Love Parade was canceled until 2022, when it returned to Berlin.

When one thinks of German electronic music, one immediately thinks of the band Kraftwerk, undoubtedly the most influential German electronic music group. Radio stations around the world played the band's albums, *Autobahn, Trans-Europe Express, The Man-Machine,* and *Computer World,* which were highly ranked on numerous charts. They often called their genre of music "industrial folk music." Florian Schneider-Esleben, one of the founding members of the group, paid homage to Sala in his introduction to *Oskar Sala: Pionier der elektronischen Musik.*[27] He and Sala, in honor of the latter's ninetieth birthday, participated in a discussion on Sala's work.[28] The group's lead singer and keyboardist Ralf Hütter said, "We always considered ourselves the second generation of electronic explorers, after Stockhausen."[29] Their song "Airwaves," from their *Radio-Activity* album, drew upon Karlheinz Stockhausen's sound experiments used in his "Studie I" and "Studie II." "Voice of Energy" from that album utilized speech synthesis as pioneered by Werner Meyer-Eppler.[30] Recall that Trautwein's experiments on artificial vowel production predates Meyer-Eppler's research. So, if Stockhausen and the Cologne Studio were Kraftwerk's parents, Trautwein and Sala were their grandparents.

Obviously, radio fidelity today is markedly better than it was in the 1920s and '30s. This is in part due to the improvement of broadcasting stations and the receiver components. But the invention of frequency modulation (FM) radio vastly improved radio broadcasting: FM possesses a much greater fidelity than AM. In 1928 American engineer Edwin Howard Armstrong, working in a basement laboratory at Columbia University, began researching FM. By 1933 he had developed wideband FM, and by the late 1930s a small number of radio stations in the United States were broadcasting in FM.

The popularity of FM stations soared in the 1970s as a direct result of the creation of album-oriented rock, an FM radio format that was critical to the playing of what we now call classic rock albums on the radio with superior fidelity. Growing up at the time, I truly appreciated listening to Pink Floyd and Led Zeppelin (as well as Jacqueline du Pré and Itzhak Perlman) on FM. By the end of the decade, the number of listeners of FM stations surpassed those of AM stations, which, by the late 1980s and '90s in the United States, were filled with talk radio, sports broadcasts, news, and foreign-language broadcasts.

Germany was one of the first countries in Europe to adopt FM. Much of their AM frequency was taken up by Allied forces radio programming after the war, and the Copenhagen Frequency Plan of 1948 only permitted Germany to have two medium-wave frequency bands that were not conducive to broadcasting.[31] After initially broadcasting on ultra-short wave (*Ultrakurzwellen*)—now called very high frequency, or VHF—scientists and engineers in the occupied territories that would become the Federal Republic of Germany realized that FM was much better than AM for VHF radio broadcasting.

Finally, just as Sala had difficulties finding music students to play the instrument throughout his lifetime, today there is only one professional trautonium player, Peter Pichler from Munich. Once Peter decides to stop playing, it is not clear if the instrument in its various instantiations will be ever be played again.

This book has used two pieces of material culture, early German radio and the trautonium, to piece together bits of culture that, today, many outside media studies and science and technology studies assume to be unrelated. By structuring a narrative around material objects, one can appreciate the links that exist between the histories of musical aesthetics; physiological, physical, and electrical engineering theories and practices; harmonic analyzers; broadcast fidelity; electronic music; tone color; and politics.

Improvements in radio fidelity and related research in tone color gave rise to a fascinating musical instrument. Its flexibility in both imitating sounds and tone colors of traditional acoustical instruments and creating original, modern tones and tone colors enabled it to survive nearly a century's worth of musical aesthetics and political ideologies. While it garnered fame from its appearances in numerous musical genres and the creation of sound effects for film and commercials, it was also an instrument relevant in debates among physicists, radio engineers, and physiologists about the formation of tone colors and vowel sounds.

A history of the trautonium also illuminates the historical development of synthesizers as well as electric, electroacoustic, and electronic music in the twentieth century. The trautonium belonged to electric and electroacoustic music from the very beginning, but the question remained—and was hotly contested—as to whether the instrument should belong to the new postwar genre of electronic music. The materiality of electroacoustical instruments, as embodied by the trautonium, was critical. For Sala and Trautwein, contra the Cologne Studio for Electronic Music and musique concrète, musical instruments were still important to the creation of music. The roles of the performer and tape in the process of composition were also points of contention between the groups. While the categories of electric, electroacoustic, and electronic music are often lumped together under the general rubric of electronic music in the secondary literature, such a classification tends to flatten the complex and historically contingent relationships between these musical genres. Precisely because of the trautonium's longevity, it can enrich our historical understanding of those relationships.

Computers with high-quality speakers can now reproduce sound with impressive levels of fidelity. In addition, thanks to advantages in artificial intelligence and machine learning, computers can compose music ranging from classical to jazz. Science and engineering not only create a musical aesthetic, they create music. In the eighteenth century, debates erupted as to whether musical automata built by the world's finest mechanicians could imitate musical instruments and evoke certain emotions by successfully duplicating a certain a musical aesthetic. Now, we are at a point where the human seems to be removed from the process altogether. Many of the fears expressed by critics of the trautonium—namely, that the human would eventually be replaced by circuits and vacuum tubes—uncannily seem to have come to fruition to some extent.

Rather fittingly, on April 1, 2023, Luftrum Sound Design played a clever April Fools' prank. It announced the launch of SynthGPT, which it claimed used artificial intelligence and machine learning to render all synthesizers obsolete. The company later acknowledged that SynchGPT "doesn't actually exist (at least not yet!) . . . The intention was simply to create a playful prank that was timely and relevant."[32] Should SynthGPT become available one day, my hunch is that the various responses to it will be similar to those elicited by the trautonium in particular and modernism in general.

Such a possibility is not as far-fetched as it seems. Back in June of 2023, Paul McCartney famously reported that he used AI to help him create "the final Beatle's record."[33] The music industry has recently been turning to AI. For example, AI singing-voice generators, such as Lyrebird, Uberduck, and Voicemod, can analyze recordings of a singer's voice and replicate it. They can even create new songs or entire albums.[34] This, of course, raises interesting questions for intellectual property law.

There are more dire ethical and legal ramifications of voice replication. In April 2023, a daughter called her mother to inform her that she had been kidnapped and that the two men responsible demanded a $1 million ransom. The mother told reporters that "the voice sounded just like [my daughter's], the inflection, everything."[35] It was a scam, and the scammers used AI-driven voice cloning to deceive the mother. Voice cloning draws upon AI techniques to train a machine-learning voice model. It takes portions of a recording of someone's voice and determines the sound spectrum that has near perfect fidelity to the original voice. The political, ethical, social, and legal implications of voice cloning, which dates back to 1998 but has only been in the news for the past few years, are enormous.

A history of radio and the trautonium belongs not only to a much more general history of the ways in which the natural sciences and engineering have contributed to the creation of musical aesthetics, it also offers us important insights on the historically contingent relationships between science, technology, and music, on the one hand, and society, culture, economics, and politics on the other.[36] To insist on disciplinary forms of knowledge, or to ignore the content of scientific and technological knowledge, is to offer an incomplete history. It prevents us from participating intelligently in current sociocultural and political debates involving the human.

NOTES

Introduction

1. While German inventors during the 1920s and '30s referred to the instruments as electric instruments, many of these instruments were what we would now call electronic musical instruments. The genre was referred to as electric music, electro-music, and electroacoustic music. Today, an electrical musical instrument is defined as one that draws upon electricity to amplify the sound. Examples include the electric organ, electric piano, and electric violin. Electronic musical instruments, on the other hand, use electronic means to generate and control the sound. Examples of electronic musical instruments include the theremin, trautonium, and Hammond organ. That said, I am sticking to the actors' categories by referring to these instruments as electric musical instruments. I shall also be discussing the historically contingent relationships between electric music, electroacoustic music, and electronic music throughout this work. See, for example, Brilmayer, "Das Trautonium: Prozesse des Technologietransfers," 7–17.

2. Peter Pichler of Munich is the sole remaining trautonium virtuoso. I thank him for taking the time to show me his various trautoniums and permitting me to play them.

3. Thérberge, *Any Sound You Can Image*. See also Sterne, *The Audible Past*, and Patteson, *Instruments for New Music*.

4. Mager, *Eine neue Epoche*.

5. Wittje, *The Age of Electroacoustics*, and Yeang, *Transforming Noise*.

6. See, for example, Patteson, *Instruments for New Music*, and Holmes, *Electronic and Experimental Music*, 3–40.

7. Matthews, "The Evolution of Real-time Music Synthesizers": S74.

8. Modern synthesizers are composed of transistors, which, of course, did not exist during the early years of the trautonium. Modern electronics replaced the old-fashioned vacuum tubes.

9. Donhauser, *Oskar Sala*, 84.

10. *Sonderweg*, or separate path, refers to a view of history that stresses the lines of continuity from the creation of the German nation-state in 1871 to the rise of fascism in 1933. It was a popular debate among historians of German from the 1960s through the '80s espoused by three leading German historians: Hans Rosenberg, Jürgen Kochka, and Hans-Ulrich Wehler. The view states that the following played critical roles in Germany's special path: the power of the military, popular militarism in the *Kaiserreich*, the parliament's inability to keep portions of the government in check, and the power of East Prussian feudal landowners over villages. For them, Hitler's rise to power was overdetermined. *Sonderweg* is often seen as a way to come to grips with the horrors of the Third Reich. It strikes me that the Weimar Republic was such a fascinating period that it should be treated on its own terms, rather than reducing it to a precursor of the Third Reich. See Rossol and Ziemann, "Introduction," in Rossol and Ziemann, eds., *The Oxford Handbook*, 2–3. During the 1980s and '90s, a number of historians criticized the various *Sonderweg* theories, arguing that Germany was actually not as unique as had been originally claimed, since its circumstances were similar to a number of other countries. The national lens might not be the best one for the analysis. For example, Detlev Peukert insists that a number of

other European countries experienced similar social, political, and economic crises that *Sonderweg* historians argued were unique to Germany. See Peukert, *Die Weimarer Republik.* David Blackbourn and Geoff Eley suggest that Germany's *Kaiserreich* was far more modern than the *Sonderweg* proponents wish to argue. See Blackbourn and Eley, *The Peculiarities of German History.* My book does not intend to engage in that debate. Rather I wish to argue that the contextual uniqueness of Germany of the 1920s resulted in the trautonium's invention. Historians, I feel, need to explain why certain events occurred in one historical context and not in another. For a nice summary of the *Sonderweg* historiography, see Everett, "The Generation of the Sonderweg"; Kochka, "German History before Hitler," 3–16; Kochka, "Looking Back at the Sonderweg," 137–42 and Smith, "When the Sonderweg Left Us," 225–40.

11. Dolan, *The Orchestral Revolution.*

12. Dolan and Rehding, eds., *The Oxford Handbook.*

13. Thompson, *The Soundscape of Modernity.*

14. Sterne, *The Audible Past.*

15. Thompson, *The Soundscape of Modernity*, 237–38.

16. Chion, *Audio-Vision*, 23, 93–95, 98–99, 107–9, 119–120. While that might be true, my actors certainly believed that one could objectively measure fidelity and thereby improve speakers, amplifiers, and loudspeakers. I thank Jonathan Sterne for pointing out Chion's work to me.

17. See for example, Trendelenburg, *Klänge und Geräusche.*

18. Sattelberg, *Dictionary of Technological Terms,* part 1, p. 93, and part 2, p. 267.

19. On this point, see Sterne, *Audible Past* (2003). Mark Katz argues that recording technologies gave rise to new forms of performing, such as violin vibrato and crooning. Katz, *Capturing Sound.*

20. Gabriel, "Opera after Optimism," Grosch, *Die Musik*, and Potter, *The Art of Suppression,* 208–11.

21. See, for example, Halliwell, *The Aesthetics of Mimesis.*

22. I thank Jonathan Sterne for pointing that out to me.

23. Mulvin, *Proxies.*

24. Ludwig, *Technik und Ingenieure.*

25. Voskuhl, "Engineering Philosophy," 725.

26. Pinch and Trocco, *Analog Days.* My work also owes much to sound studies in general. Other informative works include Pinch and Bijsterveld, eds., *The Oxford Handbook*; Braun, ed., *Creativity: Technology and Music*, "Introduction"; Bijsterveld, *Mechanical Sound*; Lee and Mills, "Vocal Features," 129–60; Mills, "Hearing Aids," 24–45; Mills and Tresch, eds., *Audio/Visual*; and Bruyninckx's *Listening in the Field.*

27. Tresch and Dolan, "Toward a New Organology," 281.

28. Wolf, "Haltbarkeit," 63–82; Wolf, "Musical Instruments," 7–11; and Rehding, "Instruments of Music Theory."

29. Patteson, *Instruments for New Music.*

30. Dettelbach, "The Face of Nature," 473–504; Jackson, *Harmonious Triads*; Brain, *The Pulse of Modernism*; Tresch, *The Romantic Machine*; and Wise, *Aesthetics, Industry and Science.*

31. Pesic, *Music and the Making of Modern Science*; Pesic, *Sounding Bodies.*

32. Hui, *The Psychophysical Ear.*

33. Kursell, "Experiments on Tone Color," 191–211; Kursell, "Carl Stumpf," 323–46; Kursell, "From Tone to Tune," 121–39; Kursell, "Listening to More Than Sounds," S39–S63; and Kursell, "'False Relations,'" 85–96.

34. Tkaczyk, *Thinking with Sound.*

35. Coen, *Vienna in the Age of Uncertainty.*

36. Brilmayer, "Das Trautonium: Prozesse des Technologietransfers," and Dörfling, *Der Schwingkreis.*

37. Donhauser, *Oskar Sala.*

38. For example, Peter Manning only mentions the instrument in passing on four pages. See Manning, *Electronic and Computer Music*, 4, 11, 40, 43. Thom Holmes also spends four pages on the trautonium: Holmes, *Electronic and Experimental Music*, 30–33, 241. Mark Vail briefly summarizes the trautonium in two pages. Vail, *The Synthesizer*, 8–9. Patteson's *Instruments for New Music*, however, has a chapter dedicated to the trautonium, 114–51.

39. Iverson, *Electronic Inspirations*.

1: Weimar Radio: The Great Experiment

1. The station was initially named Radio-Stunde on its official founding, which did not occur until December 10, 1923. It was renamed Funk-Stunde A. G. Berlin (or simply Funk-Stunde Berlin) on March 18, 1924. See, for example, Giesecke, "Wie die Organisation," 65. The initial transmitting frequency was 750.0 kHz (400 meters), subsequently changing to 697.2 kHz (430 meters) and 594.1 kHz (505 meters). See Braun, *Achtung, Achtung, Hier ist Berlin!* 81, and Weichart, "In 14 Tagen," 43–52.

2. Halefelt, "Sendegesellschaft und Rundfunkordnungen," vol. 1, 23. See also "90 Jahre Radio," https://www.deutschlandfunk.de/90-jahre-radio.724.de.html?dram:article_id=266956. For a recording of this announcement, see "Achtung, Achtung (Wort) (Hier ist die Sendestelle Berlin)," https://www.youtube.com/watch?v=C4kMLzWQ0RY.

3. See Jackson, *Harmonious Triads*, 50–53. See also Applegate and Potter, *Music and the German National Identity*, and Rehding, *Music and Monumentality*. On the notion of an invention of tradition, see Hobsbawm and Ranger, eds., *The Invention of Tradition*.

4. Weichart, "In 14 Tagen," 48.

5. Weichart, "In 14 Tagen," 49.

6. Weichart, "In 14 Tagen," 49, and Bredow, "So entstand der deutsche Rundfunk," 4–5.

7. As quoted in Ulrich, *Deutschland 1923*, 9.

8. As quoted in Ulrich, *Deutschland 1923*, 9.

9. Geyer, "The Period of Inflation," 56, 58–62.

10. For an excellent summary of Germany in 1923, see Ulrich, *Deutschland 1923*.

11. Müller, "Liberalism," 323, 333.

12. Thun, "Die Vorgeschichte," 14–15.

13. Funktechnischer Verein, ed., *Bestimmungen über den Unterhaltungs-Rundfunk*, part 3, p. 1.

14. Bredow, "So entstand der deutsche Rundfunk," 6.

15. Antoine, "Marksteine," 218. For a brief history of the Königs Wusterhausen radio station, see Gerlach, "Wie Königs Wusterhausen," 27–41. An arc transmitter is a type of spark transmitter that uses an electric arc to convert direct current into alternating current.

16. Antoine, "Marksteine," 218, and Gerlach "Wie Königs Wusterhausen," 30.

17. Gerlach, "Wie Königs Wusterhausen," 33.

18. Pohle, *Der Rundfunk*, 21–27. See also Antoine, "Marksteine," 218, and Gerlach, "Wie Königs Wusterhausen," 30.

19. Antoine, "Marksteine," 219. See also Braun, *Achtung, Achtung, Hier ist Berlin!* 80.

20. Antoine, "Marksteine," 219.

21. When it opened in 1924, the Frankfurt broadcasting station transmitted to folks within a 250-kilometer radius of the city, which included the residents of Cologne, Cassel, Eisenach, and Trier. See Lertes,"Unser Programm!" 1.

22. Magnus, "Reichs-Rundfunk-Gesellschaft," 32.

23. Bredow, "So entstand der deutsche Rundfunk," 23. See also Winzheimer, *Das musikalische Kunstwerk*, 14.

24. See Pohle, *Der Rundfunk*, 46, and Leberke, "Wirtschaftsführung im Rundfunk," 73. A significant percentage of the money collected went to radio program companies. During the

last years of the Weimar Republic, that percentage fell below 50 percent. See Führer, "A Medium of Modernity?" 727.

25. Anon., "Die Zahl der Rundfunkhörer," 28. A second source had very similar (and in some cases the identical) numbers as of January 1 of each year: 555,548 in 1925; 1,022,348 in 1926; 1,376,574 in 1927; 2,009,842 in 1928; 2,635,567 in 1929; and 3,066,6872 in 1930. See von Boeckmann, "Organisation des deutschen Rundfunks," 223. Another source had the number of radio licenses sold on April 1 of every year as 9,895 in 1924; 778,868 in 1925; 1,205,310 in 1926; 1,635,728 in 1927; 2,234,732 in 1928; 2,837,894 in 1929; and 3,238,396 in 1930. See Antoine, *Die Entwicklung*, not paginated.

26. Braun, *Achtung, Achtung, Hier ist Berlin!* 80–85. By the end of 1932, the number of licenses sold topped 4.2 million. Bredow, "So entstand der deutsche Rundfunk," 18. Note that Führer argues that the average German household totaled between two and three persons; therefore, his numbers are a tad lower. See Führer, *Wirtschaftsgeschichte des Rundfunks*, 55. According to accounts from the 1920s, authorities assumed four persons per radio license. See, for example, anon., "Zwei Millionen Hörer," 32.

27. Antoine, *Die Entwicklung*, not paginated.

28. Münster was initially chosen as the site for the Westdeutscher Funkstunde AG because the Rhineland and Ruhr region had occupying troops stationed there, and radio transmitters were therefore not permitted. On October 10, 1924, Cologne became the primary site of the radio station, with Münster hosting a relay station. Most of the broadcasting was eventually transferred to Langenberg. Braun, *Achtung, Achtung, Hier ist Berlin!* 80–82. See also Bredow, "So entstand der deutsche Rundfunk," 9, 12, 17, and von Boeckmann, "Organisation des deutschen Rundfunks," 237–38. A second broadcasting station in Berlin on Magdeburger Platz (in Tiergarten) was created on June 27, 1924, while the original Vox-Haus station closed on December 5 later that year. A third station would be erected in Berlin-Witzleben on September 25, 1925. See also Antoine, "Marksteine," 222, and Weichart, "In 14 Tagen," 52. The renowned Haus des Rundfunks began broadcasting in 1930. See Mueller, "Berlin baut sein neues Funkhaus," 219–24.

29. Antoine, "Marksteine," 222. An auxiliary transmitting station was created in Hanover on December 16, 1924. By the end of 1925, Stettin, Gleiwitz, Kassel, Dortmund, Elberfeld, Dresden, Kiel, Cologne, and Aachen all had radio stations, and Munich had a second one. Bredow, "So entstand der deutsche Rundfunk," 18.

30. Antoine, "Marksteine," 224, and Bredow, "So entstand der deutsche Rundfunk," 12.

31. Bredow, "So entstand der deutsche Rundfunk," 18. Among those stations were ones in Königs Wusterhausen, Langenberg, and Nauen, the oldest continuously broadcasting station created by Telefunken on April 1, 1906.

32. Führer, "A Medium of Modernity?" 724–25.

33. Giesecke, "Wie die Organization," 40.

34. Bredow, "So entstand der deutsche Rundfunk," 23.

35. Führer, "A Medium of Modernity?" 726.

36. Von Boeckmann, "Organisation des deutschen Rundfunks," 223, and Pohle, *Der Rundfunk*, 48.

37. Pohle, *Der Rundfunk*, 48.

38. Bredow, "Dem Deutschen Rundfunk zum Geleit," 15. See also Funktechnischer Verein, ed. *Bestimmungen*, vol. 3, p. 5, italics added by the author. Certainly the notion of radio's ability to uplift a nation's citizenry was not uniquely German. See, for example, Hilmes, *Radio Voices*.

39. As quoted in Pohle, *Der Rundfunk*, 51.

40. The use of radio as an instrument of impartiality and internationalism resonated with other European countries' view of radio during the 1920s. See Lommers, *Europe—On Air*, 196–216.

41. Bredow, "So entstand der deutsche Rundfunk," 27. See also von Boeckmann, "Organisation des deutschen Rundfunks," 221.

42. Von Boeckmann, "Organisation des deutschen Rundfunks," 221. See also Lacey, *Feminine Frequencies*, 29–30.

43. Bredow, "So entstand der deutsche Rundfunk," 7.

44. Bredow, "So entstand der deutsche Rundfunk," 3. See also Bredow, "Der Rundfunk als Kulturmacht," 124, and Führer, "A Medium of Modernity?" 729.

45. Bredow, "Beginn der Rundfunkarbeit," 45.

46. Antoine, "Marksteine," 224.

47. Reprinted in Bredow, "So entstand der deutsche Rundfunk," 27.

48. See, for example, Schütte, *Regionalität und Föderalismus im Rundfunk*.

49. Führer, "A Medium of Modernity?" 753, and Ross, *Media and the Making*, 156. The rise in nationalism and the concern that radio could be used as a tool for propaganda were not unique to Germany; those elements were true of other European countries by the early 1930s. See Lommers, *Europe—On Air*, 216–32.

50. Thierfelder, *Sprachpolitik und Rundfunk*, 18–21.

51. Magnus, "Gegen Zentralisierung," 58–61.

52. Magnus, "Gegen Zentralisierung," 59–60.

53. As quoted in Hagemann, "Die künsterlisch-kulturelle Zielsetzung," 235. See also Führer, "A Medium of Modernity?" 730.

54. Bredow, "So entstand der deutsche Rundfunk," 8.

55. Magnus, "Gegen Zentralisierung," 61.

56. Anon., "Nordischer Rundfunk," 97.

57. Anon., "Nordischer Rundfunk," 99–100.

58. KAB, "Das Stuttgarter Rundfunkprogramm," 2281.

59. Stapelfeldt, "Der Rundfunk als Träger," 233.

60. Stapelfeldt, "Der Rundfunk als Träger," 236.

61. Stapelfeldt, "Der Rundfunk als Träger," 238.

62. Bredow, "Wirtschaftliche und organisatorische Fragen," 83.

63. Bredow, "Programm-Organisation und Ultrakurzwellen," 64.

64. Bredow, "Gegenwartsfragen des deutschen Rundfunks," 3–4.

65. For the importance of isolation to radio listening, see Menzel, "Vor Mikrophon und Lautsprecher" 321–22.

66. Führer, "A Medium of Modernity?" 732, 736–7.

67. Führer has the price between 25 and 30 RM. See Führer, "A Medium of Modernity?" 736. Ross has the price between 15 to 40 RM, *Media and the Making*, 136.

68. Ross, *Media and the Making*, 136.

69. Führer, "A Medium of Modernity?" 733, 736. See also Essau, "Die Verstärkung des Rundfunksenders," 411–12, and Brandt, "Kampf den Rundfunkstörungen," 268.

70. Führer, "A Medium of Modernity?" 736

71. Führer, "A Medium of Modernity?" Also, original: *Radio Händler*, April 7, 1931, p. 361, Ross, *Media and the Making*, 136. Over half of the registered radio sets in Germany in 1931 were *Detektoren* and *Röhrenortsempfänger*. Herbert Antoine has the price of Ortsempfänger at around 110 RM in 1930. See Antoine, "Rundfunk in Zahlen," 359.

72. Antoine, "Rundfunk in Zahlen," 359.

73. Führer, "A Medium of Modernity?" 735.

74. Führer, "A Medium of Modernity?" 734. This was true in other technologically advanced nations as well. See Hollows, *Domestic Cultures*, 98–99. For the domestication of radios in the United States, see Thompson, *The Soundscape of Modernity*, 239.

75. Rühle-Gerstel, "Back to the Good Old Days?" 218.

76. Bridenthal, Grossman, and Kaplan, eds., *When Biology Became Destiny*, 43. See also Lacey, *Feminine Frequencies*, 18–29.

77. Roeseler, "Die Frau als Rundfunkhörerin," 342.

78. Dürre, "Der Kinderfunk," 347–52.

79. Lacey, *Feminine Frequencies*, 40–44.

80. Bredow, "Wirtschaftliche und organistorische Fragen," 94.

81. Führer, "A Medium of Modernity?" 736. In 1925, 1926, and 1928 the stations in Eberfeld, Stuttgart, and Cologne broadcast at 1.5 kW. By the 1930s, the Imperial Ministry of the Post Office began building much stronger transmitting stations.

82. Führer, "A Medium of Modernity?" 737, and Ross, *Media and the Making*, 138–39. It should be noted that radios were not equally distributed throughout major cities. In Berlin, for example, residents in the more exclusive districts, such as Dahlem, Charlottenburg, and Grunewald, were far more likely to possess a radio than those living in the working-class districts of the city, such as Neukölln, Wedding, and Friedrichshain.

83. Ross, *Media and the Making*, 137.

84. Führer, "A Medium of Modernity?" 739. Original: Statistisches Reichsamt, *Die Lebenshaltung*, vol. 1, 134–35.

85. Ross, *Media and the Making*, 137–38.

86. Führer, "A Medium of Modernity?" 738.

87. Ross, *Media and the Making*, 139–40.

88. Führer, "A Medium of Modernity?" 742.

89. Odendahl, "Rundfunk: Von Morgens bis Mitternacht," 345. This article summarizes the program of a typical Sunday in 1929. The total number of broadcasting hours of all German stations increased from eighty-six hours in 1923 to over 100,000 hours in 1928. Antoine, *Die Entwicklung*, not paginated.

90. Flesch, "Rundfunkmusik," 150. See also Hagen, *Das Radio*, 109.

91. Flesch, "Rundfunkmusik," 150.

92. "Protokol über die Sitzung des geschäftsführenden Kuratoriums der Rundfunkversuchsstelle, am 3. Dezember 1929," Universität der Künste–Berlin (henceforth, UdK-Berlin) Archiv, Bestand 1b, 1, folia 69, and UdK-Berlin Archiv, Bestand 1b, Nr. 3, folia 69–70, and "Umlauf bei den Herren Mitgliedern des Kuratoriums der Rundfunkversuchsstelle," February 24, 1931, ibid., folio 27.

93. Donhauser, *Elektrische Klangmaschinen*, 67.

94. Bredow, "Beginn der Rundfunkarbeit," 34–46. See also Hagemann, "Wie kommt," 301–6.

95. Bredow, "Beginn der Rundfunkarbeit," 40.

96. Antoine, *Die Geschichte* (1931), not paginated.

97. Odendahl, "Das letzte Rundfunkjahr," 92.

98. Odendahl, "Das letzte Rundfunkjahr," 92–95.

99. Antoine, "Rundfunk in Zahlen," 363. See also Antoine, *Die Entwicklung*, not paginated.

100. Guzatis, "Berliner Programmstatistik," 130. The economic situation from late 1929 through the early 1930s might explain that decline.

101. The Frankfurt station was viewed as the other progressive station.

102. For a short biography of Flesch, see Ottmann, *Im Anfang*, 25–88. See also Birdsall, "Radio Documents," S96–S128, and L. Bd., "Hans Flesch," 81–82.

103. Ottmann, *Im Anfang*, 58.

104. Flesch, "Inbetriebnahme des Frankfurter Rundfunksenders," 122. Ortmann, *Im Anfang*, 190. See also Flesch, "Zur Ausstellung des Programms im Rundfunk," 3–4.

105. Flesch, "Ein Jahr Frankfurter Programm," 155, and Ortmann, *Im Anfang*, 270.

106. Flesch, "Rundfunkmusik," in Bredow, *Aus meinem Archiv*, 239.

107. Butting, "Music of and for the Radio," 19.

108. Ortmann, *Im Anfang*, 196.

109. Flesch, "Rundfunkmusik," in Bredow, *Aus meinem Archiv*, 239.

110. Ottmann, *Im Anfang*, 192.

111. Von Boeckmann, "Grundlegende Fragen der allgemeinen Programmgestaltung," 40, as translated in Ross, *Media and the Making*, 154.

112. Ross, *Media and the Making*, 152–54.

113. On early experimentation in German radio, see Ottmann, *Im Anfang*.

114. Flesch, "Das Studio des Berliner Rundfunks," 117.

115. Flesch, "Das Studio des Berliner Rundfunks," 118.

116. Ottmann, *Im Anfang*, 221. For the feelings of radio retailers toward programming, see Ross, *Music and the Making*, 155–56.

117. Flesch, "Das Studio der Berliner Funk-Stunde," 119.

118. Brecht, "Vorschläge für den Intendanten," 122.

119. Anon., "Südwest-Deutscher Rundfunk," 75–80.

120. Butting, "Music of and for the Radio," 15.

121. Lubszynski, "Rundfunk Aufnahme," 253.

122. Lubszynski, "Rundfunk Aufnahme," 253.

123. For Flesch's programming of Funk-Stunde Berlin, see anon., "Funkstunde-Berlin," 1265–66, and Ottmann, *Im Anfang*, 278–79. For a thorough account of the music of Neue Sachlichkeit, see Grosch, *Die Musik*.

124. The 1919 essay was first published in 1923 in Bekker, *Neue Musik*, 88. See also Ottmann, *Im Anfang*, 296.

125. Bekker, *Neue Musik*, 108–9.

126. Brillouin, "Avenir de la Musique Automatique," 164–66.

127. Stuckenschmidt, "Die Mechanisierung der Musik," 1–8, reprinted in Stuckenschmidt, *Die Musik eines halben Jahrhunderts*, 9–15. See also Stuckenschmidt, "Neue Sachlichkeit in der Musik," 3–6, reprinted in Stuckenschmidt, *Die Musik eines halben Jahrhunderts*, 36–41. See also Grosch, *Die Musik*, 53, and Patteson, *Instruments for New Music* 6, 18–51.

128. Stuckenschmidt, "Die Mechanisierung," 13.

129. Composer Hanns Eisler spoke of the "wild joy of experimentation" (*wilde Experimentierfreude*). As quoted in Poirier, "Die Avantgarde in Deutschland," 61. The Welte-Mignon reproducing piano could mechanically reproduce pieces played by virtuosi. See also Patteson, *Instruments for New Music*, 26–30.

130. Poirier, "Die Avantgarde in Deutschland," 68.

131. Haas, *Forbidden Music*, 140–44.

132. Grosch, *Die Musik*, 55.

133. Grosch, *Die Musik*, 181–257.

2: High Infidelity

1. Nelson, "Radio Broadcasting Transmitters," 121, reprinted in *Proceedings of the Institute of Radio Engineers*, 1949.

2. Rider, "Why is a Radio Soprano Unpopular?" 334.

3. Anon., "Sopranos Not Always to Blame," 17.

4. Meyer, "Technische Grundlagen und Bedingungen," 97.

5. Schünemann, "Die Funkversuchsstelle," 277. See also Schünemann, "Die Aufgabe," 7. For a short and straightforward discussion of the problems of broadcasting music and voices over the radio, see Heinitz, *Klangprobleme im Rundfunk*.

6. Steidle, "Vortragskunst und große Musik," 309.

7. Lubsznynski, "Rundfunk Aufnahme," 245.

8. Meyer, "Technische Grundlagen und Bedingungen," 97.

9. Weiskopf, "Das Sphärophone," 388, as quoted in Donhauser, *Elektrische Klangmaschinen*, 31.

10. For informative accounts of the rise of electroacoustics, see Wittje, *The Age of Electroacoustics*, and Thompson, *The Soundscape of Modernity*.

11. Sterne, *The Audible Past*, 219.

12. Williams, "Base and Superstructure in Marxist Cultural Theory," 47.

13. Mumford, "Authoritarian and Democratic Technics," 1–8.

14. Winner, "Do Artifacts Have Politics?" 121–36. For a critique of Winner's account, particularly the claim that Robert Moses's bridges over New York City highways were inherently racist, see Joerges, "Do Artifacts Have Politics?" 411–31. See also Mills, "Do Signals Have Politics?" 320–46. For current examples of algorithms discriminating against women and people of color, see Garcia, "Racist in the Machine," 111–17; Buolamwini and Gebru, "Gender Shades," 1–15; Garvie and Frankle, "Facial-Recognition Software"; Simonite, "The Best Algorithms"; Bushwick, "How NIST Tested Facial Recognition"; Singer, "Amazon Is Pushing Facial Technology"; and Benjamin, *Race after Technology*.

15. Lima, Furtado, and Furtado, "Empirical Analysis," 533–38.

16. Koenecke, Nam, Lake et al., "Racial Disparities," 7684–89.

17. Chen, Li, Setlur, and Xu, "Exploring Racial and Gender Disparities in Voice Biometrics," 3723–37.

18. This is, of course, a portion of the "Forman Thesis." See Forman, "Weimar Culture, Causality, and Quantum Theory," 1–115. See also Carson, Kojevnikov, and Trischler, eds., *Weimar Culture and Quantum Mechanics*.

19. Jarausch, *The Unfree Professions*, 8. Hortleder offers a longue-durée account of the engineer's view of society. See Hortleder, *Das Gesellschaftsbild des Ingenieurs*.

20. For the ability of Technische Hochschulen to grant the degrees of Dip. Ing. and Dr. Ing. by order of Kaiser Wilhelm II, see Manegold, *Universität, Technische Hochschule und Industrie*, 282–305.

21. Jarausch, *The Unfree Professions*, 19. See also McClelland, *The German Experience*, 91, 110, 115, 123–24, 149–51.

22. Gispen, *New Profession, Old Order*, 65

23. Weber, *Gesammelte Aufsätze*, vol. I, 204.

24. On that point, see Hårt, "German Regulation," 33.

25. Mommsen, *Bürgerliche Kultur und politische Ordnung*, 72. See also McClelland, *The German Experience*, 133–34, and Sander, *Die doppelte Defensive*, 11–26.

26. Jarausch, *The Unfree Professions*, 21. See also Ludwig, *Technik und Ingenieure*, and Hortleder, *Das Gesellschaftsbild des Ingenieurs*.

27. Jarausch, *The Unfree Professions*, 43. See also Dietz, "'Technik und Kultur,'" 115–27.

28. Hårt, "German Regulation," 34.

29. Föllmer, "The Middle Classes," 457.

30. See, for example, Braun, "Technik als 'Kunsthebel' und 'Kulturfaktor,'" 35–43.

31. As translated in Jarausch, *The Unfree Professions*, 48, original Weihe, "Technik und Kultur," 2–3. See also Viefhaus, "Ingenieure in der Weimarer Republik," 296, and Voskuhl, "Engineering Philosophy," 729–32. The conscious linking of technology and culture was not new in the early twentieth century, although it was most intense then. There are examples dating back to the mid-nineteenth century. See Braun, "Technik als 'Kunsthebel' und 'Kulturfaktor,'" 35–43.

32. Dietz, "'Technik und Kultur,'" 121–27.

33. Voskuhl, "Engineering Philosophy," 721–52.

34. Mommsen, *Bürgerliche Kultur und künstlerische Avantgarde*, 8–9, and Ringer, *The Decline of the German Mandarins*.

35. Thompson, *The Soundscape of Modernity*, 237–38.

36. Thompson, *The Soundscape of Modernity*, 238.

37. Gerlach, "Wie Königs Wusterhausen," 34.

38. Gerlach, "Wie Königs Wusterhausen," 34.

39. Backhaus, "Physikalische Untersuchungen," 813.

40. Seidler-Winkler and Buschkötter, "Instrumente und Singstimmen," 251.

41. For a discussion of the boundaries of the transmissions of overtones, see Winzheimer, *Das musikalische Kunstwerk*, 19–25.

42. Anon., "Funkversuchsstelle," 1297. This is a quotation from Schünemann for the opening of the Rundfunkversuchsstelle on May 3, 1928.

43. Schünemann, "Die Funkversuchsstelle," 277. See also Schünemann, "Die Aufgaben," 7.

44. Anon., "Eine Tagung für Rundfunkmusik," 123–24, Kapeller, "Der enträtselte Rundfunk," 145–46; Band, "Musiker und Techniker über den Rundfunk," 161–62; Anon., "Musik und Technik," 225–26; and Band, "Auf dem Wege," 233–34.

45. As R. Steven Turner has argued, Helmholtz most likely got the idea that the relative amplitudes of the upper partials determine the note's timbre from August Seebeck's work on the siren. See Turner, "The Ohm-Seebeck Dispute," 10. Helmholtz took notice of Seebeck's work when analyzing the so-called Seebeck-Ohm Dispute arising from Ohm's "Ueber die Definition des Tones," 513–65. In that paper, Ohm famously drew upon Joseph Fourier's *Théorie analytique de la chaleur* of 1822, in which the Frenchman's theorem mathematically represents a periodic function of some signal (in Fourier's case, heat propagation) in terms of a converging series of sine and cosine values. Ohm applied Fourier analysis to musical tones, arguing that in the case of periodic sound waves, sine and cosine values represent amplitude, phase, and frequency of every partial tone. See Kromhout, "The Unmusical Ear," 473, 481–83. Seebeck was not convinced by Ohm, since Ohm argued that the tone in its entirety must be determined by $a \sin 2\pi (mt + p)$, if a is the intensity, m the pitch, t the time, and p has no effect on the tone. There is nothing left to determine the timbre. Seebeck reasoned, quite correctly it turns out, that the tone's timbre must be determined by the amplitude of the upper partials; therefore, the entire harmonic series must be able to be expressed mathematically. Turner, "The Ohm-Seebeck Dispute," 10.

46. Brief definitions are in order. The first harmonic is the fundamental frequency, while the first overtone is the second harmonic, the second overtone is the third harmonic, and so forth. Overtones are any resonant frequency above the fundamental: they can be both harmonic or inharmonic. All harmonics above the first (or fundamental) are overtones, but not all overtones are harmonics. A partial (or partial tone) is any sine wave that contributes to the complex tone. They need not be integer multiples (i.e., they can be inharmonic as well).

47. It should be noted that not all musical instruments' tone colors are determined by harmonics. Drums and tympany, for example, also have inharmonic partials or overtones that play a significant role in their tone colors.

48. See, for example, Winzheimer, *Das musikalische Kunstwerk*, 64–65.

49. Seidler-Winkler and Buschkötter, "Instrumente und Singstimmen," 251. See also Wagner, "Der Frequenzbereich," 455.

50. Winzheimer, *Das musikalische Kunstwerk*, 24–25. See also Herrmann-Goldap, "Über die Klangfarbe," 979–85.

51. Schünemann, "Die Aufgaben," 10.

52. Winzheimer, *Das musikalische Kunstwerk*, 32.

53. Anon., "Funkversuchsstelle," 1298.

54. Anon, "Der Unterhaltungs-Rundfunk," 185, as translated in Gilfillan, *Pieces of Sound*, 49.

55. See, for example, Marten, "Was ist Radio für die Kunst?" 34.

56. Fischer, "Reproduktion," 86.

57. Schoen, "Aufgaben und Grenzen des Rundfunk-Programs," 18.

58. Schünemann, "Die Funkversuchsstelle," 284. See also Schünemann, "Die Aufgaben," 10.

59. Rindfleisch, *Technik um Rundfunk*, 61.

60. Wiener, "Welche Instrumente eignen sich," 758–59.

61. Tedious and exact measurements of the broadcasting studio were required to determine where each instrument needed to be placed vis-à-vis the microphone. Numerous articles during the 1920s addressed the importance of determining the optimal distance. See, for example, Merten, "Was ist Radio?" 34.

62. Wiener, "Welche Instrumente," 759.

63. Wiener, "Welche Instrumente," 759.

64. Wiener, "Welche Instrumente," 759. See also Henson, "Sound Recording and the Operatic Canon," 495–508.

65. Wiener, "Die musikalische Stimme im Rundfunk," 1180–81.

66. Anon, "Der Unterhaltungs-Rundfunk," 185 as translated in Gilfillan, *Pieces of Sound*, 49.

67. Winzheimer, *Das musikalische Kunstwerk*, 28.

68. Bronsgeest, "Das Problem der Sendeoper," 331–32.

69. Wiener, "Die musikalische Stimme," 1180.

70. Wiener, "Die musikalische Stimme," 1181.

71. Throughout Germany, radio orchestras employed 248 musicians in 1926; 284 in 1927; 374 in 1928; and 439 in 1929. Radio choirs comprised 75 employees in 1926 and 1927; 76 employees in 1928; and 80 employees in 1929. Antoine, *Die Entwicklung des Deutschen Rundfunks*, not paginated.

72. Wiener, "Die musikalische Stimme," 1181.

73. Chion, *Audio-Vision*, 102.

74. On the topic of losing visual effects when listening to the radio, see Gregersen, "Zur Psychologie des Rundfunkhörens," 78.

75. On this topic, see Ettingen, "Opern für den Rundfunk," 75–76, and Ettingen, "Opernfunk—Funkoper!" 10. See also Hernried, "Geistige Probleme der Rundfunkoper," 39; Band, "Opernkrise der Gegenwart," 13–14; and Bronsgeest, "Opernsendespiel oder Übertragung?" 17–18.

76. Bronsgeest, "Die Sende-Oper," 253–55.

77. Butting, "Music of and for Radio," 17.

78. Schoen, "Aufgaben und Grenzen," 14.

79. Steidle, "Vortragskunst und große Musik," 309–31. For the question of the number of microphones and their placement in opera and concert recitals, see also Mönch, *Mikrophon und Telephon*, 147–48.

80. Steidle, "Vortragskunst und große Musik," 323.

81. Steidle, "Vortragskunst und große Musik," 326–29.

82. Steidle, "Vortragskunst und große Musik," 309.

83. Steidle, "Vortragskunst und große Musik," 309.

84. Almeida, "Die erste 'direkte Übertragung,'" 125.

85. Almeida, "Die erste 'direkte Übertragung,'" 126.

86. Nelson, "Radio Broadcasting Transmitters," 121, reprinted in *Proceedings of the Institute of Radio Engineers*, 1949.

87. Weill, "Übertragungsbedingungen für Orchesterklang" (1928) in Weill, *Musik und Theater*, 278–88. See also Grosch, *Die Musik*, 206.

88. Seidler-Winkler and Buschkötter, "Instrumente und Singstimmen," 251.

89. Szendrei, "Instrumentalmusik in Rundfunk," 209.

90. Szendrei, "Instrumentalmusik in Rundfunk," 210.

91. Butting, "Rundfunkmusik," 28. See also Grosch, *Die Musik*, 205.

92. Winzheimer, *Das musikalische Kunstwerk*. See also Winzheimer, *Übertragungstechnik*, 4–47.

93. Hernried, "Das Orchesterhören," 118.

94. Chion wishes to differentiate between "definition," which is a recording's precision and sharpness of perception in rendering its detail and which is a function of bandwidth,

and "fidelity," which requires a continuous comparison between the original and its reproduction. This definition of fidelity was not possible during the period I am studying. Indeed, none of my actors thought it was possible to compare the original sound with its reproduction throughout a length of time. Chion argues that "fidelity" does not have a precise meaning as "definition" does. My engineers and physicists defined one type of fidelity as the difference in timbre (and more quantifiably, the difference in overtones as measured by harmonic analyzers) between the original or "natural" sound and its reproduction, although, as discussed in chapter 5, a number of them began to realize that timbre changes throughout the duration of a tone. In addition, while they were certainly aware of the financial importance of fidelity—Siemens & Halske often stressed that their microphones and loudspeakers were fidelitous—it would be a mistake to claim that "high fidelity is a purely commercial" notion. Chion, *Audio-Visual*, 98.

95. Winzheimer, *Das musikalische Kunstwerk*, 23–24, and von Raman, "Musikinstrumente und ihre Klänge," vol. 8, p. 356.

96. Höpfner, "Entwicklung des Kabelwesens," 263.

97. Höpfner, "Entwicklung des Kabelwesens," 264.

98. Höpfner, "Entwicklung des Kabelwesens," 263. Hans Bredow thought in the summer of 1930 that only 6,000 kilometers of cable had been laid. See Bredow, "Der Rundfunk in Deutschland," 174.

99. Höpfner, "Entwicklung des Kabelwesens," 266. See also Antoine, *Die Geschichte des Deutschen Rundfunks*, not paginated, and Nesper, *Kompendium für Funktechnik*, 199.

100. See Van B. Roberts, "The Uses of the Three Electrode Tube," 169.

101. Van B. Roberts, "The Uses of the Three Electrode Tube," 169.

102. Leithäuser, "Theorie und Praxis der neuen Wellenverteilung," 21. See also Lommers, *Europe—On Air*, 74, 82.

103. Medium frequency radio is between 300 kHz to 3 MHz.

104. Anon., "Der neue Rundfunkwellenplan," 15.

105. See Salinger, "Physikalische Grundlagen der Empfangstechnik," 144. See also Wormbs, "Technology-dependent Commons," 94–95, and Anderson, "The Greatest Recreation Is a Test of the Receiver," 19.

106. Lommers, *Europe—On Air*, 92–93. See also Nesper, *Kompendium für Funktechnik*, 47.

107. Höpfner, "Entwicklung des Kabelwesens," 261. See also Leithäuser, "Wissenwertes über neuzeitliche Empfänger," 273.

108. Nesper, *Kompendium für Funktechnik*, 49. The Madrid/Lucerne Plan of 1932, which came into effect on January 15, 1934, also proposed 9000 Hz channel spacing, but not harmonic multiples. See "Managing the Transition to Digital Sound Broadcasting."

109. Rider, "Why Is a Radio Soprano Unpopular?" 334–37. See also Tallon, "A Century of 'Shrill.'"

110. See, for example, Tallon, "A Century of 'Shrill'"; Carson, "The Gender of Sound," 119–42; and Eidsheim, *The Race of Sound*.

111. It turns out that sopranos often suffered when various technologies were employed. See Henson, ed., *Technology and the Diva*.

112. Steinberg, "Understanding Women," 153–54.

113. Steinberg, "Understanding Women," 153, and Tallon, "A Century of 'Shrill.'"

114. K. S. Johnson of Bell Labs as cited by Rider, "Why Is a Radio?" 334.

115. Steinberg, "Understanding Women," 154.

116. As quoted in Rider, "Why Is a Radio?" 334.

117. Slotten, "Radio Engineers," 954. See also Dellinger, "Reducing the Guesswork in Tuning," 242. It should be noted that in 1925 the European radio stations were also all spaced 10,000 Hz apart.

118. Dellinger, "Reducing the Guesswork in Tuning," 241, and Lombardi and Nelson, "WWVB," 26.

119. Anderson, "The Greatest Recreation," 19.

120. Rider, "Why Is a Radio?" 337, and Tallon, "A Century of 'Shrill.'"

121. Rider, "Why Is a Radio?" 336.

122. Rider, "Why Is a Radio?" 337.

123. Rider, "Why Is a Radio?" 336.

124. Anon., "Sopranos not Always," section xx, p. 17.

125. Rider, "Why Is a Radio?" 334.

126. Rider, "Why Is a Radio?" 334.

127. Anderson, "The Greatest Recreation," 18.

128. Schneider, "The Woman Who Overcame."

129. "The Reasons Why Women's Voices are Deeper Today," BBC, June 12, 2018, https://www.bbc.com/worklife/article/20180612-the-reasons-why-womens-voices-are-deeper-today. See Pemberton, McCormack, and Russell, "Have Women's Voices Lowered Across Time?" 208–13. See also Tallon, "A Century of Shrill."

130. Clearly, German radio was not the only institution experimenting with radio dramas. The first radio broadcast was Richard Hughes's "A Comedy of Danger," transmitted by the BBC in 1924. For a good summary of Rundfunkmusik, see Stoffels, "Rundfunk als Erneuerer und Förderer," vol. 2, 847–64.

131. Butting, "Music of and for the Radio," 18. For his views on radio music, see also Butting, "Vom Wesen der Musik im Rundfunk," 201–2, and Butting, "Das Verhältnis des schaffenden Musikers," 279–98.

132. Grosch, *Die Musik*, 195–204.

133. Butting, "Music of and for the Radio," 17.

134. On that topic, see, for example, Butting, "Rundfunkmusik," 28, and Butting, "Music of and for the Radio," 16. See also Grosch, *Die Musik*, 181–257.

135. Butting, "Music of and for the Radio," 15.

136. Butting, "Das Verhältnis des schaffenden Musikers," 283.

137. Butting, "Rundfunkmusik," 28.

138. Butting, "Music of and for the Radio," 19.

139. Butting, "Rundfunkmusik," 28. Representing them well refers to "*Naturtreue*."

140. Butting, "Music of and for the Radio," 16.

141. Butting, "Rundfunkmusik," 28.

142. As cited in Adorno, "The Radio Symphony," 260.

143. Butting, "Music of and for the Radio," 16–17.

144. Hagemann, "Die künstlerisch-kulturelle Zielsetzung," 126.

145. Warschauer, "Die Zukunft der Technisierung," 432.

146. As reported in Hagemann, "Die künstlerisch-kulturelle Zielsetzung," 127.

147. Flesch, "Die neuen künsterlischen Probleme," 100–1.

148. Butting, "Music of and for the Radio," 18.

149. Brecht, "Vorschläge für den Intendanten des Rundfunks," vol. 18, 122.

150. Döblin, "Literatur und Rundfunk," 252.

151. Döblin, "Literatur und Rundfunk," 254.

152. Rosenthal, "Introduction," in Rosenthal, ed., *Radio Benjamin*, xvi.

153. Müller-Hartmann, "Klang und Stilprobleme der Rundfunkmusik," 205.

154. Levin and von der Linn, "Elements of a Radio Theory," 317.

155. Adorno, "Zum Anbruch: Exposé," 601. See also Levin and von der Linn, "Elements of a Radio Theory," 317.

156. Adorno, "The Radio Symphony," 251.

157. Adorno, "The Radio Symphony," 251. Note that the word "original" had scare quotes in the original text.

158. Adorno, "The Radio Symphony," 252.

159. Here Adorno is citing Skinner, "Music Goes into Mass Production," 487. See Adorno, "The Radio Symphony," 252.

160. Adorno, "The Radio Symphony," 252.

161. Adorno used the phrase "sound color," which would be a literal translation of *Klangfarbe*. He meant tone color. Adorno, "The Radio Symphony," 260.

162. Adorno, "The Radio Symphony," 260.

163. Adorno, "The Radio Symphony," 267–68.

164. Adorno, "The Radio Symphony," 260–61. Adorno echoed Strauss's claim about the first and second violins in Berlioz's *Grand traité d'instrumentation et d'orchestration modernes*.

165. Adorno, "The Radio Symphony," 261–62.

166. Adorno, "The Radio Symphony," 265.

167. Adorno, "The Radio Symphony," 265.

168. Adorno, "The Radio Symphony," 266.

169. Leppert, "Commentary," in Leppert, ed., *Essays on Music*, 218–28, here 219–20.

170. Thomson, "Beethoven in the House," 259.

171. Thomson, "Beethoven in the House," 259–60.

3: Analyzing Distortions and Creating Fidelity

1. Harmonic analyzers were also called frequency analyzers.

2. Von Helmholtz, "Ueber die Klangfarbe der Vocale," 280–90. On this topic, see Kursell, "Experiments on Tone Color," 202–6, and Kursell, "A Gray Box," 176–97.

3. Von Helmholtz, *On the Sensations of Tone*, 103. See also Schumann, *Akustik*, 76; Lertes, *Elektrische Musik*, 37; and Pantalony, "Seeing a Voice" (2004), 433–34.

4. Physiologists (Ernst) Wilhelm Trendelenburg and H. Wullstein made simultaneous recordings of glottal lip movement, shadowgraphs of the glottal click, and oscillograms of air variations both above and below the glottis. Their investigations showed that the vibrations generated by the vocal cords are actually not very rich in overtones; however, the vibrations produced by air existing in the glottis are. See W. Trendelenburg, "Physiologische Untersuchungen," 525–73, and W. Trendelenburg and Wullstein, "Untersuchungen über die Stimmbandschwingungen," 399–426. See also F. Trendelenburg, *Klänge und Geräusche*, 70–71, fn 3. See also Dörfling, *Der Schwingkreis*, 241–47.

5. Von Helmholtz, *On the Sensations of Tone*, 124. See also Kromhout, "The Unmusical Ear," 471–92.

6. Helmholtz used the term *Teiltöne*, or partial tones. Is this instance, the partial tones were harmonics. Helmholtz, *On the Sensations of Tone*, 127.

7. Helmholtz, *On the Sensations of Tone*, 127.

8. Hermann, ed., *Handbuch der Physiologie*, vol. 3, part 2, p. 449–50. See also Schmidgen, "The Donders Machine," 211–56.

9. See Warwick, "The Laboratory of Theory," 338–39.

10. Lertes, *Elektrische Musik* (1933), 39–40. See also Vierling, "Der Formantbegriff," 220; Trautwein, *Elektrische Musik*, 12; and Fletcher, *Speech and Hearing*, 46–49.

11. Weyer, "Probleme der Analyse und Synthese," 124. See also Trautwein, "Über elektrische Synthese," 291.

12. See for example, Wachsmuth, ed., "Klangaufnahmen," 543, and Herrmann-Goldap, "Über die Klangfarbe," 979–85.

13. Gutzmann, *Physiologie der Stimme und Sprache*, 113–18, and Lertes, *Elektrische Musik*, 35–39. Despite Gutzmann's insistence, it should be noted that several scholars argued that there was no contradiction between the two theories.

14. Fletcher, *Speech and Hearing*, 46–49, particularly 49.

15. Ferdinand Trendelenburg, "Die physikalische Eigenschaften der Sprachklänge," vol. 8, 471–72. See also F. Trendelenburg, *Klänge und Geräusche*, 72–73.

16. See, for example, Pompino-Marschall, "Carl Stumpf und die Phonetik," 131–50.

17. Stumpf, *Die Sprachlaute*, 63.

18. Stumpf, "Sprachlaute und Instrumentalklänge," 751–52, and Schumann, *Akustik*, 76–77. See also Stumpf, *Die Sprachlaute*, 63, and Vierling, "Der Formantbegriff," 221.

19. Lertes, *Elektrische Musik*, 38.

20. Révész, *Introduction to the Psychology of Music*, 48–49. It originally appeared in 1954 with the German original, *Einführung in der Musikpsychologie*. See also Stumpf, "Über die Tonlage der Konsonanten," 639; Stumpf, "Sprachlaute und Instrumentalklänge," 745–58; and A. Schneider, "Change and Continuity in Sound Analysis," 85–87.

21. Stumpf, *Die Sprachlaute*, 290–309. See also Klotz, "Timbre, *Komplexeindruck*, and Modernity," 609–40. Note that perception in general was very important to Helmholtz. It is debatable whether or not he was as reductionist as those in the 1920s maintained.

22. For notions of mechanical objectivity, see Daston and Galison, *Objectivity*.

23. Miller, *The Science of Musical Sounds*, 184–243.

24. Crandall, "The Sounds of Speech," 588. Speech sounds, also called phonemes, are the smallest part of a language that makes a difference in meaning.

25. Miller, *The Science of Musical Sounds*, 97–98. He learned the technique while he was working with Morley on light waves. See Lalli, "The Interplay of Theoretical Assumptions," 343–60.

26. Miller, "The Henrici Harmonic Analyzer," 285–322. See also Fickinger, *Physics at a Research University*, 39–42, and Case Western Reserve University, ed., "Harmonic Analyzers," https://physics.case.edu/about/history/antique-physics-instruments/harmonic-analyzer-2/.

27. Miller, *The Science of Musical Sounds*, 221–32.

28. Miller, *The Science of Musical Sounds*, 193.

29. Miller, *The Science of Musical Sounds*, 171, 198, 201.

30. See, for example, Kromhout, *The Logic of Filtering*, 23.

31. Meyer, "Technische Grundlagen," 97. On the history of electroacoustics, see Wittje, *The Age of Electroacoustics*, and Thompson, *The Soundscape of Modernity*, 90–107.

32. His first major work drew upon the works of James Clerk Maxwell, Lord Rayleigh, and Horace Lamb to argue for the analogy between mechanical and electrical properties. Crandall, *Theory of Vibrating Systems and Sound*.

33. Fagen, ed., *A History of Engineering and Science*, 928.

34. Fagen, ed., *A History of Engineering and Science*, 935.

35. Lane, "Minimal Sound Energy for Audition," 492–97, and Fletcher and Wegel, "The Frequency-Sensitivity," 553–65.

36. Wegel and Lane, "The Auditory Masking," 266–85.

37. Wente, "The Sensitivity and Precision," 498–503, and Steinberg, "The Relationship," 507–23.

38. An early version of frequency analysis was developed by American physicist Preston Hampton Edwards in his doctoral dissertation at Johns Hopkins University. He analyzed musical tones by sending the sound through a series of seventeen resonators of specific pitches and then measuring the vibrational amplitude by means of a Rayleigh disc. See Edwards, "A Method for a Quantitative Analysis," 23–37.

39. Fletcher, *Speech and Hearing*, 27.

40. Crandall and Sacia, "A Dynamical Study," 232–37.

41. Sacia, "Speech Power and Energy," 627–41.

42. While there clearly was radio research going on throughout the United States, the majority of the critical research was being carried out by Bell Telephone Laboratories, even though their primary focus was telephony, telegraphy, and talking pictures.

43. Arnold, "Introduction," in Fletcher, *Speech and Hearing*, xii–xiii.

44. Fagen, ed., *A History of Engineering and Science*, 928.

45. One example of the research on musical instruments conducted in Bell Telephone Laboratories work that was similar to the research that the Germans were doing involved Clyde Snook's use of an oscillograph and harmonic analyzer from General Electric to investigate the acoustics of the Steinway Company's (NY office) studios, audience rooms, and pianos. Clyde Snook to Paul Bilhuber, March 30, 1926, in "Development of Audiometer," Box 444-05-01, Folder 08, AT&T Archives. Yeang, *Transforming Noise*, 81.

46. Wegel and Moore, "An Electrical Frequency Analyzer," 299–323. Less than a year later, Cockcroft, Coe, Tyacke, and Walker announced their invention of an electric harmonic analyzer, "An Electric Harmonic Analyser," 69–119. In 1928 H. Tinsley & Co. of London created a portable electric harmonic analyser based on the earlier model invented by Cockcroft, Coe, Tyacke, and Walker. See Anon., "A Portable Electric Harmonic Analyser," 320–23.

47. Wegel and Moore, "An Electrical Frequency Analyzer," 299–323. Note their technique was the electrical analogue of an earlier technique of sound analysis invented by S. Garten and F. Kleinknecht for the study of sung vowel sounds. See Garten and Kleinknecht, "Beiträge zur Vokallehre III," 1–43.

48. Heterodyning is the process by which a signal is combined with another in order to produce a third frequency, which is the difference of the initial two frequencies. Moore and Curtis, "An Analyzer for Voice Frequency Range," 217–29. Another harmonic analyzer was developed by A. G. Landeen of Bell Telephone Laboratories that same year. It was also based on the heterodyning method. See Landeen, "Analyzer for Complex Electric Waves," 230–47.

49. Meyer, "Analysis of Noises and Musical Sounds," 53.

50. This summary of Karl Willy Wagner's biography is based on Wagner, "Der Lichtbogen als Wechselstromerzeuger," 121–22. For the biographical information after 1910, see Wittje, "Karl Willy Wagner," 253–54.

51. Wagner, "Der Frequenzbereich," 451–56. He offered a somewhat updated lecture on the topic four years later at the First Conference on Radio Music held in Göttingen on May 7, 1928. See Wagner, "Sprache und Musik im Rundfunk," 1–6. Some of Wagner's research was summarized in Meyer, "Technische Grundlagen und Bedingungen," 97–114.

52. Stumpf, "Die Struktur der Vokale," 333–58; Stumpf, "Zur Analyse geflüsterter Vokale," 234–54,; Stumpf, "Zur Analyse der Konsonanten," 151–81; Stumpf, "Veränderungen des Sprachverständnisses," 182–90; and Stumpf, "Über die Tonlage der Konsonaten," 636–40.

53. Both Miller and Stumpf simultaneously conducted similar research in the 1910s and '20s. While Stumpf cited Miller's research, Miller does not cite Stumpf. Miller's editions of *The Science of Musical Sounds* (1916, 1922) predate Stumpf's *Die Sprachlehre*.

54. For a history of filters, see Kromhout, *The Logic of Filtering*.

55. Wagner, "Der Frequenzbereich," 453. In telephones today, the usable voice frequency band ranges between 300 and 3400 Hz. In 1930, Prussian ministerial councilor Karl Höpfner said that the ideal range for speech and language for a radio broadcast was between 30 and 10,000 Hz, although 100 to 5000 Hz was acceptable for a musical broadcast, while 200 to 3000 Hz was sufficient for speech. See Höpfner, "Entwicklung des Kabelwesens," 261.

56. Wagner, "Der Frequenzbereich," 454.

57. A number of those diagrams subsequently appeared in Aigner, "Über die Schwingungen," 18–38. Similar diagrams of the intensities of the fundamental tone and its overtones for various musical instruments can be found in Lertes, *Elektrische Musik*, 47, 49, 51, 53–55, and in Wagner, "Der Frequenzbereich," 455.

58. Wagner, "Der Frequenzbereich," 455. As discussed in the ensuing chapter, such research proved crucial to the construction of the trautonium. Trautwein drew upon Wagner's technique for electrically generating the upper partials in his work on the trautonium.

59. In 1933 the institute changed its name to the Institute for Oscillations Research, omitting the name, since Hertz was a Jew. In the fall of 1945, it was renamed the Heinrich Hertz Institute for Oscillations Research and specialized in acoustics, telecommunications, high-frequency engineering, information processing, and control engineering. In 1975 the name was changed to the Heinrich Hertz Institute for Communications Engineering (*Nachrichtentechnik*). Finally, in 2003 it became the Fraunhofer Institute for Telecommunications, Heinrich Hertz Institute. They now specialize in photonic networks, multimedia using electronic imaging, and mobile broadband systems. See https://www.hhi.fraunhofer.de/en/fraunhofer-hhi/about-us/history-of-hhi.html. See also Wittje, *The Age of Electroacoustics*, 158–63.

60. Anon., "An alle Funkfreunde," 417–18.

61. Wagner, "Die Kunstaufgabe des Rundfunks," 4, and Wagner, "Vorwort," in "Der Lichtbogen," i.

62. Wagner, "Das Heinrich-Hertz-Institut," 296.

63. After the conversion of the Reich Post Office into the Reich Post Ministry, the Reich's Office of Telegraph Technology (Telegraphentechnisches Reichsamt) was founded as a subdepartment in 1920. It was created by merging several Reichspost facilities, which included the telegraph test office, telegraph equipment office, telephone line office, and radio operations office. In 1928 the Telegraphentechnisches Reichsamt was renamed the Reichspostzentralamt (Reich's Central Post Office).

64. Later Leithäuser became professor at the TH Berlin and leader of the High Frequency (Radio) Department of the Heinrich Hertz Institute.

65. Wagner, "Das Heinrich-Hertz-Institut," 298. For subsequent research on the production of vowel sounds via electric circuits, see anon., "Elektrisch erzeugte Vokale," 206.

66. F. Trendelenburg, "Objektive Klangaufzeichnung" (1924), 43–66, and (1925), 1–13. See also F. Trendelenburg, *Klänge und Geräusche*, 78–92.

67. F. Trendelenburg, "Objektive Klangaufzeichnung" (1924), 43. It is interesting to note that Trendelenburg uses the term *Klangtreue*, or "faithful to sound" a decade later. See F. Trendelenburg, *Klänge und Geräusche*.

68. F. Trendelenburg, "Objektive Klangaufzeichnung" (1924), 43.

69. F. Trendelenburg, "Objektive Klangaufzeichnung" (1924), 66.

70. F. Trendelenburg, "Objektive Klangaufzeichnung" (1924), 66.

71. F. Trendelenburg, "Objektive Klangaufzeichnung" (1925), 1–13.

72. F. Trendelenburg, "Objektive Klangaufzeichnung" (1925), 13. Schumann, who also worked on tone color, followed Stumpf's and Trendelenburg's research. His view of timbre was very Helmholtzian and reductionistic. See Schumann, *Akustik*, 47–86; Schumann, *Die Physik der Klangfarben*; and Reuter, "Erich Schumann's Laws of Timbre as an Alternative," 185–200.

73. Meyer, "Analysis of Noises and Musical Sounds," 53.

74. Meyer, "Die Klangspektren der Musikinstrumente," 606–11. See also Grützmacher, "Eine neue Methode der Klanganalyse," 216–28; Grützmacher, "Zur Analyse von Geräuschen," 570–72; Grützmacher, "Klanganalyse mit einem Einfadenelektrometer," 572–73; and Gerlach, "Über einen registrierenden Schallmesser und seine Anwendungen," 515–19.

75. "Beats," heterodynes, or difference tones are similar to acoustical beats and combination tones. Difference tones are a type of combination tone that sound at a frequency that is the difference between the two frequencies. Summation tones, another type of combination tone, are ones that sound at a frequency that is the sum of the two frequencies. For example, two tones sounding at 600 Hz and 400 Hz could produce a difference tone at 200 Hz, while a summation tone could be produced at 1000 Hz.

76. See, for example, Salinger, "Zur Theorie der Frequenzanalyse," 293–302. See also Meyer and Buchmann, "Die Klangspektren der Musikinstrumente," 735–78.

77. F. Trendelenburg, "Klangspektren von Musikinstrumenten," 11.

78. For a biography of Erwin Meyer, see Guicking, *Erwin Meyer*, available at http://www.guicking.de/dieter/Erwin-Meyer.pdf.

79. Guicking, *Erwin Meyer*, available at http://www.guicking.de/dieter/Erwin-Meyer.pdf.

80. Meyer and Buchmann, "Die Klangspektren der Musikinstrumente," 735–78.

81. Meyer and Buchmann, "Die Klangspektren der Musikinstrumente," 736–41..

82. The contrabassoon has fifty overtones between 30 and 15,000 Hz. Meyer and Buchmann, "Die Klangspektren der Musikinstrumente," 765, and Meyer, "Die Klangspektren," 611.

83. Meyer and Buchmann, "Die Klangspektren der Musikinstrumente," 750, and Meyer, "Die Klangspektren," 607.

84. Meyer was not the first to offer such a study for the violin in Germany. For example, Hermann Backhaus, the Siemens & Halske engineer who was researching at the University of Greifswald, used a Siemens oscillator and a Riegger high-frequency capacitor microphone to compare and contrast the timbres of various violins, some of which were crafted by Stradivarius and Guarneri, by analyzing the resonant frequency bands of the overtones. See Backhaus, "Physikalische Untersuchungen an Streichinstrumenten," 811–18, 835–39.

85. The fundamental pitch in this instance is literally a product of the human ear.

86. Meyer and Buchmann, "Die Klangspektren der Musikinstrumente," 760–61, and Meyer, "Die Klangspektren," 609.

87. Meyer and Buchmann, "Die Klangspektren der Musikinstrumente," 772–76 and Meyer "Die Klangspektren," 610.

88. Meyer and Buchmann, "Die Klangspektren der Musikinstrumente," 777, and Meyer "Die Klangspektren," 610.

89. See Meyer, "Das Gehör," 489. Meyer cites Harvey Fletcher's work at Bell Telephone Laboratories. The nonlinearity of the human ear, as discussed in chapter 5, refers to the fact that humans hear some frequencies much more easily than others. A sound at 100 Hz must have a far greater amplitude to be heard at the same volume as a frequency at 3000 Hz. Recall that Carl Stumpf also made that argument. In addition to the pioneering work of Erwin Meyer on harmonic analysis, in 1927 Martin Grützmacher developed a method similar to that of Meyer's for determining the overtones of a frequency mixture by using modulation via a pure sinusoidal voltage. Grützmacher, "Eine neue Methode der Klanganalyse," 506–9, reprinted in *Elektrische Nachrichtentechnik* 4 (1927): 533–55. See also Kalähne, "Schwallwiedergabe durch Lautsprecher," 333–34. While Kalähne is listed as the general author on the section "Schallaufnahmen und Klanganalyse," this portion of chapter 5 was actually written by Erwin Meyer. See Kalähne, "Schwallwiedergabe durch Lautsprecher," x. Walther Gerlach also invented a method for harmonic analysis and recommended it for the measurement of distortion caused by microphones, loudspeakers, and amplifiers. See also Gerlach, "Über einen registrierenden Schallmesser und seine Anwendungen," 519, and Kalähne, "Schwallwiedergabe durch Lautsprecher," 333–34. Again, while Kalähne is listed as the general author on the section "Schallaufnahmen und Klanganalyse," this portion of chapter 5 was actually written by Erwin Meyer.

90. Jeans, *Science & Music*, 104. See also Vierling, "Der Formantbegriff," 219 and Mol, *Fundamentals of Phonetics*, II, 19.

91. Jeans, *Science & Music*, 148.

92. Vierling, "Der Formantbegriff," 219.

93. Thadeusz, "Nazi-Labor."

94. Vierling, *Das elektroakustische Klavier*.

95. Deutsches Reichspatent no. 626,179, granted on July 24, 1928. See Vierling "Der Formantbegriff," 224.

96. Vierling, "Der Formantbegriff," 221, 223.

97. Weichart, "Das Mikrophon auf der Bühne," 1–4.

98. Winzheimer, *Das musikalische Kunstwerk*, 16.

99. Nesper, *Kompendium der Funktechnik*, 196

100. It was known at the time that piccolos could reach above 4000 Hz, so Nesper's value is somewhat low.

101. Nesper, *Kompendium der Funktechnik*, 196. Nespser stated the soprano's voice could reach 1290, Hz, which is approximately E6.

102. Nesper, *Kompendium der Funktechnik*, 196–97. For the definitive study on the relationship between loudness and frequency for human hearing for this period, see Fletcher, *Speech and Hearing*, 132–44.

103. See, for example, Meyer, *Electro-Acoustics*, 48.

104. The standard of microphones was Western Electric's condenser microphone, which deviated by 5 dB from 100 to 8000 Hz, or 3 dB between 100 and 5000 Hz. See, for example, Frederick, "The Development of the Microphone," 24, reprinted in *Bell Telephone Quarterly* 10 (1931): 187.

105. Nelson, "Radio Broadcasting Transmitters," 121–22.

106. Meyer, "Über eine einfache Methode," 398–403, and Gerlach, "Ueber einen registrierenden Schallmesser und seine Anwendung," 519. In 1933 Hans-Joachim von Braunmühl and Walter Weber developed a heterodyne oscillator with three (rather than two) high frequencies, which generated two (rather than one) low-frequency tones. The frequency difference remained constant if two of the high frequencies were fixed, while the third slid through the entire frequency range. The two low-frequency tones resulted in the creation of a difference tone (heterodyne) generated in the microphone being tested. It remained constant, was recorded, and could be filtered out. See Braunmüller und Weber, "Nichtlinearische Verzerrungen von Mikrophonen," 1068–70. See also Meyer, *Electro-Acoustics*, 58–59.

107. See Meyer and Neumann, *Physical and Applied Acoustics*, 273–74.

108. Waetzmann, "Die Entstehungsweise von Kombinationstönen," 729–44.

109. Kalähne, "Schwallwiedergabe durch Lautsprecher," 333–34. While Kalähne is listed as the general author on the section "Schallaufnahmen und Klanganalyse," this portion of chapter 5 was actually written by Erwin Meyer.

110. Meyer and Neumann, *Physical and Applied Acoustics*, 277. See also Goucher, "The Carbon Microphone," 163–94, and Frederick, "The Development of the Microphon," 11, reprinted (1931): 174.

111. Frederick, "The Development of the Microphone," 14, 23–24, reprinted (1931), 177–78, 185–86.

112. Nesper, *Kompendium der Funktechnik*, 62–63. See also, Kalähne, "Schwallwiedergabe durch Lautsprecher," 315–16. While Kalähne is listed as the author, this section of his contribution was actually written by Erwin Meyer.

113. Meyer and Neumann, *Physical and Applied Acoustics*, 280. See also Frederick, "The Development of the Microphone," (1931), 22, reprinted (1931), 185.

114. Temmer, "In Memoriam of Georg Neumann," 708.

115. Wente, "A Condenser Transmitter," 39–63. See Thompson, *The Soundscape of Modernity*, 95, for the diagram. See also H. J. Braun, "Introduction," 20; Williams, "Piezo-electric Loudspeakers and Microphones," 166–67; and Frederick, "The Development of the Microphone," 20–22, reprinted (1931), 183.

116. Frederick, "The Development of the Microphone" (1931), 22, 183. See also Olson, "A History of High-Quality," 798–807.

117. Kalähne, "Schwallwiedergabe durch Lautsprecher," 320–21. This portion was actually written by Erwin Meyer. See also Lockheart, "A History of Early Microphone Singing," 368.

118. Temmer, "In Memoriam of Georg Neumann," 708

119. Kalähne, "Schwallwiedergabe durch Lautsprecher," 320–21. That portion of the entry was written by Erwin Meyer.

120. Kalähne, "Schwallwiedergabe durch Lautsprecher," 321–22. See also Riegger, "Zur Theorie des Lautsprechers," 67–100, and Nesper, *Kompendium*, 63.

121. Kalähne, "Schwallwiedergabe durch Lautsprecher," 322–24, 346–47. That portion was written by Erwin Meyer. Later they created the ribbon loudspeaker. See also Olson, "Mass Controlled Electrodynamic Microphones," 56–68.

122. Nesper, *Kompendium*, 63.

123. Meyer and Neumann, *Physical and Applied Acoustics*, 290.

124. Merten, "Was ist Radio für die Kunst?" 35.

125. Weichart, "In 14 Tagen," 50.

126. Olson, "Mass Controlled," 56–68. See also Weinberger, Olson, and Massa, "A Uni-Directional Ribbon Microphone," 139–47.

127. Siemens, "New 'Velocity' Microphone," 406, 431.

128. Wente and Thuras, "Moving-Coil Telephone Receivers and Microphones," 55.

129. Wente and Thuras, "An Improved Form of Moving Coil Microphones," 8.

130. A final type of microphone being developed in the 1920s was the kathodophone. They were never popular for recording sound as they had a short lifespan and could be volatile. Kalähne, "Schwallwiedergabe durch Lautsprecher," 323–24. This portion of the entry was written by Erwin Meyer.

131. Hunt, *Electroacoustics*, 80. See also Beranek, "Loudspeakers and Microphones," 618–29.

132. Kalähne, "Schwallwiedergabe durch Lautsprecher," 347–48. This portion of the entry was written by Erwin Meyer.

133. Rice and Kellogg, "Notes on the Development," 461–75. Edward Wente at Bell Telephone Laboratories applied for a patent on a loudspeaker based on the same properties on April 1, 1925. It was granted on June 30, 1931, US patent 1,812,389.

134. Rice and Kellogg, "Notes on the Development," 474. Intensity means volume in this instance.

135. Hunt, *Electroacoustics*, 80.

136. Hunt, *Electroacoustics*, 82.

137. Meyer, "Über die nichtlineare Verzerrung," 509–15.

138. Meyer, *Electro-Acoustics*, 59–64.

139. Wente and Thuras, "A High Efficiency Receiver," 140–53.

140. Beranek, "Loudspeakers and Microphones," 622.

141. Hunt, *Electroacoustics*, 88. See also Wente and Thuras, "Moving-Coil Telephone Receivers," 44–55.

142. Beranek, "Loudspeakers and Microphones," 622–23.

143. Beranek, "Loudspeakers and Microphones," 623.

144. Wente and Thuras, "Loudspeakers and Microphones," 259–77. See also Fletcher, "Symposium on Wire Transmission of Symphonic Music," 239–44. Note that stereoscopic technology was invented by the British engineer Alan Blumlein of Electric and Musical Industries (EMI) in the early 1930s.

145. Wente and Thuras, "Loud Speakers and Microphones," 259.

146. For the classic analysis of the importance of the shape of a loudspeaker horn, see Olson and Wolff, "Sound Concentration for Microphones," 410–17.

147. The optimal range was from 30 to 10,000 Hz. Kalähne, "Schwallwiedergabe durch Lautsprecher," 338. This portion of the entry was written by Erwin Meyer.

148. Meyer, "Die Prüfung von Lautsprechern," 290, 293–95. See also Grützmacher and Meyer, "Eine Schallregistriervorrichtung," 203.

149. Grützmacher and Meyer, "Eine Schallregistriervorrichtung," 203–11.

150. Kalähne, "Schwallwiedergabe durch Lautsprecher," 338–39. This portion of the entry was written by Erwin Meyer.

151. Meyer, "Über die nichtlineare Verzerrung," 509.

152. Meyer, *Electro-Acoustics*, 48–50. By the late 1930s, the frequency range of German radio was extended on both ends, from 80 to 8000 Hz to 30 to 12,000 Hz. The upper end of this increase in frequency range could not be satisfactorily reproduced by dynamic loudspeakers of the period. See also Williams, "Piezo-electric Loudspeakers and Microphones," 166–67.

153. For the work on microphones and loudspeakers at Bell Telephone Laboratories, see Wente and Thuras, "Loud Speakers and Microphones," 259–77, and Wittje, *The Age of Electroacoustics*, 141. For a review of the leading types of amplifiers and loudspeakers featured at the German Radio Exhibition of 1934, see Schwandt, "Kraftverstärker—Lautsprecher—Elektroakustische Geräte," 649–56. See also Nestel, "Der gegenwärtige Stand der Rundfunktechnik," 489–91.

154. Thomas, "The Performance of Amplifiers," 253–78.

155. Some of the most popular vacuum tubes for radios in Germany included Telefunken's RE 074, RE 084, RE 704, RE 054, RE 124, RE 114, RE 134, RE 604, REN 1004, REN 1104, RE 124, RE 114, and RE 134. Other popular vacuum tube suppliers included Röhrenfabrikante Valvo (Radioröhrenfabrik, Hamburg) and TeKaDe (Süddeutsche Telefonapparate, Kabel- und Drahtwerke A. G., Nuremberg). Nesper, *Kompendium*, 100.

156. See Scriven, "Amplifiers," 278.

157. Reich, "A New Method of Testing," 401–15.

158. On the history of objectivity, see Daston and Galison, *Objectivity*, 115–90.

4: The RVS: Radio Experiments

1. "Bericht des Rundfunkkommissars des Reichpostministers vom 31. 12. 1928," as reprinted in Rindfleisch, *Technik im Rundfunk*, 54. For a history of the Reich's Broadcasting Corporation, see Pohle, *Der Rundfunk*, 47–56. While the Reich's Broadcasting Corporation had private investors, the majority of its shares was held by Reich's Ministry of the Post Office.

2. From 1919 to 1933 it was called the Preußisches Ministerium für Wissenschaft, Kunst und Volksbildung, or the Prussian Ministry for Science, Art, and Education

3. For histories of the RVS, see Schenk, *Die Hochschule*, 257–72; Tkaczyk, "Radio Voices," 85–109; Tkaczyk, *Thinking with Sound*, 199–207; Brilmayer, "Das Trautonium: Prozesse des Technologietransfers," 217–30, and Dörfling, *Der Schwingkreis*, 175–216.

4. Schünemann, "Die Funkversuchsstelle," 278.

5. Grosch, *Die Musik*, and Potter, *Art of Suppression*, 209.

6. It should be noted that Paul Hindemith rejected atonality.

7. Strobel, "Neue Sachlichkeit in der Musik," 3–4. See also Patteson, *Instrument for New Music*, 36.

8. Raven-Hart, "Radio, and a New Theory," 380.

9. This is more of a difference of degree rather than kind. As we shall see, the Germans were also interested in the transmission of voices, and there were a number of US scientists and engineers working at Bell Telephone Laboratories who ran radio stations and were interested in music. See, for example, Silvan, Dunn, and White, "Absolute Amplitudes," 1–42. This study analyzed the power and frequency range in music, focusing specifically on the extreme values. And we know that the Steinway Company of New York asked Bell Telephone Laboratories to test their pianos using harmonic analysis.

10. UdK-Berlin Archiv, Bestand 1b, Nr. 1, folia 132 and 135, and Schünemann's response of June 26, 1926, ibid., folia 133–134.

11. Kapeller, "Die verwaiste Kultur in Rundfunk," 178.

12. Mendelssohn, "Das Kompromiß," 294. See also Mendelssohn, "Die Versuchsanstalt für den Rundfunk," 385–86.

13. Georg Schünemann to Preußisches Ministerium für Wissenschaft, Kunst, und Volksbildung, September 21, 1926, UdK-Berlin Archiv, Bestand 1b, Nr. 1, folia 130–131, here 130.

14. Ibid.

15. Ibid.

16. The meeting was attended by K. W. Wagner; Leithäuser; Herbert Schubotz of *Deutsche Welle*; ministerial adviser Otto von Rottenburg; minister Leo Kestenberg of the Prussian Ministry of Culture and member of the Independent Social Democrat Party; and Schünemann.

17. "Sitzung des Preußischen Ministeriums für Wissenschaft, Kunst und Volksbildung, am 20. November 1926," UdK-Berlin Archiv, Bestand 1b, Nr. 1, folia 126–127.

18. Frank Warschauer, "Bemerkungen zum Plan einer Rundfunkversuchsstelle," April 19, 1927, UdK-Berlin, Bestand 1b, Nr. 1, folia 122–24.

19. Ibid., folio 123. His name was also written Reiss and Reiß.

20. Anon., "Endlich eine akustische Rundfunk-Versuchsstelle," 14.

21. Georg Schünemann to Preußisches Ministerium für Wissenschaft, Kunst und Volksbildung, Berlin, January 25, 1928, UdK-Archiv Bestand 1b, Nr. 1, folia 100–101, here 101. He specifically referred to Termin, who invented the theremin, and Mager, who invented the spherophone.

22. On September 1, 1929, Fischer successfully conducted an orchestral concert in Berlin featuring musicians playing in Berlin, London, Paris, Milan, Vienna, and Zurich. See Anon., "Ferndirigiertes Völkerbundkonzert," 1115.

23. UdK-Berlin Archiv, Bestand 1b, Nr. 16. See also Ikl., "Fernmeldetechnik," 954–55.

24. Anon, "Funkversuchsstelle," 1298. See also Dörfling, *Der Schwingkreis*, 194–95.

25. Lichtenthal, "Der Chorgesang im Rundfunk," 138, and Epstein, "Chorgesang im Rundfunk," 150.

26. Ikl., "Fernmeldetechnik," 954. See also Georg Schünemann to Minister/Präsidenten der Preußischen Bau- und Finanzdirektion, December 6, 1932, UdK-Berlin Archiv Bestand 1b, Nr. 2, folio 94.

27. Schenk, "Der Berliner Rundfunkversuchsstelle," 124.

28. Anon, "Funkversuchsstelle," here 1297.

29. Weill, "Über die Möglichkeiten einer Rundfunkversuchsstelle," (1927) reprinted in *Weill: Musik und Theater*, 245. The Funk-Stunde "Studio" worked closely with the RVS, particularly the courses on performance in front of a microphone offered by composer Max Butting. Anon., "'Studio' und 'Funkakademie,'" 146.

30. UdK-Berlin Archiv Bestand 1b, Nr.1, Folio 70, 5. Dezember 1929, "Protokol über die Sitzung des geschäftsführenden Kuratoriums der RVS, am Dienstag, den 3. Dezember ds. Js. [1929] in der RVS," and Bestand 1b, Nr. 2, Folia 47, 48, 51, 100, 103, 108–111, and 117. In fiscal year 1932, for example, the Prussian ministry paid 43,940 RM, while the Reich's Broadcasting Corporation paid 14,646 RM, totaling 58,586 RM, which was equivalent to just under $250,000 or $5.4 million in 2022 US dollars. Hence, the Prussian ministry increased its subsidy to 75 percent. In 1933 the Prussian ministry gave the RVS 30,000 RM. That would be equivalent to about $2.8 million in 2022. From October 7, 1927 to February 3, 1928, the Prussian ministry transferred 35,000 RM to the RVS account, UdK-Berlin Archiv, Bestand 1b, Nr. 1, folia 92, 96, and 106; 40,000 RM in 1929, UdK-Berlin Archiv, Bestand 1b, Nr. 1, folia 66 and 74; and just under 21,500 RM in 1930, UdK-Berlin Archiv, Bestand 1b, Nr. 1, folia 44 and 52; and 25,000 RM in 1931, UdK-Berlin Archiv, Bestand 1b, Nr. 1, folia 2 and 23. As a comparison, Germany's Physikalisch-technische Reichsanstalt had a budget of 1.5 million RM in 1931. See Cochrane, *Measures for Progress*, 310.

31. UdK-Berlin Archiv Bestand 1b, Nr. 1, Georg Schünemann, "Pressenotiz," March 8, 1930, folia 63–64. See also Schünemann, "Die Arbeit der Rundfunksversuchsstelle," 41–42.

32. MZL, "Ein enttäuschender Vortragsabend," 120.

33. Schünemann, "Von der Arbeit der Rundfunkversuchsstelle," 130–31.

34. Butting, "Music of and for Radio," 19.

35. Anon., "Berlin eröffnet seine Funk-Akademie," 188.

36. Weill, "Über die Möglichkeiten" (1927/1990), 243, italics added by the author.

37. Weill, "Über die Möglichkeiten" (1927/1990), 244.

38. Anon., "Arbeitsplan der Berliner 'Funkakademie,'" 176.

39. Stoffels, "Rundfunk als Erneuerer und Förderer," vol. 2, 857.

40. Anon, "Funkversuchsstelle," 1297. See also Schünemann, "Die neue Funkversuchsstelle," 153–54.

41. Anon, "Funkversuchsstelle," 1298.

42. Kapeller, "Die Schwierigkeiten der Rundfunk-Aufnahme," 41.

43. UdK-Berlin Archiv Bestand 1b, Nr. 1, folio 127. Leithäuser served on the RVS board of directors.

44. Schünemann, "Die Funkversuchsstelle," 292–93. See also anon., "Funkversuchsstelle," 1297.

45. As quoted in Schenk, *Die Hochschule*, 259.

46. Schünemann, "Die neue Funkversuchsstelle," 153.

47. The cost for the equipment and setup was 32,506.05 RM. See UdK-Berlin Archiv Bestand 1b, Nr. 4, folia 115–118, November 25, 1927. Some of the equipment was also rented annually by the RVS. See also UdK-Berlin Archiv Bestand 1b, Nr. 11 (G^1, Bd. II), folio 195, January 13, 1928. As the exchange rate between US dollars and German RM on January 1, 1928, was 4.21 to 1, that would be the equivalent of around $137,000 in January 1928, or just over $2 million in December 2018. For the original electrical diagrams of Siemens & Halske for the project, see UdK-Berlin Archiv, Bestand 1b, Nr. 17.

48. Ikl., "Fernmeldetechnik," 954.

49. UdK-Berlin Archiv, Bestand 1b, Nr. 11 (G^1 Bd II), folia 2 and 4, April 15 and July 28 1931. The cost for renting the large loudspeaker unit for one year (January 1 to December 31, 1931) was 1,500 RM according to the contract of February 28 1930, ibid., folio 4. See also the Siemens & Halske bill to the RVS for the installation of the large loudspeaker unit, UdK-Berlin Archiv, Bestand 1b, Nr. 11 (G^1, Bd. II), December 31, 1927, and January 13 and 18, 1928, folia 190–192 and 196–197. See also UdK-Berlin Archiv, Bestand 1b, Nr. 14, folia 1–21, October 27, 1928–July 5, 1930, for a list of equipment rentals. On August 19, 1932, Siemens & Halske transferred responsibility for the large loudspeaker unit to their "daughter company," Telefunken, which rented it to the RVS for 3,000 RM per year. See UdK-Berlin Archiv, Bestand 1b, Nr. 13, January 18, 1932, folia 59–60, and folio 32. On December 13, 1933, Telefunken decided to reduce the rental fee to 1,350 RM per year for two years. See UdK-Berlin Archiv, Bestand 1b, Nr. 13, folio 5. On April 11, 1935, Telefunken gifted the equipment to the RVS. See UdK-Berlin Archiv, Bestand 1b, Nr. 13, folio 3.

50. UdK-Berlin Archiv, Bestand 1b, Nr. 1, Georg Schünemann, "Pressenotiz," March 8, 1930, folia 63–64, January 2, 1930, folio 67, Georg Schünemann to Preußisches Ministerium, September 21, 1926, folia 130–131; UdK-Berlin Archiv, Bestand 1b, Nr. 11, (G^1, Bd II), folia 40–41, 129, 177; UdK-Berlin Archiv, Bestand 1b, Nr. 13 (22.7.30-16.10.31 G^2, Bd II), folio 218; UdK-Berlin Archiv, Bestand 1b, Nr. 1, folia 64, 127, and 130. See also Manning, *Electronic and Computer Music*, 13.

51. UdK-Berlin Archiv, Bestand 1b, Nr. 11, (G^1, Bd II), folia 41 and 129. Otto Feussner of Hanau sent the RVS a Stille teletype machine, folio 41.

52. Ibid., folia 41–43.

53. UdK-Berlin Archiv, Bestand 1b, Nr. 7, folia 10, 14–18, 42–43 and 47 and UdK-Archiv Bestand 1b, Nr. 1, folia 27 and 30. See also Schenk, *Die Hochschule*, 263. See also UdK-Archiv Bestand 1b, Nr. 8, folia 3, 4, 6 and 9, dated January 1934.

54. UdK-Berlin Archiv, Bestand 1b, Nr. 7, folio 43. See also Tkaczyk, "Radio Voices," 100–1, and Tkaczyk, *Thinking with Sound*, 205.

55. Schenk, *Die Hochschule*, 164.

56. UdK-Berlin Archiv, Bestand 1b, Nr. 11, folia 128–133.

57. Dietmar Schenk, "Der Berliner Rundfunkversuchsstelle," 124. From 1929 to 1930 a department within the RVS worked on talkies, UdK-Archiv 1b, No. 5, folia 1–89.

58. Friedrich Trautwein, patent application "Procedure and Contrivance for the Reception and Reproduction of Pictures and Sounds" (written in English), UdK-Berlin Archiv, Bestand 1b, Nr. 10 Technisches (Akustik), folia 174–198, here 174–178 and 184–187. Much like radio, the early years of German television, from the mid-1920s to the late 1920s, fell under the purview of the Reich's Post Office. Hempel, "German Television Pioneers," 126.

59. "Tätigkeitsberichte der Dozenten für die Zeit vom I. X. 29–30. IX. 1930," Paul Hindemith, "Filmmusikkurs im Rahmen der Rundfunkversuchsstelle (Schuljahr 1929/30)," September 1930, UdK-Berlin Archiv, Bestand 1b, Nr. 15, folio 9. It should be noted that French composer Darius Milhaud experimented with vocal transformations produced by varying a phonograph's rotational velocity during the 1920s. See Cros, "Electronic Music," 35.

60. As quoted in Schenk, *Die Hochschule*, 263.

61. Raz, "The Lost Movements," 37–59.

62. Schünemann, "Produktive Kräfte," 246–47. See also Katz, "Hindemith, Toch, und *Grammophonmusik*," 162.

63. Katz, "Hindemith, Toch, und *Grammophonmusik*," 162–63. See also Band, "Rundfunk im Rahmen der 'Neuen Musik,'" 133, and Schünemann, "Neue Musik Berlin 1930," 121–22.

64. "New Music Berlin 1930," *World-Radio*, June 27, 1930 in Deutsches Museum (Munich) Archiv (henceforth, DMA), Friedrich Trautwein Nachlass 187 / 114 (not numbered).

65. As quoted in Katz, "Hindemith, Toch, und *Grammophonmusik*," 164. Original: Schünemann, "Produktive Kräfte," 246–47.

66. Raz, "The Lost Movements," 41–43.

67. Schünemann, "Die Funkversuchsstelle," 283, and Schünemann, "Die Aufgaben," 11.

68. Wagner, "Der Frequenzbereich," 454.

69. Schünemann, "Die Aufgaben," 7.

70. Buchwald, "Die technischen Einrichtungen," 16.

71. Schünemann, "Die neue Funkversuchsstelle," 154.

72. Interference tubes, invented by German physicist Georg Hermann Quincke in 1866, are two glass tubes configured in a Y-shape connected to a rubber tubing. The difference in the lengths of the tubes causes the wavelengths traveling through one of the tubes to cancel out the waves traveling through the other. As a result of interference, overtones can be eliminated. See Quincke, "Ueber Interferenzapparate für Schallwellen," 177–92. See Kursell, "Coming to Terms with Sound," 35–59; Tkaczyk, "Radio Voices," 94; and Tkaczyk, *Thinking with Sound*, 201.

73. See Kursell, "Experiments on Tone Color," 191–211, and Kursell, "Musikwissenschaft am Berliner Institut für Psychologie," 73–90.

74. Anon., "Funkversuchstelle," 1297.

75. Georg Schünemann to Preußisches Ministerium für Wissenschaft, Kunst und Volksbildung, January 25, 1928, UdK-Berlin Archiv, Bestand 1b, Nr. 1, folia 100–101, here 100. See also Schünemann, "Die neue Funkversuchsstelle," 154.

76. Georg Schünemann to Preußisches Ministerium für Wissenschaft, Kunst und Volksbildung, January 25, 1928, UdK-Archiv, Bestand 1b, Nr. 1, folia 100–101, here 100.

77. Schünemann, "Die Aufgaben," 9, and Schünemann, "Die Funkversuchsstelle," 278.

78. Schünemann, "Die Aufgaben," 9 and Schünemann , "Die Funkversuchsstelle," 280–81.

79. Schünemann, "Die Aufgaben," 9. Note that he is drawing upon Stumpf's technique of formant analysis here.

80. Anon, "Funkversuchsstelle," 1298

81. Schünemann, "Die Aufgaben," 9, and Schünemann, "Die Funkversuchsstelle," 284.

82. This argument is reminiscent of Mark Katz's work on the importance of vibrato when recording violinists during the 1920s. See Katz, *Capturing Sound*, 94–108.

83. Sterne, *The Audible Past*, 215–86.

84. Schünemann, "Die Aufgaben," 9. The same experimentation with placing various instruments in a room was true when horns were used in recording. See Thompson, *The Soundscape of Modernity*, 294.

85. Schünemann, "Tonfilm und Rundfunk im Musikunterricht," 421–22 (reprinted in *Jahresbericht der Hochschule für Musik* 50 [1929]: 7–8), and Friedrich Trautwein, "Über akustische Probleme aud Grenzgebieten der Musik, Physik und der Physiologie," in UdK-Archiv, Bestand 1b, Nr. 10 Technisches (Akustik), folia 314–316.

86. Georg Schünemann to Preußisches Ministerium für Wissenschaft, Kunst und Volksbildung, December 6, 1932, UdK-Berlin Archiv Bestand 1b, Nr. 2, folio 94. Schünemann drew attention to the document stating the origins of the RVS, namely the letter from Secretary of State Lammers to Schünemann of June 11, 1926, UdK-Archiv Bestand 1b, Nr. 1, folia 132 and 135. See also Ikl., "Fernmeldetechnik" (1928), 954.

87. Raven-Hart, "Radio, and a New Theory," 380. In 1929 the RVS established the first course of study for radio speaking. See anon., "Die Schule für Rundfunkredner," 72. See also anon., "Arbeitsplan der Berliner 'Funkakademie,'" 176, and anon., "Berlin eröffnet seine Funk-Akademie," 187–90.

88. While improvements in cable were also crucial to radio—as we saw in the previous chapter—there is no evidence to suggest that the physicists and engineers at the RVS researched cable technology.

89. Schünemann to the Minister für Wissenschaft, Kunst und Volksbildung, January 25, 1928, UdK-Archiv Bestand 1b, Nr. 1, folia 100–101, here 100.

90. There are numerous examples of the importance of learning how to use the microphone. See, for example, anon., "Funkversuchsstelle," 1297. Fritz Stern to the Reichsministerium für Wissenschaft, Kunst und Volksbildung, September 12, 1934, UdK-Berlin Archiv, Bestand 1b, Nr. 2, folio 22b. See also Georg Schünemann to Preußisches Ministerium für Wissenschaft, Kunst und Volksbildung, December 6, 1932, UdK-Berlin Archiv, Bestand 1b, Nr. 2, folio 94 and Georg Schünemann to I. G. Farben A. G. (Agfa) Berlin, UdK-Berlin Archiv, Bestand 1b, Nr. 6, folio 70, May 20, 1931.

91. Mn., "Hörspielkursus," 126.

92. UdK-Berlin Archiv, Bestand 1b, Nr. 15. "Tätigkeitsberichte der Dozenten für die Zeit vom 1. X. 29–30. IX. 1930," folia 2, 3, 4, and 7. See also Braun, "Rundfunk—Bühne—Tonfilm," 72, and anon., "Berlin eröffnet seine Funk-Akademie," 189.

93. Anon., "Berlin eröffnet seine Funk-Akademie," 189.

94. Ikl. "Fernmeldetechnik," 954–55, and Buchwald, "Die technischen Einrichtungen," 17. See also UdK-Berlin Archiv, Bestand 1b, Nr. 11, (G^1, Bd II), folio 145–158, November 12, 1928, and folio 174, May 10, 1928. See also Dörfling, *Der Schwingkreis*, 208–12.

95. UdK-Berlin Archiv, Bestand 1b, Nr. 13 (22.7.30-16.10.31 G^2, Bd II), folia 145 and 190; UdK-Berlin Archiv, Bestand 1b, Nr. 14, folia 1–21; UdK-Berlin Archiv, Bestand 1b, Nr. 4, folia 115–118; and UdK-Berlin Archiv, Bestand 1b, Nr. 9, folio 5; UdK-Berlin, Archiv Bestand 1b, Nr. 11, (G^1, Bd II), folia 145–158, November 12, 1928, folio 190, December 31, 1927; UdK-Berlin Archiv, Bestand 1b, Nr. 9, folio 5, January 7, 1938; UdK-Berlin Archiv, Bestand 1b, Nr. 3, folio 4, September 10, 1934; UdK-Berlin Archiv, Bestand 1b, Nr. 4, folia 1–10, July 26 to October 27, 1929) and folia 111, 115–118, November 25, 1927.

96. Georg Schünemann to Eugen Reisz, May 12, 1928, UdK-Berlin Archiv, Bestand 1b, Nr. 11, (G^1, Bd II), folio 173. See also Georg Schünemann to Pama G.m.b.H. für Papier-Machè-Erzeugnisse, Munich, January 22, 1929, ibid., folio 135, and Georg Schünemann, "Pressenotiz," March 8, 1930, UdK-Berlin Archiv, Bestand 1b, Nr. 1, folia 63–64. A radio engineer named

Dr. Sell of the RVS developed a staff microphone, which the academy used during musical performances.

97. "Abschrift des Protokols: Versuche über 'Phasenmodulation,'" UdK-Berlin Archiv, Bestand 1b, Nr. 11 Technisches (Rundfunk), (G[1], Bd II), folio 77 and Alfons Kreichgauer, "Aktennotiz 2," folio 129; UdK-Berlin Archiv, Bestand 1b, Nr. 1, folia 123 and 127; UdK-Archiv Bestand 1b, Nr. 18, folio 2.

98. Eugen Reisz to Georg Schünemann, May 10, 1928, UdK-Berlin Archiv, Bestand 1b, Nr. 11, (G[1], Bd II), folia 171–175, here 174. He called RCA "Radio-Corporation" and Western Union, "Western Corporation." Ibid. See also folia 161–166 and 170. As stated in the previous chapter, a "flat response" was a distortion of a maximum of 5 dB over that frequency range.

99. UdK-Berlin Archiv, Bestand 1b, Nr. 11, (G[1], Bd II), folio 174. For other references to Reisz microphones, see UdK-Berlin Archiv, Bestand 1b, Nr. 18, December 4, 1929, folio 2; UdK-Berlin Archiv, Bestand 1b, Nr. 1, April 19, 1927, folia 93–94, 122–124, and November 20, 1926, folio 127.

100. UdK-Berlin Archiv, Bestand 1b, Nr. 13 (22.7.30-16.10.31 G[2], Bd II), folio 178, December 11, 1930.

101. UdK-Berlin Archiv, Bestand 1b, Nr. 11, Technisches (Rundfunk) (G[1], Bd II), folio 77.

102. UdK-Berlin Archiv, Bestand 1b, Nr. 11, Technisches (Rundfunk) (G[1], Bd II), folia 2, 4, 43, 128, and 197. The RVS rented a large loudspeaker from Siemens & Halske for 1,500 RM for the calendar year of 1931 as per the contract of February 28, 1930, ibid., folio 4; UdK-Berlin Archiv, Bestand 1b, Nr. 3, folia 3, 5, and 59; UdK-Berlin Archiv, Bestand 1b, Nr. 1, folia 59–62, 127, and 142 and UdK-Berlin Archiv, Bestand 1b, Nr. 10 Technisches (Akustik), folio 316; UdK-Berlin Archiv, Bestand 1b, Nr. 9, folia 46–49; UdK-Berlin Archiv, Bestand 1b, Nr. 11, (G[1], Bd II), folio 190; UdK-Berlin Archiv, Bestand 1b, Nr. 14, folio 1–21.

103. UdK-Berlin Archiv, Bestand 1b, Nr. 11, (G[1], Bd II), folia 111–112. In May 1929, AEG delivered four Rice-Kellogg loudspeakers to the RVS for 2,000 RM, ibid., folio 112.

104. Wittig, *The Age of Electroacoutics*, 141.

105. "Die Kollegen von damals," Oskar Sala, interview on September 1, 1989. http://www.klangspiegel.de/trautonium/kollegen-damals.

106. Werbeprospekt SH 3981: "Der bewährte Lautsprecher Siemens 072 (Protos), 1926." Reproduced at https://www.medienstimmen.de/chronik/1926-1930/1926-siemens-halske-protos-lautsprecher-siemens-072/. See also Dörfling, *Der Schwingkreis*, 209–12. Clearly an advertisement announcing their product needs to be taken with a grain of salt; however, it does inform us on how loudspeakers were judged.

107. Anon., "Was verlangt man vom Lautsprecher?" Reproduced at https://www.medienstimmen.de/chronik/1926-1930/1926-siemens-halske-protos-lautsprecher-siemens-072/. Originally published: anon., *Dortmunder Zeitung*, Nr. 572, December 7, 1929.

108. Ikl., "Fernmeldetechnik," 955. See also Buchwald, "Die technischen Einrichtungen," 18–19.

109. Georg Schünesmann, "Pressenotiz," March 8, 1930, UdK-Berlin Archiv, Bestand 1b, Nr. 1, folia 63–64.

110. UdK-Berlin Archiv, Bestand 1b, Nr. 11, (G[1], Bd II), folia 116–120, 128, 135, 161–164 and 167–169 and UdK-Berlin Archiv, Bestand 1b, Nr. 9, folio 244. Tempel of Krefeld had the German license to sell Temple Loudspeakers of Chicago, Illinois. UdK-Berlin Archiv, Bestand 1b, Nr. 11, (G[1], Bd II), folia 167–169.

111. UdK-Berlin Archiv, Bestand 1b, Nr. 11, (G[1], Bd II), folio 119, April–May 1929.

112. UdK-Berlin Archiv, Bestand 1b, Nr. 11, (G[1], Bd II), folio 102 (September 19, 1929) and folio 133 (March 2, 1929). Whether or not combinations tones are only subjective, or whether they can also be objective, i.e., independent of human hearing, has been debated for over 150 years. It was a topic of interest during the 1920s and '30s, including among scientists and engineers at the RVS. Recent studies demonstrate that some musical instruments—such as violins—produce objective combination tones.

113. Friedrich Trautwein, "Über akustische Probleme aus Grenzgebieten der Musik, der Physik und der Physiologie," UdK-Berlin Archiv, Bestand 1b, Nr. 10 Technisches (Akustik), folia 314–316, here 316.

114. Buchwald, "Die technischen Einrichtungen," 17–18.

115. UdK-Berlin Archiv, Bestand 1b, Nr. 14, folia 1–21; UdK-Berlin Archiv, Bestand 1b, Nr. 4, folia 111, 115–118; UdK-Berlin, Archiv Bestand 1b, Nr. 17; UdK-Berlin, Archiv Bestand 1b, Nr. 18, folia 2, 4–6, 20–23; UdK-Berlin, Archiv Bestand 1b, Nr. 11, (G^1, Bd II), folia 1, 38–39, 111, 145–148; UdK-Berlin, Archiv Bestand 1b, Nr. 9, folia 5, 19, and 51; UdK-Berlin Archiv, Bestand 1b, Nr. 1, folia 59–62; UdK-Berlin Archiv, Bestand 1b, Nr. 1, folia 93–94; and UdK-Berlin Archiv, Bestand 1b, Nr. 11, (G^1, Bd II), folia 32, 38–39, and 95.

116. Georg Schünemann to the Physikalisch-Technische Reichsanstalt, UdK-Berlin Archiv, Bestand 1b, Nr. 10 Technisches (Akustik), folia 124 (April 10, 1933), 125, (April 8, 1933), 135 (February 17, 1933), and 280–281; UdK-Berlin Archiv, Bestand 1b, Nr. 12, folia 5 (October 27, 1931), folia 11–12 (October 22, 1931).

117. Hans-Georg Görner to Georg Schünemann, "Abschrift des Protokolls. Versuche über 'Phasenmodulation' an der Rundfunkversuchsstelle bei der Staatl. Akad. Hochschule für Musik-Berlin. Über Arbeiten vom 1. Oktober 1929 bis Dezember 1929," UdK-Berlin Archiv, Bestand 1b, Nr. 11 (G^1 Bd II), folia 75–82, here folio 76.

118. Ernst Finking D.J. of Funking Reinton, Leipzig to Georg Schünemann, April 10, 1930, UdK-Berlin Archiv, Bestand 1b, Nr. 11 Technisches (Rundfunk) (G^1 Bd II), folio 43.

119. Meyer and Just, "Messung von Nachhalldauer und Schallabsorption," 293–300, UdK-Berlin Archiv, Bestand 1b, Nr. 25; Meyer, "Beiträge zur Untersuchung des Nachhalles," 135–39; Meyer, "Über eine einfache Methode," 398–403; Grützmacher, "Eine neue Methode der Klanganalyse," 216–28; Grützmacher and Meyer, "Eine Schallregistriervorrichtung zur Aufnahme der Frequenzkurven," 203–11. See also "Tätigkeitsberichte der Dozenten für die Zeit vom 1.X. 29–30 IX. 1930," UdK-Berlin Archiv, Bestand 1b, Nr. 15, folio 10.

120. UdK-Berlin Archiv, Bestand 1b, Nr. 13 (22.7.30-16.10.31 G^2, Bd II), folio 228.

121. UdK-Berlin Archiv, Bestand 1b, Nr. 2, Folio 92; UdK-Berlin Archiv, Bestand 1b, Nr. 1, folia 27 and 80–81.

122. UdK-Berlin, Archiv Bestand 1b, Nr. 15, folio 10 and UdK-Archiv Bestand 1b, Nr. 3, folio 69. See also anon., "Berlin eröffnet seine Funk-Akademie." 187.

123. Werner Trautwein to Siegfried Goslich, September 8, 1964, DMA, Friedrich Trautwein Nachlass 187 028, 1–3, here 2–3. This biography was given to Goslich by Trautwein's son, Werner. Apparently, it was based on his father's own autobiographical sketch. See also Dvorak, "Friedrich Trautwein," 690, which adds more to the autobiography and has somewhat different dates for studying law.

124. Dvorak, "Friedrich Trautwein," 690.

125. Trautwein, "Die Elektronenröhre in der elektrischen Meßtechnik," 101–4, 119–23, and Trautwein, "Über Verlustmessung bei hohen Frequenzen," 235–64.

126. Dörfling, *Der Schwingkreis*, 184 and Beckh, *Blitz & Anker*, 354.

127. Friedrich Trautwein, "Memorandum über das Trautonium mit besonderer Berücksichtigung seiner Bedeutung für die Filmmusik," DMA, Friedrich Trautwein Nachlass 187/013, 3. See also Goldhammer, Laufer, and Lehr, "'Achtung! Achtung! Hier ist die Sendestelle Berlin,'" 74.

128. Werner Trautwein to Siegfried Goslich, September 8, 1964, DMA, Friedrich Trautwein Nachlass 187 / 028, 1–3, here 2. See also Dvorak, "Friedrich Trautwein," 690.

129. Modulation is the addition of information to an electronic signal, for example, by altering its amplitude, frequency, or phase.

130. Trautwein, "Modulation und Übertragungsgüte in der Hochfrequenztechnik," 343–52, and Dvorak, "Friedrich Trautwein," 690.

131. Trautwein, *Drahtlose Telephonie und Telegraphie.*

132. Viefhaus, "Ingenieure in der Weimarer Republik," 292.

133. Viefhaus, "Ingenieure in der Weimarer Republik," 303.

134. Viefhaus, "Ingenieure in der Weimarer Republik," 305.

135. Gispen, *New Profession, Old Order*, 194.

136. Gispen, *New Profession, Old Order*, 202. See also König, "Die Ingenieure und der VDI als Großverein," 235–87, and Viefhaus, "Ingenieure in der Weimarer Republik," 311–12.

137. Gispen, *New Profession, Old Order*, 238.

138. Viefhaus, "Ingenieure in der Weimarer Republik," 317.

139. Gispen, *New Profession, Old Order*, 208. See also Herf, *Reactionary Modernism*, 152–88.

140. As cited in Donhauser, *Elektrische Klangmaschinen*, 42–43, original: "Elektrische Musik," broadcast on Radio Bremen on February 18,1953.

141. Friedrich Trautwein, "Memorandum über das Trautonium mit besonderer Berücksichtigung seiner Bedeutung für die Filmmusik," DMA, Friedrich Trautwein Nachlass, 187 / 013, 3.

142. Friedrich Trautwein, "Einrichtung zur Schwingungserzeugung mittels Elektronenröhren," Deutsches Reichpatent 462,980, granted October 22, 1922.

143. Trautwein, *Elektrische Musik*, 11.

144. Donhauser, *Elektrische Klangmaschinen*, 44

145. Trautwein, "Die technische Entwicklung der elektrischen Musik," 134, Deutsches Reichpatent 469,775, granted on March 4, 1924. Trautwein later wrote that the date of the patent is April 3, 1924, and April 4, 1924. See Trautwein, "Elektronische Klangerzeugung und Musikästhetik," 176, and Trautwein, *Elektrische Musik*, 11. He was awarded US Patent 2,039,201 on March 24, 1924.

146. Friedrich Trautwein, "Memorandum über das Trautonium mit besonderer Berücksichtigung seiner Bedeutung für die Filmmusik," DMA, Friedrich Trautwein Nachlass 187/013, June 1, 1949, p. 7. See also Patteson, *Instrument for New Music*, 118, and Donhauser, *Elektrische Klangmaschinen*, 69–70.

147. Dörfling, *Der Schwingkreis*, 197, and Schenk, *Die Hochschule für Musik*, 163–64.

148. Oskar Sala, "Nachruf für Prof. Dr. Ing. Friedrich Trautwein," January 24, 1957, DMA, Friedrich Trautwein Nachlass 187/066. See also DMA, Oskar Sala Nachlass 218/AV-T0420. Sala states that Trautwein was still working in electrical industry in early 1930. See also *Staatliche Akademische Hochschule für Musik in Berlin von 1. Oktober 1929 bis 30. September 1930: 51. Jahresbericht*, p. 45 in UdK-Berlin Archiv 1/D10. It should be noted that his biographical entry has him starting a lectureship in October 1930, Dvorak, "Friedrich Trautwein," 61.

149. Trautwein, "Elektrische Klangbildung und elektrische Musikinstrumente," 123.

150. Trautwein, "Elektrische Klangbildung und elektrische Musikinstrumente," 123.

151. Trautwein, "Elektrische Klangbildung und elektrische Musikinstrumente," 123. He often spoke of the unique sounds the instrument could generate; however, he also stressed the importance of imitating traditional acoustical instruments since that would increase the chances of the trautonium becoming economically successful.

152. UdK-Berlin Archiv, Bestand 1b, Nr. 11, folio 3, April 11, 1931. Hindemith used Trautwein's disc-recording and synchronous-transmission methods in his lectures at the RVS. See "Filmkurs im Rahmen der RVS (Schuljahr 1929/30)," September 1930, UdK-Berlin, Archiv Bestand 1b, Nr. 15, folio 9.

153. Friedrich Trautwein, "Über akustische Probleme aus Grenzgebieten der Musik, Physik und der Physiologie," in UdK-Berlin Archiv, Bestand 1b, Nr. 10 Technisches (Akustik), folia 314–316.

154. Friedrich Trautwein, "Über akustische Probleme aus Grenzgebieten der Musik, Physik und der Physiologie," in UdK-Berlin Archiv, Bestand 1b, Nr. 10 Technisches (Akustik), folio 314.

155. Ibid.

156. Ibid.

157. Ibid., folio 315.

158. Ibid.

159. Ibid.

160. It turns out that Trautwein was angry with the publisher of his book, Weidmann of Berlin. Three years after the book's publication, over 1,100 copies of the initial print run had still not been sold, and Schünemann's honorarium was paid one year later than agreed upon. Friedrich Trautwein to Georg Schünemann, January 13, 1933, UdK-Berlin Archiv, Bestand 1b, Nr. 10 Technisches (Akustik), folio 166.

161. Trautwein, *Elektrische Musik.*

162. As translated in Patteson, *Instrument for New Music,* 116, original: as quoted in Mersmann, "Dr. Trauteins elektrische Musik," 229.

163. Friedrich Trautwein, "Zur Ästhetik der Tonkunst im Hinblick auf elektronische Musik," DMA, Friedrich Trautweins Nachlass 187/014 (not dated, around 1954), 13. Note that the term "electronic" was widely used after World War II.

164. Trautwein, "Wesen und Ziele," 694–96.

165. Oskar Sala, "Arbiturfeier, 15.03.1929," DMA, Oskar Sala Nachlass 218/0197. See also Brilmayer, "Das Trautonium: Prozesse des Technologietransfers," 230–33.

166. DMA, Oskar Sala Nachlass 218/0437. See also Beecroft, "Oskar Sala: 1910–2002," 215–27.

167. Schenk, *Die Hochschule,* 264–65.

168. Oskar Sala, "Neue Möglichkeiten," 136.

5: The Original Trautonium

1. For an excellent account of a number of electronics musical instruments in Germany at this time, see Patteson, *Instruments for New Music.*

2. Weissmann, "Mensch und Maschine," 103

3. Leers, "Elektrische Musik in der Kritik," 178. Despite that important point, apparently the "trautonium was praised for the rich choices of sound 'that could replace an entire orchestra.'" See Weissmann, "Mensch und Maschine," 103

4. Lion, "Die technischen Grundlagen," 357–58.

5. Huth, "Elektrische Tonerzeugung," 45. As translated in Patteson, *Instruments for New Music,* 68.

6. Lion, "Die technischen Grundlagen," 358.

7. As Tresch and Dolan illustrate, researching the materiality of instruments illuminates the historically contingent relationship between music and science. Tresch and Dolan, "Toward a New Organology," 278–98.

8. Frieß and Steiner, eds., "Oskar Sala im Gespräch," 233.

9. Pinch and Trocco, *Analog Days.*

10. To view video of Leon Theremin playing the theremin and of Theremin virtuoso, Clara Rockmore, playing Chopin's Nocturen in C# Minor, see the "Resources" tab at https://press.princeton.edu/books/broadcastingfidelity.

11. Much of these opening paragraphs are based on Donhauser, *Elektrische Klangmaschinen,* 23–46. For an outstanding account of Mager in English, see Patteson, *Instruments for New Music,* 52–113.

12. Mager, *Eine neue Epoche.*

13. As translated in Patteson, *Instruments for New Music,* 60. Original: Mager, "Eine Rundfunkprophezeiung," 2952.

14. For a description of the instrument, see Patteson, *Instruments for New Music,* 61–64.

15. Battaglia, "Ondes Martenot: An Introduction," March 6, 2014. https://daily.redbullmusicacademy.com/2014/03/ondes-martenot-introduction.

16. For the engineering diagrams of a number of these electric instruments, see Vierling, "Elektrische Musik," 155–59; Janovsky, "Elektrische Musikinstrumente," 675–78, 727–30; and Lertes, "Zur historischen Entwicklung," 297–98.

17. Donhauser, *Elektrische Klangmaschine*, 121.

18. The nonlinearity of human hearing had been known since the early 1920s. See Fletcher and Wegel, "The Frequency-Sensitivity of Normal Ears," 553–65; Fletcher, "Physical Measurements of Audition," 154–58; and Fletcher, "The Physical Criterion," 435–37. Nonlinear distortions of the ear can occur with an increase in a signal's volume, which can lead to the creation of overtones and difference tones. See von Békésy, "Über die nichtlinearen Verzerrungen," 809–27.

19. Fletcher, "Loudness, Pitch and the Timbre," 61, emphasis is in the original. See also Fletcher, "Loudness and Pitch," 130–35. As Dolan and Rehding argue, timbre requires the ears as well as the eyes. See Dolan and Rehding, "Histories and Possible Futures," 3–19.

20. Jones and Knudsen, "Facts of Audition," 1016.

21. Trautwein, *Elektrische Musik*, 19. As mentioned in chapter 3, Erwin Meyer argued that the volume of the entire sound affects the perception of timbre due to the nonlinearity of the ear. Meyer, "Das Gehör," 489.

22. Apparently Trautwein returned to the theme of demonstrations involving higher order (i.e., third or fourth order) nonlinear subjective and objective combination tones. He referred to Hindemith's work at the RVS in 1932 on difference tones. See Friedrich Trautwein's letter to Oskar Sala, Deutsches Museum (Munich) Archive (henceforth DMA), Oskar Sala Nachlass 218/0053/1955/318, not dated, by most likely mid-to-late November 1955. And Sala produced subjective combination tones with two sine generators and headphones. See Oskar Sala's letter to Friedrich Trautwein, November 28, 1955, DMA, Oskar Sala Nachlass/0053/1955/317 a-b.

23. Trautwein, "Über elektrische Analogien," 40.

24. Trautwein, "Toneinsatz und elektrische Musik," 246.

25. Friedrich Trautwein, "Über akustische Probleme aus Grenzgebieten der Musik, der Physik und der Physiologie," around 1932, Universität der Künste (henceforth UdK)-Archiv Berlin, Bestand 1b, Nr. 10 Technisches (Akustik), folia 314–316, here 316. The italics are in the original text. It should be noted that Stumpf's notion of timbre was dynamic.

26. Trautwein, "Toneinsatz und elektrische Musik," 244, and Backhaus, "Über die Bedeutung," 31–46.

27. See Trautwein, "Die technische Entwicklung der elektrischen Musik," 134. He spoke of how the ear picks up on the swelling and the fading away of the sound. Engineers needed to take this into consideration when attempting to imitate sounds electrically. See also DMA, Friedrich Trautwein Nachlass 187/006. It should be noted that Carl Stumpf also believed that a static representation of overtones was insufficient in defining timbre. On the contrary, timbre was critical to his psychological research program. See Klotz, "Timbre, *Komplexeindruck*, and Modernity," 609–40.

28. Trautwein, *Elektrische Musik*, 8–10. For the basic workings of a glow discharge tube, see Lertes, *Elektrische Musik*, 83–87.

29. Impulse (or shock) excitation is a process by which oscillations are produced in a circuit by generating a signal (a stimulus) for a short period of time compared to the duration of the oscillation it produces.

30. Trautwein, "Über elektrische Analogien," 40.

31. For the original description of the trautonium, see Trautwein, *Elektrische Musik*, 28–36. Trautwein, "Elektrische Klangbildung und elektrische Musikinstrumente," 123–24; Janovsky, "Elektrische Musikinstrumente," 727–28; and Gradenwitz, "The Trautonium," 254. For a media studies' account of the circuitry and components of the trautonium, see Dörfling, *Der Schwingkreis*, 217–68. To view a video of Oskar Sala playing the original trautonium with Friedrich Trautwein looking on, see the "Resources" tab at https://press.princeton.edu/books/broadcasting fidelity.

32. Trautwein used neon tubes between 100 and 120 volts from the Gral Company of Berlin. Osram Licht AG, a lighting company created in 1919 by the merger of Auergesellschaft, Siemens & Halske, and AEG, also produced neon lamps; however, since they were made for lighting purposes, they did not produce vibrations as efficiently. If one wished to use an Osram tube, one needed to remove the series resistor. The requisite values for resistance and capacitance for generating audible frequencies were 0.5 to 4 megohms and 1,000 to 10,000 centimeters. See Trautwein, *Elektrische Musik*, 28–29.

33. A physicist or electrical engineer would say that the circuit's frequency depends on the time constant of R and C.

34. Gradenwitz, "The Trautonium," 254.

35. Trautwein, *Elektrische Musik*, 12.

36. Trautwein, *Elektrische Musik*, 36. For the June 1930 recording of Sala playing the original trautonium at the RSV, see the "Resources" tab at https://press.uprinceton.edu/books/broadcastingfidelity.

37. Trautwein, *Elektrische Musik*, 36.

38. Trautwein, *Elektrische Musik*, 31–32.

39. Georg Schünemann to the Regional Post Office Administration (*Oberpostdirektion*), July 24, 1930, UdK-Berlin Archiv Bestand 1b, Nr. 11 Technisches (Rundfunk), folio 30. See also Raven-Hart, "Neon Musical Oscillator," 650; Herbert Eimert, Radio Broadcast of the Nordwest Deutscher Rundfunk, 18 Oktober 1961: "Zehn Jahre elektronische Musik (Die Klangwelt der elektronischen Musik von 18. 10. 1951)," West Deutscher Rundfunk Archiv, Köln (Cologne) 6126844101.1.01.

40. Winkelmann, *Das Trautonium*, 14–16.

41. DMA, Oskar Sala Nachlass 218/0198/3.031.

42. In English this would generally be referred to as polyphonic, although the word has a different connotation in German.

43. Trautwein, *Elektrische Musik*, 35–36.

44. Interviews with Oskar Sala and Harald Genzmer in Weck, *Nur Einer kann es spielen*. https://www.srf.ch/audio/passage/nur-einer-kann-es-spielen-oskar-sala-meister-des-trautoniums?id=10144241.

45. Sala, "Anfänge," http://www.oskar-sala.de/oskar-sala-fonds/oskar-sala/interview/anfaenge/. See also Frieß and Steiner, eds., "Oskar Sala im Gespräch," 224.

46. Sala, "Neue Möglichkeit," 137.

47. Oskar Sala, "Anfänge," Oskar-Sala-Fonds, Deutsches Museum, München, available at http://www.oskar-sala.de/oskar-sala-fonds/oskar-sala/interview/anfaenge/. Original in the film, "Oskar Sala und sein Mixturtrautonium," Institut für Film und Bild in Grünwald als Schlußstück, Berlin, 1985. See also Frieß and Steiner, eds., "Oskar Sala im Gespräch," 224.

48. Ibid.

49. Georg Schünemann to Director Dr. Konrad Norden, October 30, 1931, UdK-Berlin Archiv, Bestand 1b, Nr. 12, folio 3.

50. Trautwein, "Wesen und Ziele," 698.

51. Janovsky, "Elektrische Musikinstrumente," 727–28. See also Lertes, *Elektrische Musik*, 183.

52. George Schünemann to Emil Mayer of Telefunken, January 27, 1933, UdK-Berlin Archiv, Bestand 1b, Nr. 10 (Technische Akustik), folio 153.

53. Schünemann's "Vorwort" to Trautwein, *Elektrische Musik*.

54. Germann, "Das Trautonium," 8. See also Deutsches Technikmuseum Berlin. Historisches Archiv I-2.060 P-Tfk 00409, p. 1–12, here 8.

55. Oskar Sala, "Bericht über das neue Trautonium, seine Enstehung und seine Spieltechnik," February 1936, DMA, Oskar Sala Nachlass 218/2462, 2–3. On September 30, 1930, Trautwein's patent attorney wrote to Schünemann, AEG, and Siemens & Halske outlining the details of a licensing agreement for the trautonium. Telefunken played a large role in attempting to make

the trautonium a household instrument. Starting in 1931 Sala and Trautwein collaborated on their so-called *Volkstrautonium* (or "trautonium of the people") funded by Telefunken, which provided an engineer (Walter Germann) and numerous parts. See Trautwein's patent attorney to Schünemann and Siemens & Halske, A.G. for licensing agreements, September 20, 1930, UdK-Berlin Archiv Bestand 1b, Nr. 13, folia 236–241, particularly folio 237. See also the letters exchanged between Friedrich Trautwein and Telefunken, UdK-Berlin Archiv, Bestand 1b, Nr. 10 Technisches (Akustik), folia 148–149, February 1933 and November 21, 1932, UdK-Berlin Archiv, Bestand 1b, Nr. 10 Technisches (Akustik), folio 168. See also Donhauser, *Elektrische Klangmaschinen*, p. 74. See also Donhauser, *Oskar Sala*, 32. Donhauser provides the best technical history of the development of the trautonium through its various instantiations that I have read.

56. "Sitzung des Kuratoriums der Rundfunkversuchsstelle bei der Staatlichen akademischen Hochschule für Musik im Ministerium für Wissenschaft, Kunst und Volksbildung, 15 November 1932," November 21, 1932, UdK-Berlin Archiv Bestand 1b, Nr. 2, between folia 92–93, 1–6, here 4.

57. UdK-Berlin Archiv, Bestand 1b, Nr. 12, Georg Schünemann to Director Dr. Konrad Norden, October 30, 1931, folio 3. See also Badge, *Oskar Sala*, not paginated. Of those composers, only Hindemith, in the end, composed for the trautonium.

58. Interview with Oskar Sala, "Anfänge," Oskar-Sala-Fonds am Deutschen Museum, http://www.oskar-sala.de/oskar-sala-fonds/oskar-sala/interview/anfaenge/index.html. See also Frieß and Steiner, eds., "Oskar Sala im Gespräch," 222, and Kämpfer, "Ein Leben für das Trautonium," 45.

59. Friedrich Trautwein, "Über akustische Probleme aus Grenzgebieten der Musik, Physik und der Physiologie," in UdK-Archiv, Bestand 1b, Nr. 10 Technisches (Akustik), folia 314–316, here 316.

60. Anon., "The Method of 'Shock-Excitation,'" 21.

61. Oskar Sala, "Anfänge," Oskar-Sala-Fonds, Deutsches Museum, München, available at http://www.oskar-sala.de/oskar-sala-fonds/oskar-sala/interview/anfaenge/.

62. Friedrich Trautwein, "Über akustische Probleme aus Grenzgebieten der Musik, Physik und der Physiologie," in UdK-Archiv, Bestand 1b, Nr. 10 Technisches (Akustik), folia 314–316, here 316. See also Dörfling, *Der Schwingkreis*, 241–47, and Donhauser, *Oskar Sala*, 16–20.

63. Friedrich Trautwein, "Über akustische Probleme aus Grenzgebieten der Musik, Physik und der Physiologie," in UdK-Archiv, Bestand 1b, Nr. 10 Technisches (Akustik), folia 315–316.

64. Friedrich Trautwein, "Bedeutung und Wesen der elektrischen Musik." In *Akademische Hochschule für Musik in Berlin zu Charlottenburg: 51. Jahresbericht vom 1. Oktober 1929 bis zum 30. September 1930*, 30 to 35, here 32, UdK-Berlin Archiv, 1-D10.

65. Trautwein, *Elektrische Musik*, 15.

66. Trautwein, *Elektrische Musik*, 16–18.

67. Oskar Sala, Oskar-Sala-Fonds am Deutschen Museum, "Anfänge," available at http://www.oskar-sala.de/oskar-sala-fonds/oskar-sala/interview/anfaenge/.

68. Trautwein, *Elektrische Musik*, 12.

69. Trautwein, *Elektrische Musik*, 13. Note that the claim that circuits are analogous to human speech organs was not originally Trautwein's. John Q. Stewart, for example, offered that analogy in his "An Electrical Analogue," 311–12.

70. Trautwein, "Über elektrische Analogien," 39–40. Trautwein also noted an important difference between the electrical and acoustical resonators that impaired artificial vowel production and played a role in the production of musical sounds. An electric oscillation possesses a single resonant frequency circuit by means of a capacitor and self-induction. The physical and simple acoustical resonators, such as wooden bodies for wooden instruments, tubes for woodwinds, and mouth and nasal cavities in humans, possess numerous resonant frequencies. One can put together a number of resonance frequencies, which can behave similarly to acoustical resonators; however, it was quite complicated to do in the 1920s and '30s. Trautwein, "Über elektrische Analogien," 42–43.

71. Trautwein, *Elektrische Musik*, 12–19. See also Raven-Hart, "Neon Musical Oscillator," 648–49; Trautwein, "Electronic Musical Instruments," 19; Mersmann, "Dr. Trautweins Elektrische Musik," 228; and DMA, Friedrich Trautwein Nachlaß, 187/006.

72. Trautwein, *Elektrische Musik*, 18.

73. Trautwein, *Elektrische Musik*, 19.

74. Lion, "Elektrische Musik: Die Untersuchungen von Dr.-Ing. Trautwein—Einblick in ein Newland—Die Hall-Formanten-Theorie," *New Yorker Staatszeitung*, October 5, 1930, DMA, Friedrich Trautwein, Nachlass 187/114 (not numbered). See also "Elektrische Musikinstrumente," ibid., "Elektrische Musik: Ein völlig neues Musikinstrument auf der Grundlage des Funks, für den Funk," in *Funkschau: Neues vom Funk* (July 27, 1930): 74 in ibid., and Mersmann, "Dr. Trautweins elektrische Musik," 228–29, DMA, Friedrich Trautwein Nachlass 187/006.

75. Tresch and Dolan, "Toward a New Organology," 278–98.

76. Trautwein, *Elektrische Musik*, 13–15. See also Trautwein, "Über elektrische Analogien," 39.

77. Krigar-Menzel and Raps, "Über Saitenschwingungen," 623. As they did not use electricity, the vibrations were generated mechanically.

78. Trautwein, *Elektrische Musik*, 16–17, F. Trendelenburg, "Die physikalischen Eigenschaften" vol. 8, 461–62, and Weyer, "Probleme der Analyse und Synthese," 124.

79. Trautwein, *Elektrische Musik*, 8, 14–15.

80. Mersmann, "Dr. Trautweins Elektrische Musik," 229, DMA, Friedrich Trautweins Nachlass 187/006 and Friedrich Trautwein, "Memorandum über das Trautonium mit besonderer Berücksichtigung seiner Bedeutung für die Filmmusik," DMA, Friedrich Trautwein Nachlass 187/013, June 1, 1949, 6.

81. Trautwein, *Elektrische Musik*, 8–13. Meyer, "Die Klangspektren der Musikinstrumente," 606–11.

82. Friedrich Trautwein, "Über akustische Probleme aus Grenzgebieten der Musik, der Physik und der Physiologie," to Schünemann, not dated, circa 1932, Universität der Künste (henceforth UdK)-Archiv Berlin, Bestand 1b, Nr. 10 Technisches (Akustik), folia 314–316, 316.

83. Friedrich Trautwein, "Memorandum über das Trautonium mit besonderer Berücksichtigung seiner Bedeutung für die Filmmusik," DMA, Friedrich Trautwein Nachlass 187/013, June 1, 1949, insert between 6–7. See also Friedrich Trautwein, "Zur Ästhetik der Tonkunst im Hinblick auf elektronische Musik," DMA, Munich: Friedrich Trautwein Nachlass 187/014, 15.

84. Badge, *Oskar Sala*, not paginated. The original German *Stimmritze* was erroneously translated as vocal cords. See also Sala, "My Fascinating Instrument," 83.

85. As quoted in Badge, *Oskar Sala*, not paginated.

86. He mistakenly attributes the word "formant" to Helmholtz, who actually never used the word. Recall that the word was coined by Ludimar Hermann. Trautwein also mentioned the research of Carl Stumpf and Erwin Meyer. Trautwein, "Über elektrische Analogien," 39–43.

87. Sala, "Das Trautonium, ein Instrument der Zukunft," 162–63. See also Sala, "Das Trautonium: Begriff und Aufgabe," 25–26.

88. Trautwein, "Perspektiven der musikalischen Elektronik," 103–4.

89. Vierling, "Der Formantbegriff," 219–32.

90. Donhauser, *Elektrische Klangmaschinen*, 70.

91. F. Trendelenburg, "Die physikalischen Eigenschaften," vol. 8, 471–72. See also F. Trendelenburg, *Klänge und Geräusche*, 70–73.

92. Trautwein, "Über elektrische Analogien," 41.

93. Trautwein, "Über elektrische Analogien," 46.

94. For the program of the *Neue Musik* Berlin 1930, see DMA, Friedrich Trautwein Nachlass 187/006, and Sala, "50 Jahre Trautonium," 78, DMA, Oskar Sala Nachlass 218/0193. See also Schenk, *Die Hochschule*, 228–29.

95. Badge, *Oskar Sala*, unpaginated. See also Schünemann, "Neue Musik Berlin 1930," 122, and Frieß and Steiner, eds., "Oskar Sala im Gespräch," 221. To view a video of Paul Hindemith's "Seven Pieces for Three Trautoniums," composed for the Neu Musik Berlin festival at the Berlin Academy of Music in 1930, see the "Resources" tab at https://press.princeton.edu/books/broadcastingfidelity.

96. Band, "Rundfunk im Rahmen," 134.

97. Lion, "Elektrische Musik: Die Untersuchungen von Dr.-Ing. Trautwein" (1930), Friedrich Trautwein, Nachlass 187/114 (not numbered). See also Band, "Rundfunk im Rahmen," 133–34.

98. Trautwein as quoted by Mersmann, "Dr. Trautweins elektrische Musik," 229, also found in DMA, Friedrich Trautwein Nachlass 187/006.

99. Ibid.

100. Sala, "Neue Möglichkeiten," 137. See also DMA, Friedrich Trautwein Nachlass 187/006.

101. Band, "Rundfunk im Rahmen," 134.

102. Band, "Rundfunk im Rahmen," 134, and DMA, Friedrich Trautwein Nachlass 187/114 (not numbered).

103. "New Music Berlin," *World-Radio*, June 27, 1930, DMA, Friedrich Trautwein, Nachlass 187/114 (not numbered).

104. As quoted in Donhauser, *Elektrische Klangenmaschinen*, 73, original: Abendroth, "Neue Musik Berlin 1930," 724.

105. Ibid.

106. Stege, "Neue Musik Berlin 1930," 645.

107. As quoted in Donhauser, *Elektrische Klangenmaschinen*, 73, original: *Deutsche Allgemeine Zeitung*, June 21, 1930.

108. O. Vetter, "Elektrische Musik: Das Trautonium und seine beiden Virtuosen," *Berliner Morgenpost*, November 9, 1940, DMA, Oskar Sala Nachlass 218/0199/2.464.

109. As quoted in Donhauser, *Elektrische Klangenmaschinen*, 73, original: Holl, "Elektroakustische Musik," 826.

110. Donhauser, *Elektrische Klangenmaschinen*, 73, original: Lade, "Zweite Tagung für Rundfunkmusik," 284–86.

111. Oskar Sala, "50 Jahre Trautonium," 78. See also DMA, Oskar Sala Nachlass 218/0193, and Frieß and Steiner, eds., "Oskar Sala im Gespräch," 222. To view a video of a modern performance of Paul Hindemith's "Konzert für Trautonium and Streichorchester" (Concerto for Trautonium and String Orchestra), see the "Resources" tab at https://press.princeton.edu/books/broadcastingfidelity.

112. "Neues vom Hindemith," *Deutsche Allgemeine Zeitung*, November 20, 1931, Beilage [Supplement], "Das Unterhaltungsblatt," as cited in Donhauser, *Elektrische Klangmaschinen*, 102.

113. *Allgemeine Musikalische Zeitung* 58, 801, as cited in Donhauser, *Elektrische Klangmaschinen*, 102.

114. Winkelmann, *Das Trautonium.*

115. Funkhusen, "Das Trautonium des Bastlers," 385–86, and Saraga, "Das Trautonium," 241–45. See also Stephani, "Ein Trautonium neuerer Ausführung," 167–72.

116. Henry Cowell to Georg Schünemann, October 18, 1932, UdK Berlin Archiv, Bestand 1b, Nr. 10 (Technische Akustik), folio 217. See also Cowell to Schünemann, February 29, 1932, ibid., folio 291.

117. Denton, "The Trautonium: A New Musical Instrument," 522–24.

118. Gernsback, "Electronic Music," 521.

119. Gernsback, "Electronic Music," 521.

120. Gernsback, "Electronic Music," 521.

121. Gernsback, "Electronic Music," 521.

122. Théberge, *Any Sound You Can Image.*

123. Sala, "My Fascinating Instrument," 85. Listen to the first minute: "Stürme über dem Montblanc (1930) [Abenteuer] ganzer Film(eutsch)," see the "Resources" tab at https://press.uprinceton.edu/books/broadcastingfidelity.

124. As quoted in Donhauser, *Elektrische Klangmaschinen*, 82, original: Arthur Koestler, "Abendteuer der Musik," *Vossische Zeitung*, July 11, 1931. See also Badge, *Oskar Sala*, not paginated. On African American vocalists performing in the Weimar Republic, see Thurmann, *Singing Like Germans*, 97–158.

125. DMA, Oskar Sala Nachlass 218/0037/1949-51/145, handwritten resume, undated (most likely 1951) and DMA, Oskar Sala Nachlass 218/0205, 7. The archives state 1932 to 1935; however, some sources report 1932–36 and 1931–36. See, for example, Frieß and Steiner, eds., "Oskar Sala im Gespräch," 215. Sala's *Ersatz-Studienbuch I* states his matriculation dates: May 1, 1931 to the summer semester 1935/36. *Ersatz-Studienbuch I*, DMA, Oskar Sala Nachlass 218/0001, p. 2. He paid fees for the summer semester 1935/36; however, he did not take any courses, and he had a leave of absence (*beurlaubt*), *Ersatz-Studienbuch I*, DMA, Oskar Sala Nachlass 218/0001, 10.

126. Köhler worked on the physiology and psychology of tone color, although there is no evidence that that subject was part of the course. *Ersatz-Studienbuch I*, DMA, Oskar Sala Nachlass 218/0001, 1–9.

127. Donhauser, *Oskar Sala*, 23.

128. Donhauser, *Oskar Sala*, 23.

129. UdK-Archiv, Bestand 1b, Nr. 10 Technisches (Akustik), folia 275–278, 296–299. See also Donhauser, *Elektrische Klangmaschinen*, 119–22.

130. Donhauser, *Elektrische Klangmaschinen*, 116.

131. Donhauser, *Elektrische Klangmaschinen*, 117.

132. J. Winckelmann, "Das Orchester der Zukunft?" *Rundschau* 52 (1932), as cited in Donhauser, *Elektrische Klangmaschinen*, 119.

6: The Nazis and the Trautonium

1. Goebbels, *Deutsche Technik*, 105–6, as translated in Herf, *Reactionary Modernism*, 196.

2. Siemens, "National Socialism," 397.

3. For the relationship between Nazi ideology and technology as a whole, see Ludwig, *Technik und Ingenieure*, 15–102; Herf, *Reactionary Modernism*; Hortleder, *Das Gesellschaftsbild*; Maier, "Nationalsozialistische Technikideologie"; Jarausch, *The Unfree Professions*; and Voskuhl, "Engineering Philosophy."

4. Ludwig, *Technik und Ingenieure*, 44–102.

5. Ludwig, "Vereinsarbeit im Dritten Reich 1933 bis 1945," 430.

6. Weihe, "Spengler und die Maschine," 37–38. See also Herf, *Reactionary Modernism*, 172.

7. Weihe, "Geistige Sozialisierung, Technik und Volksbildung," 86–87, as translated in Hård, "German Regulation," 44.

8. Herf, *Reactionary Modernism*, 155–56.

9. Herf, *Reactionary Modernism*, 89. A more recent work has shown that engineers, as evidenced by their income and memberships in various organizations, were by and large members of the petty bourgeoisie and were content to be so: their increasingly radical critique of the Weimar Republic attracted them to the utopian visions of the Third Reich. Such ephemeral radicalism was fueled in large part by the republic's inchoate democracy and ineffective economic policies. See Sander, *Die doppelte Defensive*.

10. Jarausch, *The Unfree Professions*, 100.

11. Jarausch, *The Unfree Professions*, 133. See also McClelland, *The German Experience*, 177, and Dietz, Fessner, and Maier, "'Der Kulturwelt der Technik' als Argument," 8.

12. McClelland, *The German Experience*, 188.

13. On the *Gleichschaltung* of engineers, see Hortleder, *Das Gesellschaftsbild des Ingenieurs,* 116–21. See also Maier, "Nationalsozialistische Technikideologie," 253–68. For the *Gleichschaltung* of other professions, see, for example, Jarausch, *The Unfree Professions.*

14. Jarausch, *The Unfree Professions,* 150.

15. Ludwig, "Der VDI als Gegenstand," 411.

16. On July 20, 1932, Paul von Hindenburg, following the advice of Chancellor Franz von Papen, dissolved the Prussian parliament and gave von Papen control as *Reichskommissar.*

17. As reprinted in Pohle, *Der Rundfunk,* 132.

18. Pohle, *Der Rundfunk,* 132.

19. Lacey, *Feminine Frequencies,* 50.

20. Pohle, *Der Rundfunk,* 133.

21. Bericht der Reichs-Rundfunk-Gesellschaft, "Sprachpflege," 184.

22. Bericht der Reichs-Rundfunk-Gesellschaft, "Sprachpflege," 184. Siebs's *Deutsche Bühnenaussprache. Hochdeutsch,* originally published in 1898, was a pronunciation dictionary that became the standard for German *Hochdeutsch,* "high" or proper German. Similarly, Hans Lebede praised the BBC's earlier action to create a standard "BBC voice" and scolded the Reichs-Rundfunk-Gesellschaft for neglecting this important function of radio. Lebede, "Mundart und deutsche Hochsprache," 97–98. See also Tkaczyk, "Radio Voices," 98–99, and Tkaczyk, *Thinking with Sound,* 202–5.

23. There were also articles written during the Weimar Republic that stressed the importance of radio to the art of speaking, although standardization was rarely the theme. See, for example, Grunicke, "Die Sprachkunst im Rundfunk," 149–50. See also Kapeller, "Die 'Rede' im Rundfunk," 470.

24. Tkaczyk, "Radio Voices," 99.

25. Pohle, *Der Rundfunk,* 133.

26. Levi, *Music in the Third Reich,* 127.

27. The Nazis did not desire to eradicate, nor would they have been able to eradicate, major artistic trends of the Weimar Republic. It was easier and more effective to seize and repurpose those trends that served their sinister aims. See, for example, Potter, *The Art of Suppression.*

28. See, of example, Kolb and Siekmeier, eds., *Rundfunk und Film.*

29. Pohle, *Der Rundfunk,* 333.

30. Levi, *Music in the Third Reich,* 132.

31. Levi, *Music in the Third Reich,* 129.

32. Hadamovsky, *Der Rundfunk im Dienste,* 28, as translated in Levi, *Music in the Third Reich,* 129–30.

33. Eckert, *Der Rundfunk als Führungsmittel,* 176, as translated in Levi, *Music in the Third Reich,* 130.

34. As cited in Schütte, *Regionalität und Föderalismus,* 189, as translated in Levi, *Music in the Third Reich,* 130.

35. As Levi has argued, there certainly were important strands of reactionary attitudes against the liberalism of the musical scene during the Weimar Republic. They went on to become important resources upon which the Nazis could draw. See Levi, *Music in the Third Reich,* 3–5. On the fact that the Nazis did not ban all of the music during the Weimar Republic, but critically incorporated it, see Potter, *Art of Suppression,* 175–214.

36. Interestingly, Goebbels was initially not as antimodern as some of his colleagues. However, after the Allgemeiner Deutscher Musikverein Festival in Weimar in 1936, his stance completely changed. See Levi, *Music in the Third Reich,* 88–89.

37. Graener, "Aufklang," 1, as translated in Levi, *Music in the Third Reich,* 87.

38. Hitler's speech to the Cultural Conference of the Party Convention of Greater Germany, 1938, as quoted by Wulf, *Musik im Dritten Reich,* 298.

39. Friedrich Trautwein, "Gutachten Trautweins vom 22.10.1934 über einen Zeitungsartikel des Herrn Franz Rector," Universität der Künste (henceforth UdK)-Berlin Archiv, Bestand 1, Nr. 142, folia 1–6, here 3.

40. Ibid.

41. Sala, "Auf den Wegen II." See also Oskar Sala Nachlass 218/0172, folia 1–3, here 1 (1.432a).

42. His membership number was 1,774,684. Dvorak, "Friedrich Trautwein," 691. For the number of engineers joining the Nazi Party, see Herf, *Reactionary Modernism*, 197–98. See also Donhauser, *Elektrische Klangmaschinen*, 68. Apparently, Oskar Sala was astonished to see Trautwein in the RVS wearing a Sturmabteilung (SA) uniform. Donhauser, *Elektrische Klangmaschinen*, 327. See also Fischer-Defoy, *Kunst. Macht. Politik*, 44.

43. Dvorak, "Friedrich Trautwein," 691.

44. UdK-Berlin Archiv Bestand 1b, Nr. 11, folia 25–30, particularly folia 27–30.

45. UdK-Berlin Archiv Bestand 1b, Nr. 3, folia 24, 27, and 28.

46. Trautwein's letter to the Rundfunkversuchsstelle, April 4, 1933, UdK-Berlin Archiv Bestand 1b, Nr. 3, folia 22–23.

47. Friedrich Trautwein, "Gutachten Trautweins vom 22.10.1934 über einen Zeitungsartikel des Herrn Franz Rector," Universität der Künste (henceforth UdK)-Berlin Archiv, Bestand 1, Nr. 142, folia 1–6, here 5.

48. Friedrich Trautwein to Fritz Stein, March 2, 1935, UdK-Berlin, Archiv Bestand 1b, Nr. 9, folia 89–93.

49. Friedrich Trautwein, "Gutachten Trautweins vom 22.10.1934 über einen Zeitungsartikel des Herrn Franz Rector," Universität der Künste (henceforth UdK)-Berlin Archiv, Bestand 1, Nr. 142, folia 1–6, here 6.

50. Donhauser, *Elektrische Klangmaschinen*, 127.

51. *Bayerische Funk-Echo*, January 22, 1933, 2. As cited in Donhauser, *Elektrische Klangmaschinen*, 127.

52. *Bayerische Funk-Echo*, February 19, 1933, 20. As cited in Donhauser, *Elektrische Klangmaschinen*, 127.

53. "Zukunftsmusik?" *Vossische Zeitung*, Nr. 26, January 26, 1933, *Abendausgabe* (evening edition), as cited in Donhauser, *Elektrische Klangmaschinen*, 127.

54. This Bruno Kittel should not be confused with the Austrian-born member of the SS, who oversaw the liquidation of the Vilnius Ghetto in 1943.

55. Badge, *Oskar Sala*, not paginated. Included in the signatories was Sala's piano instructor. See also Frieß and Steiner, eds., "Oskar Sala im Gespräch," 226.

56. Letter from Department III, Radio (Abteilung III, Rundfunk) of the Propaganda Minister to the Dean of the Academy of Music, UdK-Berlin Archiv, Bestand 1b, Nr. 2, folio 88. See also the letter from Fritz Stein to the Minister of Science, Art, and Education of the People, July 13, 1933, UdK-Berlin Archiv, Bestand 1b, Nr. 2, Folio 47b.

57. Donhauser, *Elektrische Klangmaschinen*, 129 and Fischer, "Worte werden zu Bildern . . ." 163.

58. UdK-Berlin Archiv, Bestand 1b, No. 2, folia 9, 18, and 20–24.

59. Fritz Stein to Herrn Minister für Wissenschaft, Kunst und Volksbildung, September 12, 1934, UdK-Berlin Archiv, Bestand 1b, No. 2, folia, 22a-25, here 22b, 23, and 24.

60. Fritz Stein to Ludwig Schidmeier (Munich), July 15, 1935, UdK-Berlin Archiv, Bestand 1b, Nr. 9, folio 66.

61. Levi, *Music in the Third Reich*, 107.

62. Berten, "Paul Hindemith und die deutsche Musik," 539, as translated in Levi, *Music in the Third Reich*, 108.

63. Levi, *Music in the Third Reich*, 110.

64. Paul Zschörlich, *Deutsche Zeitung*, March 17, 1934, as translated in Levi, *Music in the Third Reich*, 111.

65. Heiber, *Joseph Goebbels*, 170, as cited in Levi, *Music in the Third Reich*, 112.

66. Haas, *Forbidden Music*, 214.

67. As quoted in Paulding, "Mathis der Maler—The Politics of Music," 108–9, as cited in Levi, *Music in the Third Reich*, 114. The opera had a naked soprano sing an aria in a bathtub. Apparently, Hitler was irate. See Haas, *Forbidden Music*, 123.

68. Skelton, *Paul Hindemith*, 114.

69. Dümling and Girth, eds., *Entartete Musik*, 168, and Levi, *Music in the Third Reich*, 114.

70. As quoted in Dümling, "Norm und Diskriminerung," 110.

71. Oskar Sala to Dr. Alfred Rubeli, April 29, 1971, Deutsches Museum (Munich) Archiv (henceforth DMA), Oskar Sala Nachlass 218/0038, 1–2, here 2.

72. Oskar Sala to Harald Genzmer, February 13, 1954 DMA, Oskar Sala Nachlass 218/0050/1954/115.

73. Brilmayer, "Das Trautonium und Oskar Sala," 109.

74. Donhauser, *Elektrische Klangmaschinen*, 132.

75. "Trautonium: Das 'vollkommene Musikinstrument,'" *Berliner Westen*, August 20, 1933, DMA, Friedrich Trautwein Nachlass 187/114 (not numbered).

76. Noack, "More Information," 591. See also Trautwein, *Trautonium-Schule*, 5.

77. Trautwein, *Trautonium-Schule*, 4–6. See also Germann, "Das Trautonium." Note that Germann's depiction uses a different numbering scheme of the parts.

78. Gradenwitz, "The Trautonium," 254.

79. The range of the violin setting was from the G string (G3) to three octaves higher, while the range of the viola setting was from the C string (C3) to three octaves higher, the cello setting from the C (C2) string to three octaves higher, and the flute setting from an octave higher than the viola setting to three octaves higher (C6). Trautwein, *Trautonium-Schule*, 8 and 16.

80. Trautwein, *Trautonium-Schule*, 11.

81. Trautwein, *Elektrische Musik*, 23. See also Patteson, *Instruments for New Music*, 120.

82. Trautwein, *Trautonium-Schule*, 4–6; Trautwein, *Elektrische Musik*, 23; and Patteson, *Instrument for New Music*, 120. For a technical description of the workings of the *Volkstrautonium*, see Kotowski and Germann, "Das Trautonium," 389–99. See also Donhauser, *Oskar Sala*, 35–43.

83. Trautwein, *Trautonium-Schule*, 15.

84. Germann, "Das Trautonium," 1–12.

85. Kämpfer, "Ein Leben," 46.

86. Donhauser, *Elektrische Klangmaschinen*, 137. See also Donhauser, *Oskar Sala*, 42.

87. In July 1931, the Reichsbank took Germany out of the gold standard. Approximately six million Germans were unemployed during the winter of 1932.

88. Friedrich Trautwein to Georg Schünemann, January 13, 1933, UdK-Berlin Archiv, Bestand 1b, Nr. 10 Technisches (Akustik), folia 166–167, here 166.

89. Donhauser, *Elektrische Klangmaschinen*, 137.

90. Trautwein, *Trautonium-Schule*. Trautwein detailed the mechanics of the instrument, particularly the electric circuitry. Sala explained how one played the instrument, and musical examples were provided by Hindemith. Sala later claimed that the book was "a flop." Badge, *Oskar Sala*, not paginated.

91. Donhauser, *Elektrische Klangmaschinen*, 135.

92. "Entwicklungsverträge mit Professor Dr. Friedrich Trautwein," Deutsches Technikmuseum Archiv (Berlin), I.2.060 C.03505, folia 16 and 35. See also Donhauser, *Elektrische Klangmaschinen*, 142. For the various annual contracts between Trautwein and Telefunken, see Deutsches Technikmuseum Archiv (Berlin), I.2.060 C.03505; C.01783, folia 50ff; C.01782, fol. 9 and C.01783, fol. 113.

93. F. E., "Vom Rundfunk zur elektrischen Hausmusik," 144.

94. Such a critique was leveled by the Nazi paper, *Der Angriff*. "We do not understand experimenting with our classics . . ." See "Trautonium und Elektrochord," *Der Angriff*, June 26, 1934, as quoted in Donhauser, *Elektrische Klangmaschinen*, 138. See also "Bach und Mozart auf elektrischen Instrumenten," *Deutsche Allgemeine Zeitung*, June 26, 1934, evening edition, 2, as cited in Donhauser, *Elektrische Klangmaschinen*, 138.

95. Donhauser, *Elektrische Klangmaschinen*, 137–38, original: "Trautonium und Elektrochord," *Der Angriff*, April 26, 1934. See also Disterweg, "Elektromusikalisches Konzert," 408.

96. Chávez, *Toward a New Music*, 25.

97. Chávez, *Toward a New Music*, 165.

98. Schünemann, "Produktive Kräfte," 248–49, as translated in Patteson, *Instrument for New Music*, 132.

99. Patteson, *Instrument for New Music*, 132.

100. Donhauser, *Elektrische Klangmaschinen*, 139.

101. He does thank Trautwein for his technical advice. See Oskar Sala, "Bericht über das neue Trautonium: seine Enstehung und seine Spieltecknik," February 1936, DMA, Oskar Sala Nachlass 218/2462, 2.

102. Ibid., 9–10.

103. DMA, Oskar Sala Nachlaß 218/0205, 4. Their work on subharmonics led to an acrimonious intellectual property dispute. See Donhauser, *Oskar Sala*, 44. See German Reich Patent (DRP) 692,241, granted on March 5, 1931, and DRP 674,890, granted on March 26, 1937.

104. Sala, "Experimentelle und theoretische Grundlagen," 315, 317. See also Sala, "My Fascinating Instrument," 79–80, and Frieß and Steiner, eds., "Oskar Sala im Gespräch," 217.

105. Oskar Sala, "Bericht über das neue Trautonium: seine Enstehung und seine Spieltecknik," February 1936, Oskar Sala Nachlass 218/2462, 8. On this important point, see Donhauser, *Oskar Sala*, 44–52. See also Oskar Sala to the Notgemeinschaft der Deutschen Wissenschaft, undated, Oskar Sala Nachlass 218/0037/1949-51/170 a-b, here a. It should be noted that Sala stopped working with Trautwein in 1937.

106. Manual is meant here in the musical sense, i.e., a keyboard, although it possessed a string, not keys. To view a video of Oskar Sala playing Paganini on the trautonium, see the "Resources" tab at https://press.princeton.edu/books/broadcastingfidelity.

107. Donhauser, *Elektrische Klangmaschinen*, 139. See also Sala, "Biografie," available at http://www.oskar-sala.de/oskar-sala-fond/oskar-sala/biografie/1933-1935/grossansicht-2/.

108. For a recording of a part of the performance, see Gagarin Records, *Historische Aufnahmen/Historical Recordings*, GR 2013, no. 1, 2009.

109. Weck, *Nur Einer kann es spielen*, available at https://www.srf.ch/audio/passage/nur-einer-kann-es-spielen-oskar-sala-meister-des-trautoniums?id=10144241.

110. Trautwein, "Dynamische Probleme," 35.

111. Oskar Sala, "Musik auf dem Trautonium: Veranstaltung der Fachgruppe 'Musik und Technik,'" Staatliche akademische Hochschule für Musik Berlin, June 11, 1936, DMA, Oskar Sala Nachlaß 218/0982.

112. To view a video of Paul Hindemith, "Langsames Stück und Rondo für Trautonium" (A Slow Piece and Rondo for the Trautonium), 1935, see the "Resources" tab at https://press.princeton.edu/books/broadcastingfidelity.

113. While a number of scholars claimed the piece was first performed on the radio trautonium discussed below, after extensive research, Donhauser believes that it is far more likely that it was the one performed on the "concert instrument," the very same one that Goebbels heard. Donhauser, *Oskar Sala*, 57. Sala himself claimed it was the radio trautonium; however, he often erred on a number of dates in later interviews. See Sala, "My Favorite Instrument," 80. See also Frieß and Steiner, eds., "Oskar Sala im Gespräch," 223.

114. As quoted in Ebbeke, "Paul Hindemith und das Trautonium," 96, reprinted in Schaal-Gotthardt and Schader, eds., *Über Hindemith*, 173. See also Schenk, *Die Hochschule*, 264.

115. Donhauser, *Elektrische Klangmaschinen*, 141–42. See also Sala, "My Fascinating Instrument," 76, and Frieß and Steiner, eds., "Oskar Sala im Gespräch," 216.

116. Oskar Sala, "Bericht über das neue Trautonium: seine Enstehung und seine Spieltecknik," February 1936, Oskar Sala Nachlass 218/2462, 25.

117. Sala, "50 Jahre Trautonium," 78. See also DMA, Oskar Sala Nachlass 218/0193.

118. Donhauser, *Elektrische Klangmaschinen*, 151–52.

119. Badge, *Oskar Sala*, not paginated.

120. "1. Städtische Musikveranstaltungen Duisburg 1936 / 1937, II. Hauptkonzert, 26.10.1936," in DMA, Friedrich Trautwein Nachlass 187 / 118. For the recording, see Harald Genzmer, "Trautonium-Konzerte," Oskar Sala, Trautonium, WER 6266–2, Germany, first track.

121. "Die Tonkünstlerversammlung in Weimar," *Neues Musikblatt* 15 (1936): 5, and Büttner, "Besinnung und Bildung," 952, as cited in Donhauser, *Elektrische Klangmaschinen*, 153.

122. Trautwein, "Wesen und Ziele," 698–99.

123. Trautwein, "Dynamische Probleme," 33–44, and DMA, Friedrich Trautwein Nachlass 187/010. See also Friedrich Trautwein, "Memorandum über das Trautonium: mit besonderer Berücksichtigung seiner Bedeutung für die Filmmusik," in DMA, Friedrich Trautweins Nachlass 187/013, 11–12, and 15.

124. Dvorak, "Friedrich Trautwein," 691.

125. Friedrich Trautwein, "Memorandum über das Trautonium: mit besonderer Berücksichtigung seiner Bedeutung für die Filmmusik," in DMA, Friedrich Trautweins Nachlass 187/013, 15. See also Trautwein, "Dynamische Probleme," 39–42, and Werner Trautwein to Siegfried Goslich, September 8, 1964, DMA, Friedrich Trautweins Nachlass 187/028, 1–3, here 2.

126. Friedrich Trautwein, "Memorandum über das Trautonium: mit besonderer Berücksichtigung seiner Bedeutung für die Filmmusik," DMA, Friedrich Trautweins Nachlass 187/013, 16.

127. Donhauser, *Elektrische Klangmaschinen*, 155–56.

128. There are several dates in the secondary literature for the origins of the radio trautonium, including 1934, 1935, 1936, and 1937. According to Sala, he received the contract from Berliner Rundfunk in 1937. Oskar Sala to Paul Hindemith, January 25, 1947, Oskar Sala Nachlass 218/0038, 1–2, here 1 a-b. In a subsequent interview with Sala, he claimed that the date of the order was 1935. See Sala, "Das Rundfunktrautonium," http://www.oskar-sala.de/oskar-sala-fonds/trautonium/rundfunktrautonium/index.html. The radio trautonium, which was too large to be used in concert performances, was housed in the Berlin radio station, Haus des Rundfunks. Oskar Sala to Paul Hindemith, January 25, 1947, DMA: Oskar Sala Nachlass 218/0038, 1–4, here 1 a-b. See also Frieß and Steiner, eds., "Oskar Sala im Gespräch," 217. For the engineering diagrams of the radio trautonium, see Donhauser, *Oskar Sala*, 62–70.

129. Sala, "Experimentelle und theoretische Grundlagen," 315.

130. For the first problem dealing with intellectual property issues between the two, see Friedrich Trautwein to Frau Dr. Sala, Sala's mother, July 2, 1937, in DMA, Oskar Sala Nachlass 218/0035/1937/002 a-d and Oskar Sala to Paul Hindemith, January 25, 1947, DMA, Oskar Sala Nachlass 218/0038, 1–4, here 1 a. See also Donhauser, *Oskar Sala*, 64–65.

131. Werner Trautwein to Siegfried Goslich, September 8, 1964, DMA, Friedrich Trautwein Nachlass 187/028, 1–3, here 2.

132. Werner Trautwein to Siegfried Goslich, September 8, 1964, DMA, Friedrich Trautwein Nachlass 187/028, 1–3, here 2. See also Donhauser, *Elektrische Klangmaschinen*, 141.

133. Oskar Sala, "50 Jahre Trautonium," 79, also found in DMA, Oskar Sala Nachlass 218/0193. See also Sala, "Experimentelle und theoretische Grundlagen des Trautoniums," 315.

134. Sala, "My Fascinating Instrument," 78.

135. Sala, "My Fascinating Instrument," 78.

136. Andreas Thiele, "Oskar Sala und das Trautonium. Examenarbeit," Examiners: K. Warnke, E. Maronn (1994), 45–46, DMA, Oskar Sala Nachlass 218/0437. See also Sala, "My fascinating Instrument," 78.

137. Sala, "Ein neues elektrisches Soloinstrument," 5. Donhauser, *Oskar Sala*, 62–70.

138. Sala, "Experimentelle und theoretische Grundlagen," 317, and Sala, "My Fascinating Instrument," 80.

139. Oskar Sala to Paul Hindemith, January 25, 1947, DMA, Oskar Sala Nachlass 218 / 0038, 1–3, here 2.

140. See Kämpfer, "Ein Leben," 46. See "Oskar Sala - Konzert-Trautonium im Wandel—1940–1950," https://www.youtube.com/watch?v=iVOjSnjyevI; Electrische piano: Trautonium (1941), https://www.youtube.com/watch?v=TQ4wGucalpc; and "Oskar Sala, Mixtur-Trautonium, 100th Anniversary 2010," https://www.youtube.com/watch?v=Hh8-qTjPV9g&t=156s.

141. Donhauser, *Elektrische Klangmaschinen*, 184.

142. Erwin Kroll, *Deutsche Allgemeine Zeitung Berlin*, January 21, 1939, and Kroll, "Rundfunk: H. Genzmers neues Trautonium-Konzert," *Deutsche Allgemeine Zeitung Berlin*, April 14, 1939, DMA, Oskar Sala Nachlass 218/0198/2.441 and 2.436 and anon., "Trautonium: Das 'vollkommene Musikinstrument,'" in *Berliner Westen*, DMA, Friedrich Trautwein Nachlass 187/114 (not numbered). See also Ludwig Josef Kaufmann, "Aus dem Stadttheater Cottbus: Anders-Genzmer-Bruckner," *Cottbuser Anzeiger*, April 18, 1939, DMA, Oskar Sala Nachlass 218 / 0198 / 2.442, and Hans Scherer, "Das Trautonium, ein neues Konzertinstrument," *Cottbuser Anzeiger*, April 19–20 1939, DMA, Oskar Sala Nachlass 218/0198/2.443.

143. In 1980, Sala referred to the instrument as the concert trautonium, although he admitted that it went through a number of modifications until the official premiere of the instrument on October 28, 1940 in Berlin. See Sala, "50 Jahre Trautonium," 80, also found in DMA, Oskar Sala Nachlass 218/0193. For the recording, see Harald Genzmer, "Trautonium-Konzerte," Oskar Sala, Trautonium, WER 6266–2, Germany, tracks 1–3.

144. Hugo Heurich, "Uraufführung auf dem Trautonium" (1939), *Hannoverscher Anzeiger* in DMA, Oskar Sala Nachlass 218/0198/2.781.

145. Heinz Joachim, "Technik-schöpferisch verwandelt: Harald Genzmers 'Konzert für Trautonium' uraufgeführt," *Berliner Lokalanzeiger*, April 12, 1939, DMA, Oskar Sala Nachlass 218/0198/2.437.

146. Hans Scherer, "Das Trautonium" (1939) in DMA, Oskar Sala Nachlass 218/0198/2.443.

147. Ernst Krienitz, "Genzmers Konzert für Trautonium und Orchester: Uraufführung im Deutschlandsender," *Die Musikwoche*, vol. 16, April 22, 1939, DMA, Oskar Sala Nachlass 218/0198/2.438.

148. Ibid.

149. Heinz Joachim, "Technik-schöpferisch verwandelt: Harald Genzmers 'Konzert für Trautonium' uraufgeführt," *Berliner Lokalanzeiger*, April 12, 1939 in DMA, Oskar Sala Nachlass 218/0198/2.437.

150. As quoted in Donhauser, *Elektrische Klangmaschinen*, 186.

151. Alfons Krüll, "Düsseldorfer Reichsmusiktage 1939," *Rheinische Landeszeitung*, May 21, 1939, DMA, Oskar Sala Nachlass 218/0198/2.439.

152. Ludwig Josef Kaufmann, "Anders-Genzmer-Bruckner," *Cottbuser Anzeiger*, April 18, 1939, DMA, Oskar Sala Nachlass 218/0198/2.442. See also Hans Scherer, "Das Trautonium, ein neues Konzertinstrument: Oskar Sala-Berlin spielt es am Donnerstag in Cottbus," *Cottbuser Anzeiger*, April 19–20, 1939, DMA, Oskar Sala Nachlass 218/0198/2.443, and Curt Loewe, "Das IV. Städtische Sinfoniekonzert: Neue Wege der Musikwiedergabe durch das Trautonium," *1. Beilage zum Cottbuser Anzeiger*, Nr 93, April 21, 1939 in DMA, Oskar Sala Nachlass 218/0198/3.214.

153. Hans Scherer, "Das Trautonium, ein neues Konzertinstrument: Oskar Sala-Berlin spielt es am Donnerstag in Cottbus," *Cottbuser-Anzeiger*, April 19–20, 1939 in DMA, Oskar Sala Nachlass 218/0198/2.443.

154. Curt Loewe, "Das IV. Städtische Sinfoniekonzert: Neue Wege der Musikwiedergabe durch das Trautonium," *1. Beilage zum Cottbuser Anzeiger*, Nr. 93, 21. April 1939 in DMA, Oskar Sala Nachlass 218/0198/3.214.

155. Karl Laux, *Dresdener Neueste Nachrichten* 1939, DMA, Oskar Sala Nachlass 218/0198 (not numbered).

156. Wilhelm Jung, "Ein neues Musikinstrument: Das Trautonium," *Leipziger Neueste Nachrichten*, August 29, 1940, Nr. 242, Edition A, DMA, Oskar Sala Nachlass 218/0199/2.446 and 2.447, Ludwig Lade, "Vom Trautonium," *Münchner Zeitung*, November 15, 1940, DMA, Oskar Sala Nachlass 218/0199/2.471, and Andreas Thiele, "Oskar Sala und das Trautonium. Examenarbeit" Examiners: K. Warnke, E. Maronn (1994), 9, DMA, Oskar Sala Nachlass 218/0437.

157. Donhauser, *Oskar Sala*, 71–78, here 71. See also Oskar Sala, "Klänge," available at http://www.oskar-sala.de/oskar-sala-fonds/oskar-sala/interview/klaenge/.

158. Badge, *Oskar Sala*, not paginated. See also Frieß and Steiner, eds., "Oskar Sala im Gespräch," 227.

159. Brilmayer, "Das Trautonium und Oskar Sala," 112.

160. "Einführungstext für Presse: Was ist ein Trautonium?" March 1942. DMA, Oskar Sala Nachlass 218/0173. For a description of later instantiations of the concert trautonium, see Sala, "Experimentelle und theoretische Grundlagen," 315–321.

161. This was similar to the pedal construction in the radio trautonium. See Sala, "My Fascinating Instrument," 78.

162. Sala, "Experimentelle und theoretische Grundlagen," 316–17.

163. Sala, "Experimentelle und theoretische Grundlagen," 315–16.

164. Sala, "Experimentelle und theoretische Grundlagen," 321.

165. Sala, "Experimentelle und theoretische Grundlagen, II. Teil," 13–19.

166. Wilhelm Jung, "Ein neues Musikinstrument: Das Trautonium," *Leipziger Neueste Nachrichten*, August 29, 1940, Nr. 242, Edition A, DMA, Oskar Sala Nachlass 218/0199/2.446 and 2.447.

167. Oskar Sala, "50 Jahre Trautonium," 80, and DMA, Oskar Sala Nachlass 218/0193.

168. Walter Steinhauser, "Zeitgenössiches unter Schuricht: Ein erfolgreiches Sonderkonzert der Philharmoniker," *Berliner Zeitung*, October 29, 1940, DMA, Oskar Sala Nachlass 218/0199/2.454. Hermann Killer also viewed the concert as an experiment. See Hermann Killer, "Symphoniekonzert der Gegensätze: Carl Schuricht in der Philharmonie," *Völkischer Beobachter Berliner Ausgabe*, October 30, 1940, DMA, Oskar Sala Nachlass 218/0199/2.456.

169. Alfred Burgartz, "Trautonium und Hamlet-Symphonie: Interessantes Novitäten-Konzert—Harald Genzmer und Boris Blacher," *Berliner Illustrierte Nachtausgabe*, October 29, 1940, DMA, Oskar Sala Nachlass 218/0199/2.453.

170. "Erläuterungen zum Programm der Konzerte von Oskar Sala und Harald Genzmer in Wien und Dresden," DMA, Oskar Sala Nachlass 218/0173; Gerhard Schultze, "Konzert auf dem Trautonium," *Völkischer Beobachter Berlin*, November 13, 1940, DMA, Oskar Sala Nachlass 218/0199/2.466; and O. Vetter, "Berliner Tagebuch: Das Trautonium gesiegt," *Zittauer Morgenzeitung*, November 16–17, 1940, DMA, Oskar Sala Nachlass 218/0199/2.474. See also Oskar Sala, "50 Jahre Trautonium," 80, also found in DMA, Oskar Sala Nachlass 218/0193.

171. "Erläuterungen zum Programm der Konzerte von Oskar Sala und Harald Genzmer in Wien und Dresden," DMA, Oskar Sala Nachlass 218/0173.

172. Ibid.

173. Gerhard Schultze, "Konzert auf dem Trautonium," *Völkischer Beobachter Berlin*, November 13, 1940, DMA, Oskar Sala Nachlass 218/0199/2.466. See also Oskar Sala, "Was ist ein Trautonium?" (1942) in DMA, Oskar Sala Nachlass 218/0173.

174. Gertrud Runge, "Musikalische Hexenkünste: Wunder des Trautoniums," *Deutsche Allgemeine Zeitung*, November 14, 1940, DMA, Oskar Sala Nachlass 218/0199/3.212. See also DMA, Friedrich Trautwein Nachlass 187/114 (not numbered).

175. Wilhelm Jung, "Ein neues Musikinstrument: Das Trautonium," in *Leipziger Neueste Nachrichten*, August 29, 1940, Nr. 242, Edition A, DMA, Oskar Sala Nachlass 218/0199/2.446 and 2.447. It should be noted that Sala also insisted that his trautonium was not meant to imitate

other instruments. Oskar Sala, "Was ist ein Trautonium?" (1942), DMA, Oskar Sala Nachlass 218/0173. See also Heinrich Funk, "Spiel auf dem Trautonium," *Thüringer Allgemeine Zeitung* (Erfurt), November 7, 1942, DMA, Oskar Sala Nachlass 218/0201/2.542.

176. Ludwig Lade, "Vom 'Trautonium,'" *Münchner Zeitung*, November 15, 1940, DMA, Oskar Sala Nachlass 218/0199/2.471. See also S[iefried] K[allenberg], "Das Trautonium: Ein neues Musikinstrument," *Münchner Neueste Nachrichten*, November 16, 1940, DMA, Oskar Sala Nachlass 218/0199/2.473. Recall that from 1930, Sala and Trautwein both underscored the fact that the trautonium did not simply imitate more conventional instruments, but could generate unique sounds. See DMA, Oskar Sala Nachlass 218/0173.

177. Wilhelm Zentner, "Trautonium—ein neues Instrument," *Neues Münchner Tagblatt*, November 20, 1940, DMA, Oskar Sala Nachlass 218/0199/2.481. See also Rudolf Hofmüller, "Oskar Salas Trautonium: Konzert mit dem neuen Instrument," *Völkischer Beobachter München*, November 20, 1940, DMA, Oskar Sala Nachlass 218/0199/2.479.

178. Fred Hamel, *Deutsche Allgemeine Zeitung*, October 29, 1940, DMA, Oskar Sala Nachlass 218/0199/2.490. See also DMA, Friedrich Trautwein Nachlass 187/114 (not numbered).

179. Walter Steinhauer, *Berliner Zeitung* am Mittag, October 29, 1940 in DMA, Oskar Sala Nachlass 218/0199/2.489 and Friedrich Trautwein Nachlass 187/114 (not numbered).

180. Heinz Joachim, "Neuer Klang auf neuem Instrument: Erstaufführung auf dem Trautonium," *Berliner Lokalanzeiger*, October 29, 1940, DMA, Oskar Sala Nachlass 218/0199 /2.452.

181. See, for example, Hermann Killer, "Symphoniekonzert der Gegensätze: Carl Schuricht in der Philharmonie," in *Völkischer Beobachter Berliner Ausgabe*, October 10, 1940, DMA, Oskar Sala Nachlass 218/0199/2.456.

182. See, for example, Lothar Band, "Ein elektrisches Instrument in der Philharmonie," *BVZ Berlin*, November 2, 1940, Ludwig Lade, "Vom 'Trautonium,'" *Münchener Zeitung*, November 15, 1940, O. Vetter, "Berliner Tagebuch: Das Trautonium hat gesiegt," *Zittauer Morgenzeitung*, November 16–17, 1940, DMA, Oskar Sala Nachlass 218/0199/2.491/2.471 and 2.474.

183. See, for example, Heinrich Edelhoff, "Heimisches Kulturleben: Das Trautonium-Konzert," *Mitteldeutsche National-Zeitung*, January 1942, DMA, Oskar Sala Nachlass 218/0201 (not numbered).

184. Heinz Kirschninck, "Das Trautonium klingt: Oskar Sala führte es gestern vor Greifswald," *Greifswalder Zeitung*, December 2, 1940, DMA, Oskar Sala Nachlass 218/0199/2.486.

185. Lothar Band, "Ein elektrisches Instrument in der Philharmonie," *BVZ Berlin*, November 2, 1940. DMA, Oskar Sala Nachlass 218/0199/2.491. Similarly, Franz Hauschild emphasized that the trautonium "was not some mere mechanical contraption." Hauschild, "Trautonium, virtuos gemeistert," *Greizer Zeitung und Tageblatt*, December 3, 1941, DMA, Oskar Sala Nachlass 218/0200/2.520.

186. "Das 'Trautonium': Instrument der tausend Möglichkeiten," *Münchner Abendblatt*, November 15, 1940, HFM, "Das Trautonium: Ein neues Instrument," *Völkischer Beobachter*, November 14, 1940, A. Würz, "Das Trautonium," *Münchner Abendblatt*, November 19, 1940, Siegfried Kallenberg, "Neue Klänge," *Münchner Neueste Nachrichten*, November 20, 1940, DMA, Oskar Sala Nachlass 218/0199/2.467 and 2.469/2.468/2.477 and 2.480. See also "Auszug aus den Pressestimmen zum 1. "Konzert auf dem Trautonium," im Herkulessaal in München am 18.11.1940," in ibid., 2.759.

187. O. Vetter, "Berliner Tagebuch: Das Trautonium hat gesiegt," *Zittauer Morgenzeitung*, November 16–17, 1940, DMA, Oskar Sala Nachlass 218/0199/2.474.

188. Hans Jenkner, ". . . hat seine zwei Saiten: Oskar Sala breichtet von Trautonium," *Der Angriff*, November 26, 1940. DMA, Oskar Sala Nachlass 218/0199/2.484.

189. A. Würz, "Das Trautonium," *Münchner Abendblatt*, November 19, 1940, DMA, Oskar Sala Nachlass 218/0199/2.477.

190. Rudolf Hofmüller, "Oskar Salas Trautonium: Konzert mit dem neuen Instrument," *Völkischer Beobachter München*, November 20, 1940. DMA, Oskar Sala Nachlass 218/0199/2.479.

191. Erwin Bareis, "Trautonium-Konzert," January 15, 1941, DMA, Oskar Sala Nachlass 218/0200/2.501.

192. Alexander Eisenmann, "Nochmals das Trautonium," *Württembergische Zeitung* (Stuttgart), January 13, 1941, DMA, Oskar Sala Nachlass 218/200/2.495. See also *Neues Stuttgarter Tagblatt*, DMA, Oskar Sala Nachlass 218/0200/2.570.

193. Weickert, "Neue Darbietungen auf dem Trautonium," 22. See also Donhauser, *Elektrische Klangmaschinen*, 190.

194. Sala, "Experimentelle und theoretische Grundlagen," 315.

195. Utrechtsche Muziekreferent, "Utrechtsch Stedeljik Orchest," *Niewe Rotterdamsche Crt*, February 24, 1941, DMA, Oskar Sala Nachlass 218/0199/2.761. See also DMA, Friedrich Trautwein Archiv 187/114 (not numbered). On February 23 and 25, 1941, General Music Director Fritz Zaun served as guest conductor to the Utrecht City Orchestra and to the Radio Orchestra in the Hilversum Broadcasting Station for Harald Genzmer's Dutch premiere. See DMA, Oskar Sala Nachlass 218/0200/2.761.

196. *Utrechtsch Nieuwsblad* and *Het Centrum Utrecht*, DMA, Oskar Sala Nachlass 218/0200/2.761. To view a video of Oskar Sala on the Konzerttrautonium in the Netherlands in 1942, see the "Resources" tab at https://press.princeton.edu/books/broadcastingfidelity.

197. "Maggio Musicale Fiorentino: Il 'Trautonium' al Verdi," *La Nazione Florenz*, May 18–19, 1941, DMA, Oskar Sala Nachlass 218/ 0200/2.762. See also DMA, Friedrich Trautwein Nachlass 187/114 (not numbered).

198. Frank Wohlfahrt, "Trautonium zum erstenmal in Florenz: Oscar [sic] Sala auf dem Maggio Musicali," *Berliner Zeitung*, May 20, 1941, DMA Oskar Sala Nachlass 218/0200/2.504.

199. See, for example, Kurt Mandel, "Zauberklänge auf dem Trautonium," *Der Oberschlesische Kurier*, November 21, 1941, and Werner Hübschmann, "Sinfoniekonzert in Beuthen," *Ostdeutsche Morgenpost*, November 21, 1941, DMA, Oskar Sala Nachlass 218/0200/2.509.

200. Oskar Sala, "50 Jahre Trautonium," 80, and DMA, Oskar Sala Nachlass 218/0193.

201. Theodor Staar, "Von Bruckner zum Trautonium," *Kurier Tageblatt*, November 20, 1941, and Erich Limmert, "Bruckner und das Trautonium," *Hannoverscher Anzeiger*, November 29–30, 1941, DMA, Oskar Sala Nachlass 218/0200/2.512.

202. Franz Hauschild, "Trautonium, virtuos gemeistert," *Greizer Zeitung und Tageblatt*, December 3, 1941, DMA, Oskar Sala Nachlass 218/0200/2.520.

203. Otto Witzke, "Ein Abend unter dem Zauber des Trautoniums," *Thüringer Gauzeitung*, December 3, 1941, DMA, Oskar Sala Nachlass 218/0200/2.518.

204. Horst Garbe, "Interessant und ergreifend: Das 3. Symphoniekonzert des Städtischen Orchesters für KdF," *Wilhelmshavener Zeitung*, December 4, 1941. DMA, Oskar Sala Nachlass 218/0200/3.215.

205. Hanns Reich, "Junge deutsche Musik: Erstes Konzert im Arbeitskreis für neue Musik," *Strassburger Neueste Nachrichten*, December 10, 1941, DMA, Oskar Sala Nachlass 218/0200/2.516.

206. Friedrich Herzfeld, "Japanische Klanggeheimnisse in Berlin," *Deutsche Allgemeine Zeitung Berlin*, January 3, 1942, DMA, Oskar Sala Nachlass 218/0201/2.765.

207. Oskar Sala, "50 Jahre Trautonium," 80, and DMA, Oskar Sala Nachlass 218/0193.

208. Oskar Sala, "50 Jahre Trautonium," 80, and DMA, Oskar Sala Nachlass 218/0193. Recall that the ondes Martenot was an early electric musical instrument invented in 1928 by Maurice Martenot.

209. Stuckenschmidt, "Die Ordnung der Freiheit," (1961), republished in H. H. Stuckenschmidt, *Die Musik eines halben Jahrhunderts*, 193.

210. See the various reviews of the concerts in DMA, Oskar Sala Nachlass 218/0202, particularly 2.773, *Ujság*, March 29, 1942.

211. Heinrich Funk, "Spiel auf dem Trautonium," *Thüringer Allgemeine Zeitung*, November 7, 1942 in DMA, Oskar Sala Nachlass 218/0201/2.542.

212. J. Wellenreuther, "Trautonium—was ist denn das?" in *Fuldaer Zeitung*, November 19, 1942, DMA, Oskar Sala Nachlass 218/0201/2.544. Wilhelm Hambach too felt that the instrument possessed a strong musical personality and a soul. Wilhelm Hambach, "Instrument der 'stählernen Romantik': Das Trautonium in dritten städtlichen Simfoniekonzert." *Mitteldeutsche National-Zeitung* (Halle) January 17, 1942, DMA, Oskar Sala Nachlass 218/0201 /2.529.

213. E. Freund, "Musikbericht: Das Trautonium, ein neues Musikinstrument," in *Oberhessische Zeitung Marburg*, November 26, 1942, DMA, Oskar Sala Nachlass 218/0201/2.548.

214. J. Wellenreuther, "Ein neuer Klang im Konzertsaal: Oskar Sala und Harald Genzmer musizierten in Fulda," *Amtliches Kreisblatt*, November 23, 1942, DMA, Oskar Sala Nachlass 218/0201/2.543.

215. Martin Koegel, "Oskar Sala musiziert auf dem Trautonium," *Braunschweigische Landeszeitung*, November 26, 1942, DMA, Oskar Sala Nachlass 218/0201/2.564.

216. Willi Wöhler, "Musik auf elektro-akustischem Wege: Eine neue Epoche des Musikinstrumentenbaues? Vorführung in der Niedersächsischen Musikgesellschaft Braunschweig," *Tageszeitung Braunschweig*, November 26, 1942, DMA, Oskar Sala Nachlass 218/0201/2.549.

217. See, for example, DMA, Oskar Sala Nachlass 218/0201/2.550, 2.551, 2.552, 3.073, 3.071, 3.072, 3.074, 3.076.

218. Franz Clemens Gieseking, "Begegnung mit Oskar Sala: Eine MZ-Besprechung mit dem Meister des Trautoniums zu seinem Konzert am 19. Januar in der Aula der Hermann-Löns-Schule," *Münsterische Zeitung*, January 19, 1943, DMA, Oskar Sala Nachlass 218/0203/2.571.

219. Max Thomas Heinrich, "Konzert auf dem Trautonium," *Vogtländische Anzeiger* (Plauen), February 22, 1943 and Kurt Spindler, "Oskar Sala auf dem Trautonium," *NS-Tageszeitung* (Plauen), February 22, 1943, DMA, Oskar Sala Nachlass 218/0203/2.562.

220. "Konzertzeit 1942/43 ging zu Ende: Begegnung mit dem Trautonium," *Westfälischer Beobachter (Gelsenkirchener Zeitung)*, May 8, 1943, DMA, Oskar Sala Nachlass 218/0203/1.494.

221. Donhauser, *Elektrische Klangmaschinen*, 191.

222. Frieß und Steiner, eds., "Oskar Sala im Gespräch," 225; Sala, "Historische Übersicht," 94–96; and Donhauser, *Elektrische Klangmaschinen*, 190–91.

223. Oskar Sala to Paul Hindemith, January 25, 1947, DMA, Oskar Sala Nachlass 218/0038, 1–3, here 2.

224. Wilhelm Hambach, "Instrument der 'stählernen Romantik': Das Trautonium in dritten städtlichen Simfoniekonzert," *Mitteldeutsche National-Zeitung* (Halle), January 17, 1942, DMA, Oskar Sala Nachlass 218/0201/2.529. See also Donhauser, *Elektrische Klangmaschinen*, 189.

225. Ibid. See also Patteson, *Instruments for New Music*, 133–41, and Donhauser, *Elektrische Klangmaschinen*, 190. Goebbels first used the phrase "steely Romanticism" in a speech to the Berlin Philharmonic in 1933.

226. Patteson, *Instruments for New Music*, 6.

227. Fritz Skorzeny, "Das Trautonium als Konzertinstrument," *Neues Wiener Tagblatt*, April 11, 1942, DMA, Oskar Sala Nachlass 218/0201/1.4689. See also Donhauser, *Elektrische Klangmaschinen*, 189.

228. As cited in "Einführungstext für Presse: Was ist ein Trautonium?" March 1942, originally in a review in V. B. München, DMA, Oskar Sala Nachlass 218/0173 (not numbered). See also Otto Steinhagen, "Schuricht dirigiert: Vorführung des verbesserten Trautoniums," *Berliner Börsenzeitung*, October 29, 1940, DMA, Oskar Sala Nachlass 218/0199/2.450, and Hermann Killer, "Symphoniekonzert der Gegensätze: Carl Schuricht in der Philharmonie," *Völkischer Beobachter Berliner Ausgabe*, October 30, 1940, DMA, Oskar Sala Nachlass 218/0199/2.456. See also *Dresdner Neueste Nachrichten*, March 23, 1942, DMA, Oskar Sala Nachlass 218/0173 (not numbered), and the *Völkischer Beobachter* review of the concert featuring Sala playing Genzmer

in the Herkulessaal in Munich on November 18, 1940, DMA, Friedrich Trautwin Nachlass 187/114 (not numbered).

229. O. V. [Vetter], "Elektrische Musik: Das Trautonium und seine beiden Virtuosen," *Berliner Morgenpost*, November 9, 1940, DMA, Oskar Sala Nachlass 218/0199/2.464.

230. Werner Kobes, "Bei Oskar Sala und seinem Trautonium," *Altenburger Zeitung für Stadt und Land*, January 26, 1942, DMA, Oskar Sala Nachlass 218/0201/2.522. See also "Personalfragebogen, Magistrat der Stadt Berlin," March 8, 1948, DMA, Oskar Sala Nachlass 218/0001/1949-51/189 a-d, here b.

231. Oskar Sala to Mr. Philipp, March 3, 1947, DMA, Oskar Sala Nachlass 218/0036 (not numbered). See also Oskar Sala to Paul Hindemith, January 25, 1947, DMA, Oskar Sala Nachlass 218/0038, 1–2, here 1 a.

7: The Trautonium after the War

1. Oskar Sala to Paul Hindemith, January 25, 1947, Deutsches Museum (Munich) Archiv (henceforth DMA), Oskar Sala Nachlass 218/0038, 1–2, here 1 a.

2. Werner Trautwein to Oskar Sala, August 31, 1948, DMA, Oskar Sala Nachlass 218/0037/1949-51/ 425 a-b and 426 a-b and August 25, 1948, 427 a-b. See also, "Das erste Turmmusik-Spiel," *Rhein-Neckar-Zeitung*, June 26, 1948, DMA, Oskar Sala Nachlass 218/0037/1949-1951/428 a-b and DMA, Friedrich Trautwein Nachlass 187/021.

3. Werner Trautwein to Siegfried Goslich, September 8, 1964, DMA, Friedrich Trautwein Nachlass 187/028, 1–3, here 2–3. See also Dvorak, "Friedrich Trautwein," 691.

4. Friedrich Trautwein, "Memorandum über das Trautonium mit besonderer Berücksichtigung seiner Bedeutung für die Filmmusik," DMA, Friedrich Trautwein Nachlaß 187/013, June 1, 1949, 16–17.

5. Ibid., 19–20.

6. Werner Trautwein to Siegfried Goslich, September 8, 1964, DMA, Friedrich Trautwein Nachlass 187/028, 1–3, here 3.

7. JBA, "Der Komponist der Zukunft ein Techniker? Professor Trautweins 'gegenstandslose' Malerei in Tönen—Das musikalische Brot der Vielen," *Der Mittag* (Düsseldorf), February 25, 1954, DMA, Oskar Sala Nachlass 218/0211/1.443. See also S-b, "Junge Tonmeister bewährten sich: Schüler Professor Trautweins betreuten Niederrhein-Film," *Neue Rhein Zeitung* (Düsseldorf), February 25, 1954, DMA, Oskar Sala Nachlass 218/0211/1.400.

8. Trautwein, "Das Elektronische Monochord," 345–52.

9. Donhauser, *Elektrische Klangmaschinen*, 226–28. He also built a "Radio-piano," or "Radioklavier," a keyboard instrument with vibrating metal reeds.

10. Oskar Sala to Friedrich Trautwein, June 6, 1953, DMA, Nachlass Oskar Sala 218/0048/1953/316 a-b, here b.

11. DMA, Friedrich Trautwein Nachlass 187/025 and DMA, Oskar Sala Nachlass 218/0048/1953/333. For the correspondence between the various attorneys from 1951 to 1953, see DMA, Oskar Sala Nachlass 218/0039 and 218/0045. One of the earliest letters we have on the topic is from Friedrich Trautwein to Oskar Sala's mother, the singer Annemarie, dated July 2, 1937 in DMA, Oskar Sala Nachlass 218/0035/1937/002 a-d.

12. Oskar Sala to Dr. Fromm (attorney), December 13, 1953, DMA, Oskar Sala Nachlass 218 / 0046 / 1953 / 041.

13. Oskar Sala to Friedrich Trautwein, June 6, 1953, DMA, Oskar Sala Nachlass 218/00481953/316 a-b, here a.

14. Oskar Sala to Siegfried Goslich, January 7, 1954, DMA, Oskar Sala Nachlass 218/0046/1953/047. The planned lecture series was titled "Music and Technology."

15. Trautwein, "Elektronische Klangerzeugung und Musikästhetik," 180, also found in DMA, Friedrich Trautwein Nachlass 187/009. See also Friedrich Trautwein, "Zur Ästhetik der

Tonkunst im Hinblick auf elektronische Musik," DMA, Friedrich Trautwein Nachlaß 187/014 (not dated, circa 1954), 1–20, here 1 and 18 and Trautwein, "Prospektiven der musikalischen Elektronik," 103–4, also found in Friedrich Trautwein Nachlass 187/010.

16. Trautwein, "Elektronische Klangerzeugung," 181.

17. Friedrich Trautwein, "Zur Ästhetik der Tonkunst im Hinblick auf elektronische Musik," DMA, Friedrich Trautwein Nachlass 187/014 (circa 1954), 2. Trautwein, "Besinnung auf das Gehör," 93, also found in DMA, Friedrich Trautwein Nachlass 187/009, and Trautwein, "Elektronische Klangerzeugung und Musikästhetik," 180.

18. Friedrich Trautwein, "Zur Ästhetik der Tonkunst im Hinblick auf elektronische Musik," DMA, Friedrich Trautwein Nachlass 187/014 (circa 1954), 2.

19. Trautwein, "Besinnung auf das Gehör," 93, also found in DMA, Friedrich Trautwein Nachlass 187/009.

20. Friedrich Trautwein, "Zur Ästhetik der Tonkunst im Hinblick auf elektronische Musik," DMA, Friedrich Trautwein Nachlass 187/014 (circa 1954), 3.

21. As quoted in Donhauser, *Elektrische Klangmaschinen*, 226, original: radio broadcast of "elektrische Musik" (Electric Music) on Radio Bremen on February 18, 1953.

22. Friedrich Trautwein, "Zur Ästhetik der Tonkunst im Hinblick auf elektronische Musik," DMA, Friedrich Trautwein Nachlass 187/014 (circa 1954), 3–4.

23. Ibid., 4.

24. As translated in Jarausch, *The Unfree Professions*, 161, original, Bundesarchiv (Koblenz) NS 14, No. 8.

25. Hårt, "German Regulation," 40.

26. As quoted in Herf, *Reactionary Modernism*, 60–61.

27. Herf, *Reactionary Modernism*, 169.

28. Voskuhl, "Engineering Philosophy," 721–52. As for natural scientists as *Kulturträger*, one thinks of Helmholtz, Planck, and Einstein.

29. Ibid., and Hårt, "German Regulation," 33–67.

30. Friedrich Trautwein, "Mensch und Maschine in der Tonkunst: Vortrag in der Volkhochschule Düsseldorf am 30. 11. 1956," DMA, Friedrich Trautwein Nachlass 187/015, 3–4, emphasis in the original.

31. Ibid., 4, emphasis in the original.

32. Herf, *Reactionary Modernism*, 1. See also Dietz, Fessner and Maier, "Der 'Kulturwert' der Technik," 22–26.

33. On José Ortega y Gasset, see Steege, *An Unnatural Attitude*, 69–93.

34. Ortega y Gasset, *Betrachtungen über die Technik*, 105. See also Mitcham, *Thinking through Technology*, 48–49.

35. Friedrich Trautwein, "Mensch und Maschine in der Tonkunst: Vortrag in der Volkshochschule Düsseldorf am 30. 11. 1956," DMA, Friedrich Trautwein Nachlass 187/015, 1–16, here 5. Clearly the quote is not the most admirable or elegant way of disputing Ortega y Gasset's point.

36. Herf, *Reactionary Modernism*, 159.

37. See also Dietz, Fessner, and Maier, "Der 'Kulturwert der Technik,'" 29–30.

38. Hortleder, *Das Gesellschaftsbild des Ingenieurs*, 139–43.

39. Hortleder, *Das Gesellschaftsbild des Ingenieurs*, 143–44. See also Verein deutscher Ingenieure, ed., *Der deutscher Ingenieur*.

40. Hortleder, *Das Gesellschaftsbild des Ingenieurs*, 143–45.

41. Hortleder, *Das Gesellschaftsbild des Ingenieurs*, 145–49.

42. Hortleder, *Das Gesellschaftsbild des Ingenieurs*, 157.

43. Hortleder, *Das Gesellschaftsbild des Ingenieurs*, 157–58, original: Koeßler, "Das Gespräch über die Technik," 70.

44. Oskar Sala to Werner Meyer-Eppler, October 22, 1951, DMA, Oskar Sala Nachlass 218/0042 (not numbered), 1–3, here 2.

45. Oskar Sala to Werner Meyer-Eppler, November 8, 1951, DMA, Oskar Sala Nachlass 218/0037 (not numbered), 1–2, here 2.

46. Friedrich Trautwein to Oskar Sala, October 27, 1954, DMA, Oskar Sala Nachlass 218/0052/1954/362 a-b, here a.

47. Friedrich Trautwein, "Elektro-akustische Mittel in der aktiven Tonkunst, (Vorläufige Inhaltsangabe)" March 1953, DMA, Oskar Sala Nachlass 218/0048/1953/329 a-c here a. He uses the terms "electric and electronic musical instruments." By "economic advantages" he means replacing numerous musicians with electronic musical instruments.

48. Mulvin, *Proxies*.

49. Oskar Sala to Mr. Philipp, March 3, 1947, DMA, Oskar Sala Nachlass 218/0036 (not numbered) and Oskar Sala to Paul Hindemith, January 25, 1947, DMA, Oskar Sala Nachlass 218/0038, 1–2, here 1 a. See also DMA, Oskar Sala Nachlass 218/0437.

50. Oskar Sala to Paul Hindemith, January 25, 1947, DMA, Oskar Sala Nachlass 218/0038, 1–2, here 1 a.

51. Julius Weismann to Mr. [Siegfried] Goslich, November 14, 1947, DMA, Oskar Sala Nachlass 218/0036/1947/001 and Ursel Küppers-Weismann to Oskar Sala, November 24, 1947, ibid., 002.

52. To view a video of Oskar Sala performing the Konzerttrautonium for the Landessendung Weimar, 1946, see the "Resources" tab at https://press.princeton.edu/books/broadcasting fidelity.

53. Oskar Sala, "Vorlesungsplan: Musikalische Akustik und ihre allgemeinen Grundlagen," April 15, 1947, DMA, Oskar Sala Nachlass 218/0001.

54. Siegfried Goslich, "Bescheinigung," Mitteldeutscher Rundfunk Landessender Weimar, May 27, 1948, DMA, Oskar Sala Nachlass 218/0001/1949-51/215 a (dated June 22, 1948).

55. Oskar Sala to Paul Hindemith, January 25, 1947, DMA, Oskar Sala Nachlass 218/0038, 1–2, here 1 b.

56. Donhauser, *Oskar Sala*, 70.

57. "Anmeldung bei der polizeilichen Meldebehörde," DMA, Oskar Sala Nachlass 218/0001/1949-21/188 a, dated June 19, 1948. He lived at Hektorstrasse 4, Berlin-Halensee.

58. Playbill of *Jeanne d'Arc auf dem Scheiterhaufen*, Arthur Honegger and Paul Claudel, Städtische Oper Berlin, December 9, 1947, DMA, Oskar Sala Nachlass 218/0931. He received an honorarium of 1,000 RM. Letter from Diederichs of the City Opera of Berlin to Oskar Sala, December 15, 1947, DMA, Oskar Sala Nachlass 218/0037/1949-5/409 and 410a. See also Robert Heger to Oskar Sala, October 22, 1947, DMA, Oskar Sala Nachlass 218/0037/1949-51/411 a, DMA, Oskar Sala Nachlass 218/0204, and Sala, "My Fascinating Instrument," 78–79.

59. Stuckenschmidt, "Die Ordnung der Freiheit," (1961), republished in Stuckenschmidt, *Die Musik eines halben Jahrhunderts*, 193.

60. Oskar Sala, "Zur 'Faust'-Musik," Playbill of *Faust der Tragödie Erter Teil, Deutsches Theater* in DMA, Oskar Sala Nachlass 218/0420.

61. Ernst Koster, "Das Trautonium: Interview mit einem neuen Instrument in der Staatsoper," *Hamburger Freie Presse*, January 27, 1950, DMA, Oskar Sala Nachlass 218/0206/1.476.

62. Dr. H. Weiher-Waege, "Johanna auf dem Scheiterhaufen," *Hamburg Freie Presse*, 23. January 1950, DMA, Oskar Sala Nachlass 218/0206/1.475.

63. Hans Hauptmann, "'Johanna auf dem Scheiterhaufen': Beispielhafte Hamburger Erstaufführung des Oratoriums von Claudel und Honegger," *Hamburger Allgemeine Zeitung*, January 23, 1950 in DMA, Oskar Sala Nachlass 218/0206/2.584.

64. See, for example, DMA, Oskar Sala Nachlass 218/0036/1948/E 003a-039a; 1949/E 018a-051 a and DMA, Oskar Sala Nachlass 218/0037/1949-51/445. See also Siegfried Goslich,

"Bescheinigung," Mitteldeutscher Rundfunk Landessender Weimar, May 27, 1948, DMA, Oskar Sala Nachlass 218/0001/1949-51/215 a, dated June 22, 1948. See also Kämpfer, "Ein Leben für das Trautonium," 46; Sala, "Mixtur-Trautonium" *Melos* (1950), 249; and Sala, "50 Jahre Trautonium," 80, also found in DMA, Oskar Sala Nachlass 218/0193.

65. Sala, "Das Trautonium," *Der Rundfunk*: 21. See also DMA, Oskar Sala Nachlass 218/0176. Other articles propagandizing the trautonium include Sala, "Mixtur-Trautonium," *Melos* (1950), 247–51; Sala, "Das Trautonium, ein Instrument der Zukunft," 162–63; and Sala, "Das Trautonium: Begriff und Aufgabe," 25–28.

66. Sala, "Das Trautonium," *Der Rundfunk*, 21. See also DMA, Oskar Sala Nachlass 218/0176.

67. Friedrich Trautwein to Oskar Sala, May 29, 1948, DMA, Oskar Sala Nachlass 218/0036/1948/017 a and b, here b.

68. Friedrich Trautwein to Oskar Sala, August 25, 1948, DMA, Oskar Sala Nachlass 218/0036/1948/022 a and b, here a.

69. Oskar Sala to Werner Egk, not dated, but certainly 1948, DMA, Oskar Sala Nachlass 218/0036/1949-1951/042 a-b.

70. Oskar Sala, "Elektro-akustisches Musikstudio," July 9, 1948, DMA, Oskar Sala Nachlass 218/0036/1949-1951/043 a-b. In 1943 Julius Weismann composed "Variationen und Fugue mit Orchester" for the concert trautonium. Other composers for the instrument included Hermann Ambrosius, Joseph Ingenbrand, Georg Häntzschel (Haentzschel), and Gustav Adolf Schlemm. See Oskar Sala to Paul Hindemith, January 25, 1947, DMA, Oskar Sala Nachlass 218/0038, 1–2, here 2 b.

71. "Protokoll der 1. Sitzung der Arbeitsgruppe für Elektro-Akustik," DMA, Oskar Sala Nachlass 218/0036/1949-1951/044 a-e, here b, undated, but definitely 1948, before July 9.

72. Ibid., 044 e.

73. *Nordwestdeutscher Rundfunk* broadcast in West Berlin from September 22, 1945 to December 31, 1955.

74. Letters from Hans-Martin Majewski to Oskar Sala, May 1 and 26, 1949, DMA, Oskar Sala Nachlass 218/0037/1949-1951/167 and 168. See also Werner Fiedler, "'Dergleichen hab' ich nie vernommen: Neuer Versuch des Faust-Dramaturgie," in *Der Tag*, April 13, 1949, DMA, Oskar Sala Nachlass 218/0205/2.583. It was also broadcast on September 25, 1949 on radio by the Deutschlandsender station. See DMA, Oskar Sala Nachlass 218/0205/2.578, and Beecroft, *Conversations*, 220, and Sala, "My Fascinating Instrument," 86.

75. Werner Fiedler, "'Dergleichen hab' ich nie vernommen: Neuer Versuch des Faust-Dramaturgie," in *Der Tag*, April 13, 1949, DMA, Oskar Sala Nachlass 218/0205/2.583.

76. Program and newspaper clipping of *Faust I*, August 28–30, 1949, DMA, Oskar Sala Nachlass 218/0420.

77. B., "Es tönt nach neuer Weise . . . Gespräch mit Oskar Sala, dem einzigen Trautoniumspieler der Welt," *Berliner Zeitung*, September 28, 1948, DMA, Oskar Sala Nachlass 218 0205/2.579. See also Richard Prätorius, "Genuß an der Hölle: Faust I. Teil im Deutschen Theater," *Berliner Kurier*, August 29, 1949, DMA, Oskar Sala Nachlass 218/0205/1.483, Oskar Sala, "50 Jahre Trautonium," 80, also found in DMA, Oskar Sala Nachlass 218/0193, and Rö, "Das Trautonium: Oskar Sala und sein elektro-akustisches Wunder," *Allgemeine Zeitung* (Mainz), December 6, 1949, DMA, Oskar Sala Nachlass 218/0205/2.582.

78. Oskar Sala to Schuricht, December 5, 1949, DMA, Oskar Sala Nachlass 218/0037/1949-1951/241.

79. Friedrich Luft, "Faust der Tragödie erster Teil in Max Reinhardts Deutschem Theater, Schumannstrasse [Berlin]," *Die neue Musikzeitung*, August 30, 1949, DMA, Oskar Sala Nachlass 218/0205/2.576.

80. B., "Es tönt nach neuer Weise . . ." *Berliner Zeitung*, September 28, 1949, DMA, Oskar Sala Nachlass 218/0205/2.579.

81. Oskar Sala, "Zur 'Faust'-Musik," Playbill of *Faust der Tragödie Erter Teil, Deutsches Theater,* DMA, Oskar Sala Nachlass 218/0420.

82. Ibid.

83. Rö, "Das Trautonium: Oskar Sala und sein elektro-akustisches Wunder," *Allgemeine Zeitung* (Mainz), December 6, 1949, 8 and B., "Es tönt nach neuer Weise . . . Gespräch mit Oskar Sala, dem einzigen Trautoniumspieler der Welt," *Berliner Zeitung,* September 28, 1949, DMA, Oskar Sala Nachlass 218/0205/2.579. See also letter to Oskar Sala from the director of the City Opera of Berlin (Städtische Oper Berlin) about the role of the trautonium in Werner Egk's *Abraxas,* DMA, Oskar Sala Nachlass 218/0037/1949-1951/185 and 218/0205/2.580. See Oskar Sala to Werner Meyer-Eppler, October 22, 1951, DMA, Oskar Sala Nachlass 218/0042 (not numbered), 1–2 and Hans Borgelt, "Spiel mit dem Feuer: Erfolgreiche Erstaufführung des Balletts *Abraxas* in der Städtischen Oper," October 11, 1949, DMA, Oskar Sala Nachlass, 218/0205/2.580. It should be noted that Heine did his best to distance himself from certain aspects of Goethe's classic.

84. Sala, "50 Jahre Trautonium," (1980), 81, also found in DMA, Oskar Sala Nachlass 218/0193. See also Oskar Sala, "Das Mixturtrautonium," *Die Musik-Woche* 30–31 (1951), 245–46, DMA, Oskar Sala Nachlass 218/0182. Sala repeated his performance in Naples in 1955. For the 1955 performances of Wagner at Bayreuth and in Naples, Sala used his mixture trautonium as discussed below.

85. Oskar Sala to Mr. Urban, Orchesterinspektor der Städtischen Oper, Berlin, February 28, 1951, DMA, Oskar Sala Nachlass 218/0037/1949-1951/269. See also Oskar Sala to Harald Genzmer, not dated, but circa March 1951, DMA, Oskar Sala Nachlass 218/0036/1949-1951/052 a and b, here b.

86. See, for example, the various reviews of "Puck" in DMA, Oskar Sala Nachlass 218/0207.

87. For a detailed account of the history of the radio drama and its metamorphosis into an opera, see Preuß, *Brechts "Lukullus" und seine Vertonungen,* 31–81.

88. Obermayer, "'Yes, to Nothingness!'" 226.

89. Obermayer, "'Yes, to Nothingness!'" 229–30.

90. Calico, *Brecht at the Opera,* 111.

91. Oskar Sala to Harald Genzmer, not dated, but around March 1951, DMA, Oskar Sala Nachlass 218/0036/1949-1951/052 a and b, here b. Sala writes that he hopes there will be no scandal. See also Sala, "50 Jahre Trautonium," 80–81, and DMA, Oskar Sala Nachlass 218/0193.

92. The opera was performed again in the Berlin State Opera in 1960. Fritz Hennenberg, "Zur Musik des 'Lukullus,' program note in *Deutsche Staatsoper, Premiere, Die Verurteilung des Lukullus,*" February 10, 1960, DMA, Oskar Sala Nachlass 218/0936.

93. Dessau, *Die Verurteilung des Lukullus,* measures 26 to 68, 65–66.

94. In the first two measures, the trautonium just plays a D flat, followed by ten measures of B. One measure is a half note of A flat, followed by a triplet of G flat, A flat, G flat, and six measures of A flat, measures 286 to 304. Dessau, *Die Verurteilung des Lukullus,* 78–79.

95. The trautonium plays D3, D2, and D1; see measures 1 to 3 and 19 to 20. Dessau, *Die Verurteilung des Lukullus,* 86.

96. The trautonium initially plays an A minor chord, rests for a measure, and then plays an inverted C# minor sixth, measures 133 to 137. Dessau, *Die Verurteilung des Lukullus,* 92–93.

97. It plays an A minor chord, see measures 212 to 214. Dessau, *Die Verurteilung des Lukullus,* 98.

98. Measures 1 to 17, Dessau, *Die Verurteilung des Lukullus,* 129. To view a video of Scene 11 of Bertolt Brecht and Paul Dessau "Die Verurteilung des Lukullus," see the "Resources" tab at https://press.princeton.edu/books/broadcastingfidelity.

99. Measures 11 and 12 in Dessau, *Die Verurteilung des Lukullus,* 129.

100. Such an effect was created by playing f sharp and c whole notes in measure 105, Dessau, *Die Verurteilung des Lukullus,* 134.

101. In measures 126 and 127 the trautonium plays an A flat major triad, Dessau, *Die Verurteilung des Lukullus*, 135.

102. The trautonium plays a whole note comprising a E flat, F sharp, and G, Dessau, *Die Verurteilung des Lukullus*.

103. The trautonium starts at the end of the preceding sentence, "There are still hungry mouths, of which you have so many up there," measures 68 to 71, Dessau, *Die Verurteilung des Lukullus*, 148–49.

104. Measures 72 to 83, Dessau, *Die Verurteilung des Lukullus*, 149–51.

105. Measures 157 to 158, Dessau, *Die Verurteilung des Lukullus*, 158.

106. Measures 163 to 201, Dessau, *Die Verurteilung des Lukullus*, 162–72.

107. I thank the musician and trautonium player Peter Pichler for making me aware of this opera.

108. Petersen, "Nachwort," I.

109. Dessau, *Deutsches Miserere*, 44–61, 67–82, 159–69.

110. Lf, "Tage zeitgenössischer Musik: Erschütternde Anklage," *Leipziger Volkszeitung*, September 22, 1966, p. 6, as reproduced in Lehmann et al., eds., *Fokus* Deutsches Miserere, 160 and anon, "Erschütterndes Monument, Paul Dessaus 'Deutsches Miserere'" uraufgeführt, in *Säschsisches Tageblatt*, September 24, 1966, reproduced in ibid, 161. Note that the part played by the trautonium featured chords at slow tempi, many of which were dissonant. Since there is no evidence that there were any trautoniums in the German Democratic Republic at the time, one not should assume that the Leipzig premiere featured one. The trautonium part was most likely played by an organ. Dessau included the trautonium in the original score of 1947 because of its ability to generate dark tone colors.

111. Rienäcker, "Fibel-Musik?" in Lehmann et al., eds., *Fokus* Deutsches Miserere, 97.

112. Anon, "Erschütterndes Monument, Paul Dessaus 'Deutsches Miserere' uraufgeführt," in *Sächsisches Tageblatt*, September 24, 1966, reproduced in Lehmann, ed., *Fokus* Deutsches Miserere, 161–62, here 162.

113. Karl Ristenpart of RIAS to Oskar Sala, September 24, 1948, DMA, Oskar Sala Nachlass 218/0037/1949-51/200.

114. Borgelt, *Das war der Frühling*, 375–76. I would like to thank Pam Potter for bringing this reference to my attention.

115. Borgelt, *Das war der Frühling*, 375–76.

116. Borgelt, *Das war der Frühling*, 376.

117. H. H. Stuckenschmidt, "Der heitere Sir John: Philharmonisches Osterkonzert im Titania Palast," *Neue Zeitung*, April 12, 1950. DMA, Oskar Sala Nachlass 218/0206/1.477.

118. Stellv[ertretender] Intendant Werner Fehling (MDR Sender Leipzig), January 5, 1951 to Oskar Sala, DMA, Oskar Sala Nachlass 218/0037/1949-1951/146.

119. *Kirche und Rundfunk*, Nr. 11, 22. Mai 1950. DMA, Oskar Sala Nachlass 218/0206/2.566.

120. Donhauser, *Oskar Sala*, 66–68 and 72–77.

121. Many of the reviews during that period simply referred to the instrument as the trautonium.

122. Sala, "My Fascinating Instrument," 79. See Frieß and Steiner, eds., "Oskar Sala im Gespräch," 217, 226.

123. Sala, "Die Geschichte des Trautoniums," 11, DMA, Oskar Sala Nachlass 218 / 0195. See also Donhauser, *Oskar Sala*, 72–73.

124. Oskar Sala to Notgemeinschaft der Deutschen Wissenschaft, Bad Godesberg (Büchelerstr. 55), "Bericht über den Stand meiner elektro-musikalischen Arbeiten und Vorschläge zur Förderung einiger im allgemeinen Interesse liegender Weiterentwicklungen," undated, DMA, Oskar Sala Nachlass 218/0037/1949-1951/170 a-c. We know that he had finished an earlier version in July 1950, but it was not ready for public performances. See DMA, Oskar Sala Nachlass

218/0036/1949-1951/ 004 a. Sala informed Genzmer in 1951 that the latest version of the mixture trautonium was not yet complete and could only create sound illustrations in the studio. See Oskar Sala to Harald Genzmer (not dated, but 1951), DMA, Oskar Sala Nachlass 218/0036/1949-1951/052 a-b. It should be noted that Sala referred to his mixture trautonium as the improved version of his concert trautonium. See, for example, Sala, "Mixtur-Trautonium," in *Melos* 17 (1950), 249.

125. Wolfgang Bührle, "Sala musiziert mit Elektronen," *Norddeutsche Zeitung* (Hannover), July 19, 1957; Bührle,, "Oskar Sala macht Elektronen konzertreif," *Westdeutsche Rundschau* (Wuppertal), May 1, 1957; and Bührle,, "Elektronen werden konzertreif: Mixtur-Trautonium-Spieler muß Mathematik[,] Physik und Musik studieren," *8-Uhr-Blatt* (Munich), August 17, 1957, DMA, Oskar Sala Nachlass 218/0219/1.235, 2.617, 2.648. See also Friedrich Trautwein to Oskar Sala, August 25, 1948, DMA, Oskar Sala Nachlass 218/0036/1948/022 a.

126. Sala, "Das Mixtur-Trautonium," *Physikalische Blätter*, 390–98; Sala, "Klanggestaltung mit dem Mixtur-Trautonium," 28–30, also found in DMA, Oskar Sala Nachlass 218/0192, Sala, "Das Mixtur-Trautonium," in *Klangstruktur der Musik*, 91–108, and Sala, "Das Trautonium: Begriff und Aufgabe," 25–28.

127. See Donhauser, *Oskar Sala*, 79–84 for the technical aspects of the instrument.

128. Note that even as late as the 1950s, Sala was using Hermann's explanation of formant theory to explain how tone colors worked on his trautoniums.

129. Sala, "Das Mixtur-Trautonium" *Melos* (1950), 390–98; Sala, "Klanggestaltung mit dem Mixtur-Trautonium," 28–30, DMA, Oskar Sala Nachlass 218/0192; and Sala, "Das Mixtur-Trautonium" in *Klangstruktur der Musik*, 91–108. For an account of the mixture trautonium for a general audience, see Oskar Sala, "Das neue Mixtur-Trautonium," 346–48.

130. Sala, "Klanggestaltung mit dem Mixtur-Trautonium," 28.

131. Sala, "Psycho-physische Konsequenzen," 13–20.

132. See, for example, Oskar Sala to Dr. U. Jetter of *Physikalische Blätter* 6, March 13, 1950, DMA, Oskar Sala Nachlass 218/0037/1949-51/134.

133. Ibid. See also Sala, "Objektive und subjektive Resonanzeffekte," 250–58, which deals with the creation of subjective and objective effects generated by the subharmonics of various frequencies. Trautwein and Sala corresponded about Békésy's work in March 1954. See Friedrich Trautwein to Oskar Sala, DMA, Oskar Sala Nachlass 218/0052/1954/414 a-b and Oskar Sala to Friedrich Trautwein, March 3, 1954, DMA, Oskar Sala Nachlass 218/0052/1954/413 a-b.

134. Sometimes the work is referred to as "Concert for Mixture Trautonium and Orchestra." To view videos of Peter Pichler playing Harald Genzmer's "Concerto for Mixture-Trautonium and Orchestra," Largo movement and Scherzo movement, see the "Resources" tab at https://press.princeton.edu/books/broadcastingfidelity.

135. Sala, "Elektronische Klanggestaltung mit dem Mixtur-Trautonium," 78–87, and Sala, "Das Mixtur-Trautonium," *Klangstruktur der Musik*, 91–108.

136. H. H. Stuckenschmidt, "Die singende Maschine: Sawallisch dirigeirte ein Sala-Trautonium," *Neue Zeitung*, July 3, 1954, in DMA, Oskar Sala Nachlass 218/0210, 3.

137. Günter Engler, "Neue Musik in Baden-Baden: Acht Novitäten beim Südwestrundfunk," *Neue Zeitung*, December 19, 1952 in DMA, Oskar Sala Nachlass 218/0208/1.480.

138. Hanns Reich, "Neue Musik im Südwestrundfunk," *Badische Zeitung*, December 23, 1952, DMA, Oskar Sala Nachlass 218/0208/2.597.

139. Willy Werner Göttig, "Neue Musik im Südwestrundfunk: Strobel und Rosbaud arbeiten bahnbrechend in Baden-Baden," *Abendpost*, December 18, 1952, DMA, Oskar Sala Nachlass 218/0208/2.595.

140. TJ, "Tübinger Musikfrage 1953: Zweimal Orchestermusik," *Schwäbisches Tagblatt*, May 19, 1953, DMA, Oskar Sala Nachlass 218/0209/2.404.

141. Fred K. Prieberg, "Musik der Zukunft: Ergebnisse eines Vortrags von Oskar Sala," *Badische Zeitung*, May 15–16, 1954, DMA, Oskar Sala Nachlass 218/0211/1.451. See also Fred K.

Prieberg, "Musik der Zukunft: Was ist elektronische Musik?" *Badische Zeitung* (Freiburg), August 13–14, 1955 in DMA, Oskar Sala Nachlass 218/0213/2.401.

142. Oskar Sala to Mr. Marschat, Bayerischer Rundfunk, Abteilung Musik, October 13, 1954, DMA, Oskar Sala Nachlass 218/0049/1954/025 and Oskar Sala to Dr. Götze, Bayerischer Rundfunk, Abteilung Musik, August 26, 1954, DMA, Oskar Sala Nachlass 218/0049/1954/017.

143. Karl Amadeus Hartmann to Oskar Sala, September 23, 1954, DMA, Oskar Sala Nachlass 218/0050/1954/168 and May 21, 1954, DMA, Oskar Sala Nachlass 218/0050/1954/174 a, Oskar Sala to Karl Amadeus Hartmann, May 24, 1954, DMA, Oskar Sala Nachlass 218/0050/1954/173, the correspondence between Sala and Will Götze of Bayerischer Rundfunk, April 1953, DMA, Oskar Sala Nachlass 218/0046/1953/026-27, and "Das einzige 'Mixtur-Trautonium' der Welt: Sala macht Musik mit Elektronen," *Eßlinger Zeitung*, May 15, 1957, DMA, Oskar Sala Nachlass 218/0219/1.158. Finally, see "Startschuß der Musica viva im Herkulessaal: Musikalische Hagelstürme," *Abendzeitung*, November 2, 1954, DMA, Oskar Sala Nachlass 218/0212/1.439, and Helmut Schmidt-Garre, "Zehnte Musica-Viva-Saison eröffnet," *Münchner Merker*, November 2, 1954, DMA, Oskar Sala Nachlass 218/0212/1.440.

144. K. H. "Mixtur-Trautonium—ein technisches Musikinstrument," *Die Südpost* (München), October 28, 1954, DMA, Oskar Sala Nachlass 218/0212/1.438.

145. DMA, Oskar Sala Nachlass 218/0212/1.407. See also 218/0212/1.413 and 2.334. Apparently, he was often called back to the stage for applause, winning new friends for the instrument as well as Genzmer's composition. Ibid., 2.334

146. Werner Oehlmann, "Zwei Welten: Elektronenklänge im Philharmoniker-Konzert," *Der Tagesspiegel* (Berlin), July 3, 1954, DMA, Oskar Sala Nachlass 218/0212/1.411.

147. Kurt Westphal, "Ton—Klang—Geräusch," *Der Kurier* (Berlin), July 2, 1954, DMA, Oskar Sala Nachlass 218/0212/1.408.

148. Anon., "Basel: Der Abschlus der Tagung für elektronische und konkrete Musik," *Basler Nachrichten*, May 24, 1955 in DMA, Oskar Sala Nachlass 218/0213/2.290.

149. Ibid.

150. Josef Eidens, "Niederrheinisches Musikfest: Orchesterwerke problematischen Inhalts," *Aachener Volks-Zeitung*, July 7, 1954, DMA, Oskar Sala Nachlass 218/0212/2.335.

151. "Begegnung mit dem Trautonium," *General-Anzeiger*, October 1, 1954, DMA, Oskar Sala Nachlass 218/0212/2.347.

152. "Mit elektronischer Musik: Philharmoniker unter W. Sawallisch," *Der Tag* (Berlin), July 2, 1954, DMA, Oskar Sala Nachlass 218/0212/1.406.

153. H. L., "Eine elekrische Tonmaschine: Anregende Zwischensaison bei den Berliner Philharmonikern," *Neue Zeit, Berlin*, July 3, 1954, DMA, Oskar Sala Nachlass 218/0212/1.410.

154. "Ein elektronisches Phantasieorchester: Das Mixtur-Trautonium," *Kölner Anzeiger*, July 9, 1954, DMA, Oskar Sala Nachlass 218/0212/1.422.

155. Friedrich Herzfeld, "Sinfonische Zwischensaison," *Berliner Morgenpost*, July 2, 1954, DMA, Oskar Sala Nachlass 218/0212/1.409.

156. "Neue Musik problematisch: Erstes Orchesterkonzert der 'Freunde neuer Musik,'" *Neue Rhein-Zeitung*, October 2, 1954, DMA, Oskar Sala Nachlass 218/0212/2.349.

157. Oskar Sala to Friedrich Trautwein, June 23, 1955, DMA, Oskar Sala Nachlass 218/0053 /1955/330.

158. Wolfgang Bührle, "Sala musiziert mit Elektronen: In Bayreuth 'lautet' und 'schmiedet' er—Das Trautonium kann alles," *Norddeutsche Zeitung*, July 19, 1957, DMA, Oskar Sala Nachlass 218/0219/1.235, Günter Bendig, "Wieland Wagners 'klassische' Inszenierung: Parsifal und 'Ring' mit geringfügigen Veränderungen," ibid., 2.646 and "Richard Wagners Welt- und Menschheitstragödie," *Fränkische Presse*, July 29, 1957, ibid., 1.220. See also DMA, Oskar Sala Nachlass 218/0934, July 26, 1957.

159. Andrew Clements, "Wagner: Das Rheingold," *The Guardian*, December 14, 2006, https://www.theguardian.com/music/2006/dec/15/classicalmusicandopera2.

160. Brown, *The Quest*, 188–221.

161. Willnauer, "Wieland Wagners Bayreuther Wirken," as quoted in Schäfer, *Wieland Wagner*, 66. It should be noted that Jörg Mager electronically recreated the Parsifal bells in Bayreuth back in 1933. He only did it once. See Donhauser, *Elektrische Klangmaschinen*, 200–3. The mixture trautonium was far better suited for Wieland Wagner's desire to shock the audience, as the critics commented.

162. Karl Ganzer, "Bayreuther Festspielkommentar," *Schwäbische Landeszeitung* (Augsburg), July 30, 1956, DMA, Oskar Sala Nachlass 218/0216/2.249.

163. Ibid.

164. Hans Schnoor, "Hat Bayreuth noch eine Zukunft?" *Westfalen-Blatt* (Biefeld, CDU), July 28, 1956, DMA, Oskar Sala Nachlass 218/0216/2.248.

165. K.U., "Götter und Menschen: Erster Zyklus endete mit dem 'Ring.'" *Westfälisches Volksblatt* (Paderborn), August 8, 1956, DMA, Oskar Sala Nachlass 218/0216/2.257.

166. A. K. Lassmann, "Ausklang der Bayreuther Festspiele," *Süddeutsche Zeitung* (Munich), September 1, 1956, DMA, Oskar Sala Nachlass 218/0216/2.263.

167. Peter Otto Schneider, "Die Zauberling von Bayreuth," *Basler National-Zeitung*, August 19, 1956, DMA, Oskar Sala Nachlass 218/0216/2.258 a-e, here e.

168. H. H. Stuckenschmidt, "Notiz über 'Lucullus,'" *Frankfurter Allgemeine Zeitung*, February 15, 1960, DMA, Oskar Sala Nachlass 218/0227/1.836.

169. Erika Wilde, "Die Verurteilung des Lukullus," *Deutsche Woche* (Berlin), February 24, 1960, DMA, Oskar Sala Nachlass 218/0227/1.841.

170. Werner Tamms, *Westdeutsche Allgemeine Zeitung*, September 9, 1960, DMA, Oskar Sala Nachlass 218/0226/3.085.

171. Hubert Schonger to Oskar Sala, March 31, 1950, DMA, Oskar Sala Nachlass 218/0037/1949-1951/237.

172. Oskar Sala to Hubert Schonger, around April 1950, DMA, Oskar Sala Nachlass 218/0037/1949-1951/236 a-b.

173. See the various reviews in DMA, Oskar Sala Nachlass 218/0218.

174. See the various reviews in DMA, Oskar Sala Nachlass 218/0220.

175. See the various reviews in DMA, Oskar Sala Nachlass 218/0221.

176. As quoted in Hans-Martin Majewski, "'Labyrinth' und seine Klangkulissen," *Der Mittag* (Düsseldorf), October 27, 1959, DMA, Oskar Sala Nachlass 218/0225/1.803.

177. See reviews in DMA, Oskar Sala Nachlass 218/0231.

178. Müller-Uri, "Oskar Sala und das Trautonium," 30.

179. Universität der Künste-Berlin Archiv Besand 1b, Nr. 6, folia 8 and 9 and 18–20. Tobis-Melofilm sent the RVS relevant samples of their sound movies, ibid., folio 10.

180. DMA, Oskar Sala Nachlass 218/0239. Film music belongs to the genre of *Gebrauchsmusik*: it is intended for amateurs rather than elite professionals.

181. DMA, Oskar Sala Nachlass 218/0237.

182. Frieß and Steiner, eds., "Oskar Sala im Gespräch," 230.

183. Oskar Sala, "50 Jahre Trautonium," 82, and DMA, Oskar Sala Nachlass 218/0193. To view a video of Oskar Sala playing the mixture trautonium for Manfred Durniok's "Eine Reise zum Mond" (A Trip to the Moon) of 1976, see the "Resources" tab at https://press.princeton.edu/books/broadcastingfidelity.

184. Oskar Sala, "Mixture-Trautonium and Studio Technique," 53–61. See also Müller-Uri, "Oskar Sala und das Trautonium," 30. For a complete list of films with Sala and his mixture trautonium until 1968, see Davies, *International Electronic Music Catalogue*. To view a video of a

German Coca-Cola commercial from the 1950s with the trautonium, see the "Resources" tab at https://press.princeton.edu/books/broadcastingfidelity.

185. "Musik der Atome," *Berliner Zeitung*, March 4, 1960, DMA, Oskar Sala Nachlass 218/0229/953. See also Remi Gassmann, "Elektronischer Orpheus," *Städtische Oper Berlin Spielzeit 1959–60, Program Heft Nr. 9*, in DMA, Oskar Sala Nachlass 218/0936.

186. DMA, Oskar Sala Nachlass 218/0936.

187. DMA, Oskar Sala Nachlass 218/0230, March 22–23, 1961.

188. To view a video of Oskar Sala playing the mixture trautonium for the ballet, "Electronics," music by Remi Gassmann, choreographed by George Balanchine, see the "Resources" tab at https://press.princeton.edu/books/broadcastingfidelity.

189. Laemmli, "A Case in Pointe," 17–20.

190. As quoted in Laemmli, "A Case in Pointe," 18, original, Balanchine, "A Word from George Balanchine," 34–35.

191. As quoted in Laemmli, "A Case in Pointe," 20, original, Horosko, "A Matter of Values," 24–25.

192. George Balanchine as quoted by Lawrence, "Electronics by Leaps and Bounds," 72.

193. Candby, "'Electronics': A Side-Review," 10.

194. Candby, "'Electronics': A Side-Review," 12.

195. Candby, "'Electronics': A Side-Review," 12, 14.

196. Gassmann as quoted by Lawrence, "Electronics by Leaps and Bounds," 70.

197. Candby, "'Electronics': A Side-Review," 14.

198. Candby, "'Electronics': A Side-Review," 78.

199. Lawrence, "Electronics by Leaps and Bounds," 70–72.

200. Lawrence, "Electronics by Leaps and Bounds," 72.

201. Martin, "The Dance," X11. See also Wierzbicki, "Shrieks, Flutters, and Vocal Curtains." 15.

202. "Wie wird das 'Trautonium' bedient? Einführungsabend mit Professor Trautwein, Oskar Sala und Günter Raphael," *Westdeutsche Rundschau*, October 1, 1954, DMA, Oskar Sala Nachlass 218/0212/1.435.

203. Ter., "Ist das die Musik der Zukunft? Oskar Sala führte Prof. Trautweins Trautonium vor," *Düsseldorfer Nachrichten*, September 15, 1954, DMA, Oskar Sala Nachlass 218/0212/1.430.

204. E. G., "Kompositionen ohne Zukunft: Ein extrem-kompromißloses Programm in der Stadthalle in Elberfeld," *Westdeutscher Rundfunk*, October 2, 1954, DMA, Oskar Sala Nachlass 218/0212/2.348.

205. Arthur van Dyck, "Sphärenmusik und Katzenjammer: Wuppertal eröffnete Konzertreihe mit Raphaels V. Sinfonie," *Westdeutsche Allgemeine* (Essen), October 2, 1954, DMA, Oskar Sala Nachlass 218/0212/1.436.

206. E. H., "Elektronische Musik—Musik der Zukunft? Zum Internationalen musikwissenschaftlichen Kongreß in Köln," *Industriekurier* (Düsseldorf), DMA, Oskar Sala Nachlass 218/0212/2.352.

207. Ff., "Das Ende der Musik?" *Sie und Er* (Zofingen), June 2, 1955, DMA, Oskar Sala Nachlass 218/0213/2.394 a-b.

208. Ibid.

209. H. H., "Die 'allermodernste' Musik: Mixtur-Trautonium gefiel nicht," *Deister- und Weserzeitung* (Hameln), July 2, 1955, DMA, Oskar Sala Nachlass 218/0213/1.384. See also H. H., "Mixtur-Trautonium und Martenot-Wellen," *Wilhelmshavener Zeitung*, July 9, 1955, DMA, Oskar Sala Nachlass 218/0213/2.303, also printed in the *Coburg Tageblatt*, July 12, 1955, DMA, Oskar Sala Nachlass 218/0213/1.386.

210. To view a video of Oskar Sala playing his "Elektronische Tanzsuite" (Electronic Dance Suite), see the "Resources" tab at https://press.princeton.edu/books/broadcastingfidelity.

211. Friedrich Siebert, "Radio Stuttgarts 'Woche der Leichten Musik,'" *Der Artist* (Düsseldorf), November 21, 1955, DMA, Oskar Sala Nachlass 218/0213/2.321.

212. Fritz Hammes, "Bilanz der 'Woche Leichter Musik': Donaueschingen im Programm—Und die Weinwerbewoche," *Die Rheinpfalz* (Ludwigshafen), October 22, 1955, DMA, Oskar Sala Nachlass 218/0213/2.318.

213. Ed., "Neue leichte Musik ist doch nicht so leicht so machen," *Untertürkheimer Zeitung* (Stuttgart-Untertürkheim), October 15, 1955, DMA, Oskar Sala Nachlass 218/0213/2.315.

214. C. O. E., "Woche der leichten Musik," *Mannheimer Morgen*, October 14, 1955, DMA, Oskar Sala Nachlass 218/0213/2.313.

215. Ho., "Pizziboogiefugato: Leichte Kammermusik beim SDR," *Stuttgarter Nachrichten*, October 13, 1955, DMA, Oskar Sala Nachlass 218/0213/1.390.

216. Dr. Gassert, "Begegnung mit dem Mixtur-Trautonium," *Südkurier*, May 19, 1954, DMA, Oskar Sala Nachlass 218/0211/1.454.

217. The original plan dates back to April 1954, Oskar Sala to Dr. Goslich, April 26, 1954, DMA, Oskar Sala Nachlass 218/0049/1954/048 a. They originally wished to play in June, but there was a conflict on the proposed days. See Friedrich Trautwein to Oskar Sala, June 6, 1954, DMA, Oskar Sala Nachlass 218/0052/1954/384 a.

218. G., "Elektro-Fantasieorchester: Prof. Trautwein und Oskar Sala demonstrieren das Mixtur-Trautonium vor Musikstudenten," *Neue Rhein-Zeitung* (Cologne), July 9, 1954, DMA, Oskar Sala Nachlass 218/0212/1.340. See also T.S., "Wie wird das 'Trautonium' bedient? Einführungsabend mit Professor Trautwein, Oskar Sala und Günter Raphael," *Westdeutsche Rundschau*, October 1, 1954, DMA, Oskar Sala Nachlass 218/0212/1.435.

219. Fred K. Prieberg, "Musik der ungewissen Zukunft: Die Tagung für elektronische Musik in Basel," *Badische Zeitung* (Freiburg), May 25, 1955, DMA, Oskar Sala Nachlass 218/0213/2.296. See also Prieberg, "Der Schrotthaufen und die Elektronik: Kongreß über elektronische und konkrete Musik in Basel," *Badische Neueste Nachrichten* (Karlsruhe), May 31, 1955, DMA, Oskar Sala Nachlass 218/0213/2.298.

8: Sala & Trautwein vs. the Cologne Studio for Electronic Music

1. Manning, "The Influence of Recording Technologies," 6.

2. Ungeheuer, *Wie die elektronische Musik*, 112. The summer courses on the new music took place in Darmstadt starting in 1946. They became very popular by the early 1950s.

3. Morawski-Büngeler, *Schwingende Elektronen*, 7 and 9.

4. As quoted in Holmes, *Electronic and Experimental Music*, 65. See also Kautny, "Pionierzeit der elektronischen Musik," 315–37.

5. "Elektronische Musik: Erfinder zeigen ihre Versuche," *Rheinische Post*, Düsseldorf, July 21, 1951, Deutsches Museum (Munich) Archives (henceforth DMA), Oskar Sala Nachlass 218/0212/1.423. The view that taped music did away with interpretation of the performer was repeated three years later in Enrst H. Haux, "Vor dem Abschied vom Konzertsaal: Die elektronische Musik wartet nur noch auf ihre Komponisten—Interpreten werden überflüssig," *Handelsblatt Deutsche Wirtschaftszeitung*, June 18, 1954, DMA, Oskar Sala Nachlass 218/0212/2.329.

6. Luening, "An Unfinished History," 46.

7. As quoted in Luening, "An Unfinished History," 46. See also Cross, "Electronic Music," 49.

8. "Bericht über eine Besprechung anlässlich der Vorführung des Nachtprogrammbandes über elektronische Musik am 18. 10. 51," Cologne, October 18, 1951, reprinted in Morawski-Büngeler, *Schwingende Elektronen*, 8. Northwestern German Broadcasting was split into Western German Broadcasting and Northern Germany Broadcasting on January 1, 1956. The Studio for Electronic Music belonged to Western Germany Broadcasting (Westdeutscher Rundfunk, or WDR). See also Cross, "Electronic Music," 49.

9. Holmes, *Electronic and Experimental Music*, 64.

10. Stuckenschmidt, "The Third Stage," 11.

11. Eimert, "Elektronische Musik," 5.

12. Stockhausen, "Electronic and Instrumental Music," 60, and Stockhausen, "The Origins of Electronic Music," 649–50. See also Cross, "Electronic Music," 53. The trautonium was actually called a monochord. Richard Toop has questioned Stockhausen's memory, arguing that the composer first used sine-wave generators in Cologne. See Toop, "Stockhausen and the Sine-Wave," 379–91.

13. Stockhausen, "Komposition 1953, No. 2," 46–51.

14. Morawski-Büngeler, *Schwingende Elektronen*, 15.

15. As translated in Iverson, *Electronic Inspirations*, 77, original: Stockhausen, "Arbeitsbericht 1953," 42, emphasis in the original.

16. Meyer-Eppler, *Elektrische Klangerzeugung*.

17. Herter, Schmidt-Görg, Weisgerber and Sendhoff, *In Memoriam Werner Meyer-Eppler*, 23–26.

18. Meyer-Eppler, "Fortschritte in der Akustik." See also Herter, Schmidt-Görg, Weisgerber and Sendhoff, *In Memoriam Werner Meyer-Eppler*, 23–26, Ungeheuer, *Wie die elektronische Musik*, 23, and Iverson, *Electronic Inspirations*.

19. Iverson, *Electronic Inspirations*.

20. Meyer-Eppler, "Informationstheorie," 341–47.

21. von Essen, "Werner Meyer-Eppler," 189–93.

22. Ungeheuer, *Wie die electronische Musik*, 97–99, and Kautny, "Pioneerzeit der elektronischen Musik," 316–17. Some three years later, one remarked at these conferences that the *Tonmeister* was the mediator between music and engineering. It clearly was becoming an increasingly important profession. Stauder, "Der Tonmeister," 79–81.

23. Luening, "An Unfinished History," 46. See also Ungeheuer, *Wie die elektronische Musik*, 98–99.

24. Eimert, "Der Komponist und die elektronischen Klangmittel," 242. It is interesting to note that the Eimert's claim is four years after his initial collaboration with Meyer-Eppler and Beyer.

25. As quoted in Ungeheuer, *Wie die elektronische Musik*, 99, original Varèse, "Musik auf neuen Wegen," 404.

26. Werner Meyer-Eppler to Oskar Sala, January 19, 1949, DMA, Oskar Sala Nachlass 218/0036/1949-1951/029 a. See also Oskar Sala to Werner Meyer-Eppler, November 22, 1950, DMA, Oskar Sala Nachlass 218/0037/1949-1951/149.

27. Werner Meyer-Eppler to Oskar Sala, March 1, 1949, DMA, Oskar Sala Nachlass 218/0036/1949-1951/028.

28. Werner Meyer-Eppler to Oskar Sala, November 13, 1950, DMA, Oskar Sala Nachlass 218/0037/1949-1951/150.

29. Oskar Sala to Werner Meyer-Eppler, 22 November 1950, DMA, Oskar Sala Nachlass 218/0037/1949-1951/149.

30. Oskar Sala to Werner Meyer-Eppler, October 22, 1951, DMA, Oskar Sala Nachlass 218/0042, 1.

31. Oskar Sala to Werner Meyer-Eppler, October 22, 1951, DMA, Oskar Sala Nachlass 218/0042, p. 1 and Oskar Sala to Werner Meyer-Eppler, November 8, 1951, DMA, Oskar Sala Nachlass 218/0037, 1–2, here 1. See also Oskar Sala, "Aus der Praxis der elektro-akustischen und elektronischen Musik," March 10, 1953, DMA, Oskar Sala Nachlass 218/0046/1953/161 a-e, here b-c.

32. Meyer-Eppler, "Über die Anwendung," 133.

33. Ibid. Meyer-Eppler sent Sala a copy. See DMA, Oskar Sala Nachlass 218/0046/1989/224 a-g.

34. Stuckenschmidt, "Die Mechanisierung der Musik," 1–8.

35. Meyer-Eppler, "Elektronische Kompositionstechnik," 5–9.

36. Meyer-Eppler, "Elektronische Kompositionstechnik," 5.

37. Meyer-Eppler, "Elektronische Kompositionstechnik," 5–6, 8.

38. Meyer-Eppler, "Elektronische Kompositionstechnik," 7, italics are in the original.

39. Oskar Sala to Friedrich Trautwein, DMA, Oskar Sala Nachlass 218/0048/1953/294 a-b.

40. Oskar Sala, "Aus der Praxis der elektro-akustischen und elektronischen Musik," March 10, 1953, DMA, Oskar Sala Nachlass 218/0046/1953/161 a-e, here c.

41. Ibid.

42. Ibid., d. Often with the mixture trautonium, Sala was both performer and composer.

43. Ibid.

44. "Authentische Komposition," Meyer-Eppler, "Elektronische Kompositionstechnik," 5–9. See also Friedrich Trautwein to Oskar Sala, March 19, 1953, DMA, Oskar Sala Nachlass 218/0048/1953/326 a-b, here b.

45. Friedrich Trautwein, "Mensch und Maschine in der Tonkunst: Vortrag in der Volkshochschule Düsseldorf am 30. 11. 1956," DMA, Friedrich Trautwein Nachlass 187/015, 1–16, here 15–16.

46. At least one of Sala's letters to Meyer-Eppler predates his 1953 publication. The notion of authenticity occurs in their correspondence as early as 1951. See also Oskar Sala's letter to Harald Genzmer on attacking Meyer-Eppler's view of authentic music, January 11, 1954, DMA, Oskar Sala Nachlass 218/0050/1954/162.

47. Oskar Sala to Werner Meyer-Eppler, DMA, Oskar Sala Nachlass 218/0042, October 22, 1951, 2.

48. Oskar Sala to Werner Meyer-Eppler, undated, but early 1954, DMA, Oskar Sala Nachlass 218/0050 (not numbered), 1–2, here 1. See also Oskar Sala to Werner Meyer-Eppler, October 22, 1951, DMA, Oskar Sala Nachlass 218/0042/2 and Oskar Sala to Werner Meyer-Eppler, November 8, 1951, DMA, Oskar Sala Nachlass 218/0037, 1–2, here 1.

49. Oskar Sala to Werner Meyer-Eppler, November 8, 1951, DMA, Oskar Sala Nachlass 218/0037, 1–2, here 1.

50. Oskar Sala, "Aus der Praxis der elektro-akustischen und elektronischen Musik," March 10, 1953, DMA, Oskar Sala Nachlass 218/0046/1953/161 a-e, here d.

51. Oskar Sala to Werner Meyer-Eppler, October 22, 1951, DMA, Oskar Sala Nachlass 218/0042, 2

52. Harald Genzmer to Oskar Sala, (first page and therefore date and number missing), DMA, Oskar Sala Nachlass 218/0050, 2.

53. Oskar Sala to Harald Genzmer, January 11, 1954, DMA, Oskar Sala Nachlass 218/0050/1954/162.

54. Oskar Sala to Werner Meyer-Eppler, undated, early 1954, DMA, Oskar Sala Nachlass 218/ 0050 (not numbered, but before 1954/194), 1–2, here 1.

55. See, for example, Oskar Sala to Werner Meyer-Eppler, November 8, 1951, DMA, Oskar Sala Nachlass 218/0037, 1–2, here 1, and Oskar Sala, "Aus der Praxis der elektro-akustischen und elektronischen Musik," March 10, 1953, DMA, Oskar Sala Nachlass 218/0046/1953/161 a-e, here b.

56. Oskar Sala to Werner Meyer-Eppler, undated, but early 1954, DMA, Oskar Sala Nachlass 218/0050 (not numbered), 1–2, here 1. After much inquiry, in April 1954, Enkel and Beyer finally showed Sala the Cologne Studio. Oskar Sala to Werner Meyer-Eppler, April 15, 1954, DMA, Oskar Sala Nachlass 218/0050/1954/206 a-b, here a.

57. Oskar Sala to Friedrich Trautwein, March 21, 1953, DMA, Oskar Sala Nachlass 218/0048/1953/325.

58. Friedrich Trautwein to Oskar Sala, May 2, 1953, DMA, Oskar Sala Nachlass 218/0048/1953/321 a-d, here c.

59. Fred K. Prieberg, "Zukunftsmusik oder Tonchaos? UNESCO-Kongreß für Musik und Elektro-Akustik," *Der Tag* (Berlin), August 22, 1954, DMA, Oskar Sala Nachlass 218/0212/2.343. Prieberg was referring to Trautwein's lecture and discussion at the UNESCO Congress for Music and Electro-Acoustics. Prieberg's quotation was meant as a summary: he was

ventriloquizing Trautwein. He did go on to say that the Cologne School was not the music of the future, since it was so "unproductive." He did think that the mixture trautonium could be the music of the future, as long as it was "no longer a catalog of effects and became simply and transparently modern." DMA, Oskar Sala Nachlass 218/0212/2.343.

60. Friedrich Trautwein to Oskar Sala, May 2, 1953, DMA, Oskar Sala Nachlass 218/0048/1953/321 a-d, here c.

61. Friedrich Trautwein to Oskar Sala, March 23, 1953, DMA, Oskar Sala Nachlass 218/0048/1953/324 a-b, here a, underline in the original.

62. Friedrich Trautwein's letter to Oskar Sala, October 27, 1954, DMA, Oskar Sala Nachlass 218/0052/1954/362 a-b, here a.

63. Oskar Sala to Friedrich Trautwein, April 10, 1953, DMA, Oskar Sala Nachlass 218/0048/1953/323 a-c, here a.

64. Friedrich Trautwein to Oskar Sala, September 22, 1953, DMA, Oskar Sala Nachlass 218/0048/1953/304 a-b, here a.

65. Friedrich Trautwein to Oskar Sala, May 2, 1953, DMA, Oskar Sala Nachlass 218/0048/1953/321 a-d, here c.

66. Ibid. The actual date was May 25, 1953: Trautwein erroneously wrote May 26, 1953.

67. Ibid.

68. Oskar Sala to Siegfried Goslich, May 30, 1953. DMA, Oskar Sala Nachlass 218/0046/1953/063.

69. Friedrich Trautwein to Oskar Sala, March 19, 1953, DMA, Oskar Sala Nachlass 218/0048/1953/326 a-b, here a. Trautwein was referring to an earlier letter he received from Sala in which Sala asked if he wanted a copy of an "Anti-Meyer-Eppler" tape, which he had just produced. See Oskar Sala to Friedrich Trautwein, May 5, 1953, DMA, Oskar Sala Nachlass 218/0048/1953/332. In June 1954, Trautwein reiterated that "I have always done well with Dr. Meyer-Eppler on an objective and factual basis, and that is important." Friedrich Trautwein to Oskar Sala, June 6, 1954, DMA, Oskar Sala Nachlass 218/0052/384 a-c, here b.

70. Friedrich Trautwein to Oskar Sala, June 5, 1953, DMA, Oskar Sala Nachlass 218/0048/1953/318 a-b, here b.

71. Oskar Sala to Werner Meyer-Eppler, November 16, 1954, DMA, Oskar Sala Nachlass 218/0050/1954/196 a-b, here a. Meyer-Eppler agreed with Sala that there was a fundamental problem with definitions and that electronic music could not be defined by the musical style of the composition, but rather only by the uncompromising use of purely electronic means. He felt that the only difference was between good and bad electronic music. See Werner Meyer-Eppler to Oskar Sala, November 26, 1954, DMA, Oskar Sala Nachlass 218/0050/1954/195 a-b, here a.

72. Oskar Sala to Friedrich Trautwein, November 19, 1954, DMA, Oskar Sala Nachlass 218/0052/361 a-b, here b.

73. Friedrich Trautwein to Oskar Sala, June 28, 1954, DMA, Oskar Sala Nachlass 218/0052/1954/376 a.

74. Friedrich Trautwein to Dr. Goslich (Radio Bremen) and Oskar Sala, June 6, 1954 in DMA, Oskar Sala Nachlass 218/0052/1954/384 a-c, here b.

75. Friedrich Trautwein to Dr. Goslich (Radio Bremen) and Oskar Sala, June 6, 1954 in Oskar Sala Nachlass 218/0052/1954/384 a-c, here b. For more on Trautwein's view that Eimert was attempting to usurp electronic music, see Friedrich Trautwein's letter to Oskar Sala, May 19, 1954, DMA, Oskar Sala Nachlass 218/0052/1954/390 a-e, here e. See also Oskar Sala to Friedrich Trautwein, October 30, 1953, DMA, Oskar Sala Nachlass 218/0048/1953/301.

76. Oskar Sala, "Aus der Praxis der elektro-akustischen und elektronischen Musik," DMA, Oskar Sala Nachlass 218/0046/1953/161 a-e, here b. See also Oskar Sala to Friedrich Trautwein, April 10, 1953, DMA, Oskar Sala Nachlass 218/0048/1953/323 a-c, here c.

77. Friedrich Trautwein to Oskar Sala, March 19, 1953, DMA, Oskar Sala Nachlass 218/0048/1953/326 a-b, here b.

78. Oskar Sala to Friedrich Trautwein, June 6, 1953, DMA, Oskar Sala Nachlass 218/0048/1953/316 a-b, here b. Generally, Sala agreed with Trautwein that the two groups defined terms differently, as became clear during their various collaborations. When Sala sent Meyer-Eppler a manuscript of his lecture on formants, Meyer-Eppler pointed out that Sala was using the term "formant" in two different ways: one for the predominant spectral region of overtones that determine the tone color of the sound, and the other for the filter circuit that produced that spectral region. After an exchange of views, they settled on calling the filter circuit the formant circuit. See the correspondence between Sala and Meyer Eppler from September 22, 1954 to October 8, 1954, DMA, Oskar Sala Nachlass 218/0050/1954 /199-202.

79. Werner Meyer-Eppler to Oskar Sala, November 26, 1954, DMA, Oskar Sala Nachlass 218/0050/1954/195 a-b.

80. Oskar Sala to Friedrich Trautwein, May 5, 1953, DMA, Oskar Sala Nachlass 218/0048/1953/320.

81. Friedrich Trautwein to Oskar Sala, May 19, 1954, DMA, Oskar Sala Nachlass 218/0052/1954/390 a-e, here e.

82. Friedrich Trautwein to Oskar Sala, October 27, 1954 in DMA, Oskar Sala Nachlass 218/0052/1954/362 a-b, here a. In that year, Trautwein published an article on the contemporary methods to generate formants that resembled traditional musical instruments. See Trautwein, "Elektroakustische Mittel in der aktiven Tonkunst," 258. See also DMA, Friedrich Trautwein Nachlass 187/008. This work was particularly relevant to the electronic monochord he built for the Studio of Electronic Music in Cologne.

83. Oskar Sala to Friedrich Trautwein, February 12, 1954 in Oskar Sala Nachlass 218/0052/1954/419.

84. Ibid.

85. Oskar Sala to Werner Meyer-Eppler, April 15, 1954, DMA, Oskar Sala Nachlass 218/0050/1954/206 a-b, here a. See also Werner Meyer-Eppler to Oskar Sala, March 5, 1954, DMA, Oskar Sala Nachlass 218/0050/1954/245.

86. Oskar Sala to Werner Meyer-Eppler, April 15, 1954, DMA, Oskar Sala Nachlass 218/0050/1954/206 a-b, here a.

87. Ibid., b.

88. Oskar Sala to Friedrich Trautwein, May 21, 1953, DMA, Oskar Sala Nachlass 218/0052/1954/389 a-c, here b, Sala's quotation of Beyer's statement.

89. Friedrich Trautwein to Oskar Sala, July 3, 1953, DMA, Oskar Sala Nachlass 218/0048/1953/312 a-b, here b.

90. Oskar Sala to Werner Meyer Eppler, September 11, 1953, DMA, Oskar Sala Nachlass 218/0046/1989/214 and Werner Meyer-Eppler to Oskar Sala, October 6, 1953, DMA, Oskar Sala Nachlass 218/0046/1989/215. See also Werner Meyer-Eppler to Oskar Sala, January 22, 1951, DMA, Oskar Sala Nachlass 218/0036/1949-51/006 and Werner Meyer-Eppler to Oskar Sala, October 19, 1953, DMA, Oskar Sala Nachlass 218/0046/1989/217.

91. Werner Meyer-Eppler to Oskar Sala, December 18, 1953, DMA, Oskar Sala Nachlass 218/0046/1989/221.

92. "Arbeitsplan des Kongresses 'Musik und Elektroakustik' in Gravesano," DMA, Oskar Sala Nachlass 218/0051/1954/276 a-b.

93. Friedrich Trautwein to Oskar Sala, October 27, 1954, DMA, Oskar Sala Nachlass 218/0052/1954/362 a-b, here a.

94. "Tagung für elektronische und Konkrete Musik / L'électronique dans la musique," 19–May 21, 1955, DMA, Oskar Sala Nachlass 218/0049/1954/020 and 029. See also Hans Heinz

Stuckenschmidt, "Der Klang aus der Röhre: Zur Basler Tagung 'Elektronsiche und konkrete Musik,'" *Stuttgarter Zeitung*, May 25, 1955 in DMA, Oskar Sala Nachlass 218/0213/2.292.

95. Beecroft, *Conversations*, 220.

96. Beecroft, *Conversations*, 220.

97. Frieß and Steiner, eds., "Oskar Sala im Gespräch," 231–32. Sala was referring to the music created in the Siemens' Electronic Music Studio.

98. Friedrich Trautwein to Oskar Sala, March 19, 1953, DMA, Oskar Sala Nachlass 218/0048/1953/326 a-b, here b.

9: Epilogue

1. Rösser, *Bilder zum Hören*, 283–304. To view a video of the mixture trautonium providing the sounds of birds' screeching and flapping their wings in Alfred Hitchock's "The Birds" of 1963, see the "Resources" tab at https://press.princeton.edu/books/broadcastingfidelity.

2. As quoted in Wierzbicki, "Shrieks, Flutters, and Vocal Curtains," 11, original: Gilling, "The Colour of Music," 37.

3. As quoted in Badge, *Oskar Sala*.

4. As cited in Moral, *The Making*, 157.

5. As quoted in Moral, *The Making*, 157.

6. As quoted in Moral, *The Making*, 158.

7. Deutsches Museum (Munich) Archive (henceforth DMA), Oskar Sala Nachlass 218/0238 and 218/0240.

8. As quoted in Moral, *The Making*, 158.

9. As quoted in Moral, *The Making*, 166.

10. Moral, *The Making*, 161–62.

11. Allen, "The Sound," 97–120.

12. See for example, "Ablauf der Festveranstaltung am 28.09.92," in DMA, Oskar Sala Nachlass 218/0436.

13. DMA, Oskar Sala Nachlass 218/0877/22/001. For the construction manual of the remade mixture trautonium, see DMA, Oskar Sala Nachlass 218/1519.

14. Hans-Jörg Borowicz, Dietmar Rudolph, and Helmut Zahen, "Entwicklung eines Mixtur-Trautoniums nach OSKAR SALA an der Fachhochschule Berlin der Deutschen Bundespost Telekom," DMA, Oskar Sala Nachlass 218/0436.

15. Müller-Uri, "Oskar Sala und das Trautonium," 30.

16. Sala, "My Fascinating Instrument," 75.

17. "Die Deutsche Verwaltung für Volksbildung in der Sowjetischen Besatzungszone—Generalintendanz des Rundfunksenders betr. Bau eines Quartett-Trautoniums mit Vertrag," DMA, Oskar Sala Nachlass 218/0037.

18. Donhauser, *Oskar Sala*, 77–79.

19. Donhauser, *Oskar Sala*, 103–4.

20. Böhme-Mehner, "Berlin Was Home," 34. It should be noted that a number of the dates in the article are incorrect.

21. Speech of Khruschchev, "In hohem Ideengehalt," 3–6. See also Slonimsky, ed., *Music since 1900*, 1377–78.

22. Böhme-Mehner, "Berlin Was Home," 42.

23. Böhme-Mehner, "Berlin Was Home," 42.

24. Donhauser, *Oskar Sala*, 105–8. See also Moog-Koussa, "Uncovering the Moogtonium," Bob Moog Foundation, https://moogfoundation.org/from-the-archives-moogtonium-discovered/.

25. Personal communication between Herbert Deutsch and the author, March 1, 2012.

26. Donhauser, *Oskar Sala*, 113–15. See also Spix, "The Digital Trautonium," available at https://archive.ph/plDuL.

27. Badge, *Oskar Sala*, unpaginated.

28. "VA Wöfl-Portrait mit Florian Schneider," November 22, 2000, DMA, Oskar Sala Nachlass 218/0129/2000/102.

29. Mark Dery, interview with Ralf Hütter, *Keyboard Magazine* 10 (1991), as quoted in Schütte, *Kraftwerk*, 90. See also Buckley, *Kraftwerk Publikation*, 14–15.

30. Schütte, *Kraftwerk*, 88–90.

31. European Broadcasting Convention: Copenhagen Plan Annexed, available at https://search.itu.int/history/HistoryDigitalCollectionDocLibrary/4.65.43.en.100.pdf,.

32. See "Luftrum Sound Design: SynthGPT," https://www.luftrum.com/synthgpt/,and "R.I.P. Synths: Hello SynthGPT!" https://www.youtube.com/watch?v=wcFmbjQ_mEc.

33. Savage, "Sir Paul McCartney," https://www.bbc.com/news/entertainment-arts-65881813.

34. Anon., "Top 3 AI," https://musicreviewworld.com/ai-singing-voice-generators/.

35. Karimi, "'Mom, these bad men have me,'" available at https://www.cnn.com/2023/04/29/us/ai-scam-calls-kidnapping-cec/index.html.

36. On those themes, see Katz, *Music and Technology*.

BIBLIOGRAPHY

Archives

Deutsches Museum (Munich) Archiv (abbreviated DMA), Museumsinsel 1, Munich, Germany.
Stiftung Deutsches Technikmuseum Berlin. Historisches Archiv. Trebinner Strasse 9, Berlin, Germany.
Universität der Künste-Berlin (abbreviated UdK-Berlin) Archiv, Einsteinufer 43–53, Berlin, Germany.
Westdeutscher Rundfunk-Archivhaus, Köln (Cologne), An der Rechtschule 4, Köln (Cologne), Germany.

Secondary Sources

Abendroth, W. "Neue Musik Berlin 1930." *Allgemeine Musikzeitung* 57 (1930): 724

Adorno, Theodor W. "The Radio Symphony: An Experiment in Theory." In *Essays on Music*. Translated by Susan H. Gillespie. Berkeley: University of California Press, 2002, 251–70.

———. "Zum Anbruch: Exposé." In *Musikalische Schriften IV*, edited by Rolf Tiedemann, , 601–2. *Gesammelte Schriften*, vol. 19 of 20 vols. Frankfurt: Suhrkampf, 1970–1986

Aigner, Franz. "Über die Schwingungen der Sprache und Musikinstrumente und über die Quellen der Verzerrung." In *Die wissenschaftlichen Grundlagen des Rundfunkempfangs*, edited by K. W. Wagner, 18–38. Berlin: Julius Spring, 1927.

Allen, Richard. "The Sound of *The Birds*." *October* 164 (2013): 97–120.

Almeida, Ludwig Graf. "Die erste 'direkte Übertragung' in München." *Funk: Die Wochenschrift des Funkwesens* 11 (March 13, 1925): 125–26.

Anderson, J. E. "The Greatest Recreation Is a Test of the Receiver." *Radio World* 6 (November 19, 1927): 18–19.

Anon. "An alle Funkfreunde." In *Rundfunk Jahrbuch 1929*, edited by Reichs-Rundfunk-Gesellschaft, 417–18. Berlin: Union Deutsche Verlagsgesellschaft, 1929.

Anon. "Arbeitsplan der Berliner 'Funkakademie'" *Funk: Die Wochenschrift des Funkwesens* 38 (September 20, 1929): 176.

Anon. "Berlin eröffnet seine Funk-Akademie." *Funk: Die Wochenschrift des Funkwesens* 41 (October 11, 1929): 187–90.

Anon. "Elektrisch erzeugte Vokale." *Funk: Die Wochenschrift des Funkwesens* 11 (March 9, 1934): 206.

Anon. "Endlich eine akustische Rundfunk-Versuchsstelle." *Funk: Die Wochenschrift des Funkwesens* 2 (January 7, 1927): 14.

Anon. "Ferndirigiertes Völkerbundkonzert: Ein Experiment der Rundfunkversuchsstelle in Berlin am 1. September." *Der deutsche Rundfunk: Rundschau und Programm für alle Funkteilnehmer* 7 (1929): 1115.

Anon. "Funknachrichten." *Neue Zeitschrift für Musik* 100 (1933): 528
Anon. "Funk-Stunde Berlin." *Der deutsche Rundfunk: Rundschau und Programm für alle Funkteilnehmer* 2 (1924): 1265–66.
Anon. "Funkversuchsstelle: Eröffnung einer wissenschaftlichen Forschungsstätte an der Musikhochschule in Berlin." *Der deutsche Rundfunk: Rundschau und Programm für alle Funkteilnehmer* 6 (1928): 1297–98.
Anon. "The Method of 'Shock-Excitation' in Wireless Telegraphy." *Nature* 91 (1913): 21.
Anon. "Musik und Technik: Bericht über die Rundfunkmusiktagung in München." *Funk: Die Wochenschrift des Funkwesens* 29 (July 17, 1931): 225–26.
Anon. "Der neue Rundfunkwellenplan." *Funk: Die Wochenschrift des Funkwesens* 2 (January 7, 1927): 15.
Anon. "Nordischer Rundfunk: Zur Psychologie des Noraghörers: Eine Programmbegründung." In *Rundfunk Jahrbuch 1931*, edited by Reichs-Rundfunk-Gesellschaft, 92–118. Berlin: Union Deutsche Verlagsgesellschaft, 1931.
Anon. "A Portable Electric Harmonic Analyzer." *Journal of Scientific Instruments* 5 (1928): 320–23.
Anon. "Die Schule für Rundfunkredner." *Funk: Die Wochenschrift für Funkwesen* 17 (April 26, 1929): 72.
Anon. "Sopranos Not Always to Blame If Voices Are Harsh on Radio." *New York Times*, October 14, 1928, section xx, 17.
Anon. "'Studio' und 'Funkakademie,'" *Funk: Die Wochenschrift für Funkwesen* 33 (August 16, 1929): 146.
Anon. "Südwest-Deutscher Rundfunk: Wie wir die ästhetischen Aufgaben des Rundfunks mit seinen Programmaufgaben vereinigen wollen." In *Rundfunk Jahrbuch 1931*, edited by the Reichs-Rundfunk-Gesellschaft, 75–80. Berlin: Union Deutsche Verlagsgesellschaft, 1931.
Anon. "Eine Tagung für Rundfunkmusik in Göttingen." Funk: *Die Wochenschrift des Funkwesens* 16 (April 13, 1928): 123–124.
Anon. "Top 3 AI Singing Voice Generators." *Music Review World*. Downloaded 8 May 2023. https://musicreviewworld.com/ai-singing-voice-generators/.
Anon. "Der Unterhaltungs-Rundfunk: Vortrag von Dr. Hans Flesch." *Radio Umschau: Wochenschrift über die Fortschritte in Rundfunkswesen* 1 (1924): 185.
Anon. "Was verlangt man vom Lautsprecher?" *Dortmunder Zeitung* 572, December 7, 1929. https://www.medienstimmen.de/chronik/1926-1930/1926-siemens-halske-protos-lautsprecher-siemens-072/.
Anon. "Die Zahl der Rundfunkhörer stieg über 3 ½ Millionen." *Funk: Die Wochenschrift des Funkwesens* 4 (January 23, 1931): 28.
Anon. "Zwei Millionen Hörer—acht Millionen Teilnehmer!" *Funk: Die Wochenschrift des Funkwesens* 4 (January 20, 1928): 32.
Antoine, Herbert. *Die Entwicklung des Deutschen Rundfunks in Zahlen, 1923–1930. Zum 5jährigen Bestehen der Reichs-Rundfunk-Gesellschaft*. Berlin: Reichs-Rundfunk-Gesellschaft, 1931.
———. "Marksteine." In *Rundfunk Jahrbuch 1931*, edited by Reichs-Rundfunk-Gesellschaft, 215–32. Berlin: Union Deutsche Verlagsgesellschaft, 1931.
———. "Rundfunk in Zahlen." In *Rundfunk Jahrbuch 1930*, edited by the Reichs-Rundfunk-Gesellschaft, 359–70. Berlin: Union Deutsche Verlagsgesellschaft, 1930.
Applegate, Cecilia, and Pamela Potter. *Music and the German National Identity*. Chicago: University of Chicago Press, 2002.
Backhaus, Hermann. "Über die Bedeutung der Ausgleichsvorgänge in der Akustik." *Zeitschrift für technische Physik* 13 (1932): 31–46.
———. "Physikalische Untersuchungen an Streichinstrumenten." *Die Naturwissenschaften* 27 (1929): 811–18, 835–39.
Badge, Peter. *Oskar Sala: Pionier der elektronischen Musik*. Göttingen: Satzwerk Verlag, 2000.

Balanchine, George. "A Word from George Balanchine." *Dance Magazine* 32 (1958): 34–35.
Band, Lothar. "Hans Flesch: Berlins Rundfunkintendant." *Funk: Die Wochenschrift des Funkwesens* 19 (May 10, 1929): 81–82.
———. "Musiker und Techniker über den Rundfunk: Die Göttinger, 'Erste Tagung für Rundfunkmusik.'" *Funk: Die Wochenschrift des Funkwesens* 21 (May 18, 1928): 161–62.
———. "Opernkrise der Gegenwart: Der Rundfunk als Vermittler." *Funk: Die Wochenschrift des Funkwesens* 4 (January 24, 1930): 13–14.
———. "Rundfunk im Rahmen der 'Neuen Musik.'" *Funk: Die Wochenschrift des Funkwesens* 26 (June 27, 1930): 133–34.
———. "Auf dem Wege zur Synthese Musiker und Physiker: Die Münchner 'Zweite Tagung für Rundfunkmusik.'" *Funk: Die Wochenschrift des Funkwesens* 30 (July 24, 1931): 233–34.
Battaglia, Andy. "Ondes Martenot: An Introduction." https://daily.redbullmusicacademy.com/2014/03/ondes-martenot-introduction.
Beckh, Joachim. *Blitz & Anker, Band I: Informationstechnik-Geschichte und Hintergründe.* Norderstedt: Books on Demand GmbH, 2005.
Beecroft, Norma. *Conversations with Post World War II Pioneers with Electronic Music.* Oshawa, Canada: Canadian Council for the Arts, 2015.
Benjamin, Ruha. *Race after Technology: Abolitionist Tools for the New Jim Code.* Cambridge, UK: Polity Press, 2019.
Bekker, Paul. *Neue Musik: Dritter Band der gesammelten Schriften.* Berlin and Stuttgart: Deutsche Verlags-Anstalt, 1923.
Beranek, Leo L. "Loudspeakers and Microphones." *Journal of the Acoustical Society of America* 26 (1954): 618–29.
Bericht der Reichs-Rundfunk-Gesellschaft (January 17, 1933). "Sprachpflege im deutschen Rundfunk." In *Aus meinem Archiv: Probleme des Rundfunks*, edited by Hans Bredow, 184–88. Heidelberg: Vowinckel, 1950.
Berten, Walter. "Paul Hindemith und die deutsche Musik." *Zeitschrift für Musik* 100 (1933): 537–44.
Bijsterveld, Karin. *Mechanical Sound: Technology, Culture and Public Problems of Noise in the Twentieth Century.* Cambridge, MA: MIT Press, 2017.
Birdsall, Carolyn. "Radio Documents: Broadcasting, Sound Archiving, and the Rise of Radio Studies in Interwar Germany." *Technology and Culture* 60 (2019, supplement): S96–S128.
Blackbourn, David, and Geoff Eley. *The Peculiarities of German History: Bourgeois Society and Politics in Nineteenth-Century Germany.* NY: Oxford University Press, 1984.
Böhme-Mehner, Tatjana. "Berlin Was Home to the First Electronic Studio in the Eastern Bloc: The Forgotten Years of the Research Lab for Inter-disciplinary Problems in Musical Acoustics." *Contemporary Music Review* 30 (2011): 33–47.
Borgelt, Hans. *Das war der Frühling von Berlin: Eine Berlin-Chronik.* Munich: Schneekluth, 1980.
Brain, Robert M. *The Pulse of Modernism: Physiological Aesthetics in Fin-de-Siècle Europe.* Seattle: University of Washington Press, 2016.
Brandt, Hugo. "Kampf den Rundfunkstörungen." In *Rundfunk Jahrbuch 1930*, edited by Reichs-Rundfunk-Gesellschaft, 267–72. Berlin: Union Deutsche Verlagsgesellschaft.
Braun, Alfred. *Achtung, Achtung, Hier ist Berlin! Aus der Geschichte des Deutschen Rundfunks in Berlin 1923–1932.* Berlin: Haude & Spenersche Verlagsbuchhandlung, 1968.
———. "Rundfunk—Bühne—Tonfilm: Erfahrungen an der 'Rundfunkversuchsstelle.'" *Funk: Die Wochenschrift des Funkwesens* 13 (March 28, 1930): 72.
Braun, Hans-Joachim, ed. *Creativity: Technology and Music, Studien zur Technik-, Wirtschafts- und Sozialgeschichte.* Frankfurt am Main: Peter Lang, 2016.
Braun, Hans-Joachim. "Introduction." In *"I Sing the Body Electric": Music and Technology in the 20th Century*, edited by Hans-Joachim Braun, 9–32. Baltimore, MD: Johns Hopkins University Press, 2002.

———. "Technik als 'Kunsthebel' und 'Kulturfaktor': Zum Verhältnis von Technik und Kultur bei Franz Reuleaux." In *Technische Intelligenz und 'Kulturfaktor Technik.' Kulturvorstellungen von Technikern und Ingenieuren zwischen Kaiserreich und früher Bundesrepublik Deutschland*, edited by Burkhard Dietz, Michael Fessner, and Helmut Maier, 35–43. Münster: Waxmann, 1996.

Braunmüller, Hans-Jochim, and Walter Weber. "Nichtlinearische Verzerrungen von Mikrophonen." *Elektrotechnische Zeitschrift* 54 (1933): 1068–70.

Brecht, Bertolt. "Vorschläge für den Intendanten des Rundfunks." In *Bertolt Brecht: Gesammelte Werken in 20 Bänden*, 18, 121–23. Frankfurt: Suhrkamp, 1968.

Bredow, Hans. "Beginn der Rundfunkarbeit" (1927). In *Aus meinem Archiv: Probleme des Rundfunks*, edited by Hans Bredow, 34–47. Heidelberg: Vowinckel, 1950.

———. "Dem Deutschen Rundfunk zum Geleit" (1923). In *Aus meinem Archiv: Probleme des Rundfunks*, edited by Hans Bredow, 15. Heidelberg: Vowinckel, 1950.

———. "So entstand der deutsche Rundfunk: Persönliche Erinnerungen von Hans Bredow." *Rufer und Hörer: Monatshefte für Rundfunk und Fernsehen* 8 (1953): 1–31.

———. "Gegenwartsfragen des deutschen Rundfunks." In *Rundfunk Jahrbuch 1931*, edited by Reichs-Rundfunk-Gesellschaft, 1–11. Berlin: Union Deutsche Verlagsgesellschaft, 1931.

———. *Aus meinem Archiv: Probleme des Rundfunks*. Heidelberg: Vowinckel, 1950.

———. "Programm-Organisation und Ultrakurzwellen" (1930). In *Aus meinem Archiv: Probleme des Rundfunks*, edited by Hans Bredow, 61–64. Heidelberg: Vowinckel, 1950.

———. "Der Rundfunk als Kulturmacht." *Deutschland: Jahrbuch für das deutsche Volk* 3 (1929): 124–30.

———. "Der Rundfunk in Deutschland." *Funk: Die Wochenschrift des Funkvereins* 35 (August 29, 1930): 173–74.

———. "Wirtschaftliche und organisatorische Fragen." In *Rundfunk Jahrbuch 1930*, edited by Reichs-Rundfunk-Gesellschaft, 83–94. Berlin: Union Deutsche Verlagsgesellschaft, 1930.

Bridenthal, Renate, Atina Grossman, and Marion Kaplan, eds. *When Biology Became Destiny: Women in Weimar and Nazi Germany*. NY: Monthly Review Press, 1984.

Brillouin, Jacques. "Avenir de la Musique Automatique." *Cahiers d'Art* 4 (1928): 164–66.

Brilmayer, Benedikt. "Das Trautonium und Oskar Sala." *Musik in Bayern* 78 (2013): 92–119.

———. "Das Trautonium: Prozesse des Technologietransfers im Musikinstrumentenbau." PhD diss., Department of Musicology, University of Augsburg, 2017.

Bronsgeest, Cornelis. "Opernsendespiel oder Übertragung?" *Funk: Die Wochenschrift des Funkwesens* 3 (January 16, 1931): 17–18.

———. "Das Problem der Sendeoper." *Funk: Die Wochenschrift des Funkwesens* 49 (November 30, 1928): 331–32.

———. "Die Sende-Oper." In *Aus meinem Archiv: Probleme des Rundfunks*, edited by Hans Bredow, 253–255. Heidelberg: Vowinckel, 1950.

Brown, Hilda Meldrum. *The Quest for Gesamtkunstwerk and Richard Wagner*. NY: Oxford University Press, 2016.

Bruyninckx, Joeri. *Listening in the Field: Recording and the Science of Birdsong*. Cambridge, MA: MIT Press, 2018.

Buchwald, Joachim. "Die technischen Einrichtungen der Funkversuchsstelle." In *Staatliche akademische Hochschule für Musik in Berlin zu Charlottenburg: Jahrbericht vom 1. Oktober 1927 bis 30 September 1928*, edited by Georg Schünemann, 16–20. Berlin: Hochschule für Musik, 1929.

Buckley, David. *Kraftwerk Publikation*. NY: Omnibus Press, 2012.

Büttner, H. "Besinnung und Bildung." *Zeitschrift für Musik* 30 (1936): 952

Buolamwini, Joy, and Timnit Gebru. "Gender Shades: Intersectional Accuracy Disparities in Commercial Gender Classification." *Proceedings of Machine Learning Research* 81 (2018): 1–15.

Bushwick, Sophie. "How NIST Tested Facial Recognition Algorithms for Racial Bias." *Scientific American*, December 27, 2019. https://www.scientificamerican.com/article/how-nist-tested-facial-recognition-algorithms-for-racial-bias/.
Butting, Max. "Music of and for the Radio." *Modern Music* 8 (1931): 15–19.
———. "Rundfunkmusik." *Die Sendung* 6 (1929): 28.
———. "Das Verhältnis des schaffenden Musikers zum Rundfunk." In *Kunst und Technik*, edited by Leo Kerstenberg, 279–98. Berlin: Volksverband der Bücherfreunde / Wegweiser-Verlag, GmbH, 1930.
———. "Vom Wesen der Musik im Rundfunk: Nach einem Vortrag, gehalten auf der 'Ersten Tagung für Rundfunkmusik.'" *Funk: Die Wochenschrift des Funkwesens* 26 (June 22, 1928): 201–2.
Calico, Joy H. *Brecht at the Opera*. Berkeley: University of California Press, 2009.
Candby, Edward Tatnall. "'Electronics': A Side-Review." *Audio* 45 (1961): 10, 12, 14, 78–79.
Carson, Anne. "The Gender of Sound." In *Glass, Irony and God*, 119–42. NY: New Directions, 1992.
Carson, Cathryn, Alexei Kojevnikov, and Helmuth Trischler, eds. *Weimar Culture and Quantum Mechanics: Selected Papers by Paul Forman and Contemporary Perspectives on the Forman Thesis*. London/Singapore: Imperial College Press/World Scientific, 2011.
Case Western Reserve University, ed. "Harmonic Analyzers." https://physics.case.edu/about/history/antique-physics-instruments/harmonic-analyzer-2/.
Chávez, Carlos. *Toward a New Music: Music and Electricity*. New York: Norton & Comp., 1937.
Chen, Xingyu, Zhengxiong Li, Srirangaraj Setlur, and Wenyao Xu. "Exploring Racial and Gender Disparities in Voice Biometrics." *Nature* 12 (2022): 3723–37.
Chion, Michel. *Audio-Vision: Sound on Screen*, foreword by Walter Murch, edited and translated by Claudia Gorbman. NY: Columbia University Press, 1994.
Clements, Andrew. "Wagner: Das Rheingold." *The Guardian*, December 14, 2006.
Coen, Deborah R. *Vienna in the Age of Uncertainty: Science, Liberalism & Private Life*. Chicago: University of Chicago Press, 2007.
Crandall, Irving B. *Theory of Vibrating Systems and Sound*. NY: D. Van Nostrand Company, 1926.
Crandall, Irving B. "The Sounds of Speech." *Bell System Technical Journal* 4 (1925): 586–624.
Crandall, I. B., and C F. Sacia. "A Dynamical Study of Vowel Sounds." *Bell System Technical Journal* 3 (1924): 232–37.
Cochrane, Rexmond C. *Measures for Progress: A History of the National Bureau of Standards*. Washington, D.C.: National Bureau of Standards, U.S. Department of Commerce, 1966.
Cockcroft, J. D., R. T. Coe, J. A. Tyacke, and Miles Walker. "An Electric Harmonic Analyser." *Journal of the Institution of Electrical Engineers* 63 (1925): 69–119.
Cross, Lowell. "Electronic Music, 1948–1953." *Perspectives of New Music* 7 (1968): 32–65.
Das Haus der Technik e. V. Essen veranstaltet zusammen mit der Nordwestdeutschen Ausstellungs-Gesellschaft Düsseldorf, ed. "Dialog zwischen Prof. Dr. F. Trautwein, Düsseldorf und Dr. S. Goslich, Radio Bremen: Die technische Gestaltung und musikkulturelle Bedeutung elektronischer Musikinstrumente." In *Technik und Musik*. Düsseldorf: Deutsche Musikmesse Düsseldorf, 1952.
Daston, Lorraine, and Peter Galison. *Objectivity*. NY: Zone Books, 2007.
Davies, Hugh. *International Electronic Music Catalogue*. Cambridge, MA: MIT Press, 1968.
Dellinger, J. H. "Reducing the Guesswork in Tuning." *Radio Broadcast* (July 1923): 241–45.
Denton, Clifford E. "The Trautonium: A New Musical Instrument." *Radio-Craft* 4 (1933): 522–24.
Dessau, Paul. *Deutsches Miserere für gemischten Chor, Kinderchor, Sopran-, Alt-, Tenor- und Baß-Solo, großes Orchester, Orgel und Trautonium*. Text from Bertolt Brecht. Study Score 650. Leipzig/Frankfurt: C. F. Peters Musikverlag und Suhrkamp, 1979.

———. *Die Verurteilung des Lukullus: Oper in 12 Szenen*. Text from Bertolt Brecht. Berlin: Henschelverlag, 1961.

Dettelbach, Michael. "The Face of Nature: Precise Measurement, Mapping, and Sensibility in the Work of Alexander von Humboldt." *Studies in History and Philosophy of Science, Part C: Studies in the History and Philosophy of Biological and Biomedical Sciences* 30 (1999): 473–504.

Dietz, Burkhard. "'Technik und Kultur' zwischen Kaiserreich und Nationalsozialismus." In *Technische Intelligenz und "Kulturfaktor Technik." Kulturvorstellungen von Technikern und Ingenieuren zwischen Kaiserreich und früher Bundesrepublik Deutschland*, edited by Burkhard Dietz, Michael Fessner, and Helmut Maier, 105–30. Münster: Waxmann, 1996.

Dietz, Burkhard, Michael Fessner, and Helmut Maier, eds. "'Die Kulturwelt der Technik' als Argument der Technischen Intelligenz für sozialen Aufstieg und Anerkennung." In *Technische Intelligenz und "Kulturfaktor Technik." Kulturvorstellungen von Technikern und Ingenieuren zwischen Kaiserreich und früher Bundesrepublik Deutschland*, 3–32. Münster: Waxmann, 1996.

———. *Technische Intelligenz und 'Kulturfaktor Technik.' Kulturvorstellungen von Technikern und Ingenieuren zwischen Kaiserreich und früher Bundesrepublik Deutschland*. Münster: Waxmann, 1996.

Disterweg, A. "Elektromusikalisches Konzert." In *Allgemeine Musikzeitung* 61 (1934): 408.

Döblin, Alfred. "Literatur und Rundfunk." In *Alfred Döblin, Schriften zur Ästhetik, Politik und Literatur*, edited by Erich Kleinschmidt, 251–61. Freiburg: Olten Verlag, 1989.

Dörfling, Christina. *Der Schwingkreis: Schaltungsgeschichten an den Rändern von Musik und Medien*. Leiden/Paderborn: Brill/Fink, 2022.

Dolan, Emily I. *The Orchestral Revolution: Haydn and the Technologies of Timbre*. NY: Cambridge University Press, 2012.

Dolan, Emily I., and Alexander Rehding, eds. *The Oxford Handbook of Timbre*. NY: Oxford University Press, 2021.

Dolan, Emily I., and Alexander Rehding, eds. "Histories and Possible Futures for the Study of Music." In *The Oxford Handbook of Timbre*, 3–19. NY: Oxford University Press, 2021.

Donhauser, Peter. *Elektrische Klangmaschinen: Die Pioneerzeit in Deutschland und Österreich*. Vienna: Böhlau, 2007.

———. *Oskar Sala als Instrumentenbauer: Ein Leben für das Trautonium*. Deutsches Museum Studies 11. Munich: Deutsches Museum, 2022.

Dümling, Albrecht. "Norm und Diskriminerung: Die Reichsmusiktage 1938 in Düsseldorf und die Austellung 'Entartete Musik.'" In *Das "Dritte Reich" und die Musik*, edited by Stiftung Schloss Neuhardenberg, 105–11. Berlin: Nicolai, 2006.

Dümling, Albrecht, and Peter Girth, eds. *Entartete Musik. Eine kommentierte Rekonstruktion*. Düsseldorf: Düsseldorfer Symphoniker, 1988.

Dürre, Konrad. "Der Kinderfunk: ein Freudenquell für unsere Kleinen." In *Rundfunk Jahrbuch 1929*, edited by Reichs-Rundfunk-Gesellschaft, 347–52. Berlin: Union Deutsche Verlagsgesellschaft, 1929.

Dvorak, Helga. "Friedrich Trautwein." In *Biographisches Lexikon der Deutschen Burschenschaften*, edited by Peter Kaupp, vol. 2, 690–92. Heidelberg: Universitätsverlag Winter, 2018.

Ebbeke, Klaus. "Paul Hindemith und das Trautonium." In *Hindemith-Jahrbuch* 11 (1982): 77–113. Reprinted in *Über Hindemith: Aufsätze zu Werk, Ästhetik und Interpretation*, edited by Susanne Schaal-Gotthardt und Luitgard Schader. Mainz: Schott, 1996.

Eckert, Gerhardt. *Der Rundfunk als Führungsmittel der nationalsozialistischen Bewegung*. Berlin: Kurt Vowinckel, 1941.

Edwards, Preston Hampton. "A Method for a Quantitative Analysis of Musical Tone." *Physical Review* 32 (1911): 23–37.

Eidsheim, Nina Sun. *The Race of Sound: Listening, Timbre & Vocality in African American Music*. Durham, NC: Duke University Press, 2019.
Eimert, Herbert. "Elektronische Musik." *Technische Hausmitteilungen des Nordwestdeutschen Rundfunks* 6 (1954): 4–5.
———. "Der Komponist und die elektronischen Klangmittel." *Das Musikleben* 7 (1954): 242–45.
Electronic Communications Committee (ECC). "Managing the Transition to Digital Sound Broadcasting in the Frequency Bands Below 80 MHz: Athens, February 2008/Revised Gothenburg, September 2010." Report 117, p. 7. https://docdb.cept.org/download/79cbf93a-c276/ECCRep117.pdf.
Epstein, Margot. "Chorgesang im Rundfunk." *Funk: Die Wochenschrift des Funkwesens* 30 (July 25, 1930): 150.
Essau, A. "Die Verstärkung des Rundfunksenders und ihre Rückwirkung auf den Empfang und die Empfänger." *Mitteldeutsche Monatshefte* 10 (1927): 411–12.
Ettingen, Max. "Opern für den Rundfunk." *Funk: Die Wochenschrift des Funkwesens* 3 (March 28, 1930): 75–76.
———. "Opernfunk—Funkoper!" *Funk: Die Wochenschrift des Funkwesens* 13 (March 28, 1930): 110.
European Broadcasting Convention: Copenhagen Plan Annexed to European Broadcasting Convention (Bern: General Secretariat of the International Telecommunication Union, 1948). https://search.itu.int/history/HistoryDigitalCollectionDocLibrary/4.65.43.en.100.pdf.
Everett, Annie. "The Generation of the Sonderweg." *International Social Science Review* 91 (2015). http://digitalcommons.northgeorgia.edu/issr/vol91/iss2/1.
Fagen, M. D., ed. *A History of Engineering and Science in the Bell System: The Early Years (1875–1925)*. NY: Bell Telephone Laboratories, Inc., 1975.
F. E. "Vom Rundfunk zur elektrischen Hausmusik." *Funk: Die Wochenschrift des Funkwesens* 36 (September 2, 1932): 144.
Fickinger, William. *Physics at a Research University: Case Western Reserve 1830–1990*. Cleveland, OH: Case Western Reserve University, 2006.
Finkelstein, Gabriel. *Emil du Bois-Reymond: Neuroscience, Self, and Society in Nineteenth-Century Germany*. Cambridge, MA: MIT Press, 2013.
Fischer, Wilhelm. "Reproduktion" (1924). In *Aus meinem Archiv: Probleme des Rundfunks*, edited by Hans Bredow, 83–88. Heidelberg: Vowinckel, 1950.
Fischer, Wolfgang. "Worte werden zu Bildern . . . Ein Besuch beim technischen Leiter der Rundfunk-Versuchsstelle in Berlin." *Funk: Die Wochenschrift des Funkwesens* 9 (February 23, 1934): 163.
Fischer-Defoy, Christine. *Kunst. Macht. Politik: Die Nazifizierung der Kunst- und Musikhochschulen in Berlin*. Berlin: Elefanten Press, 1988.
Flesch, Hans. "Zur Ausstellung des Programms im Rundfunk." *Die Besprechung (Beilage der Radio-Umschau)* 1 (1924): 3–4.
———. "Inbetriebnahme des Frankfurter Rundfunksenders." *Radio-Umschau* 1 (1924): 122–23.
———. "Ein Jahr Frankfurter Programm, gehalten in Frankfurter Sender am 1. April 1925 von Dr. Hans Flesch." *Radio-Umschau* 2 (1925): 154–57.
———. "Die neuen künsterlischen Probleme." In *Rundfunk Jahrbuch 1930*, edited by Reichs-Rundfunk-Gesellschaft, 100–1. Berlin: Union Deutsche Verlagsgesellschaft, 1930.
———. "Rundfunkmusik." In *Rundfunk Jahrbuch 1930*, edited by Reichs-Rundfunk-Gesellschaft, 146–53. Berlin: Union Deutsche Verlagsgesellschaft, 1930. Reprinted in *Aus meinem Archiv: Probleme des Rundfunks*, edited by Hans Bredow, 237–41. Heidelberg: Vowinckel, 1950.

———. "Das Studio der Berliner Funk-Stunde." In *Rundfunk Jahrbuch 1930*, edited by Reichs-Rundfunk-Gesellschaft, 117–21. Berlin: Union Deutsche Verlagsgesellschaft, 1930.

Fletcher, Harvey. "Loudness and Pitch." *Bell Laboratories Record* 13 (1935): 130–35.

———. "Loudness, Pitch and the Timbre of Musical Tones and Their Relation to the Intensity, the Frequency and the Overtone Structure." *Journal of the Acoustical Society of America* 6 (1934): 59–69.

———. "The Physical Criterion for Determining the Pitch of a Musical Tone." *Physical Review* 23 (1924): 427–37.

———. "Physical Measurements of Audition and Their Bearing on the Theory of Hearing." *Bell System Technical Journal* 2 (1923): 145–80.

———. *Speech and Hearing*, with an introduction by H. D. Arnold. NY: D. van Nostrand Co., 1929.

———. "Symposium on Wire Transmission of Symphonic Music and Its Reproduction in Auditory Perspective: Basic Requirements." *Bell System Technical Journal* 13 (1934), 239–44.

Fletcher, Harvey, and R. L. Wegel. "The Frequency-Sensitivity of Normal Ears." *The Physical Review* 19 (1922): 553–65.

Föllmer, Moritz. "The Middle Classes." In *The Oxford Handbook of the Weimar Republic*, edited by Nadine Rossol and Benjamin Ziemann, 454–74. NY: Oxford University, 2020.

Forman, Paul. "Weimar Culture, Causality, and Quantum Theory, 1918–1927: Adaptation by German Physicists and Mathematicians to a Hostile Intellectual Environment." *Historical Studies in the Physical Sciences* 3 (1971): 1–115.

Frederick, H. A. "The Development of the Microphone." *Journal of the Acoustical Society of America* 3 (1931): 1–25, reprinted in *Bell Telephone Quarterly* 10 (1931): 164–88.

Freke, Oli. *Synthesizer Evolution: From Analogue to Digital (and Back)*. London: Velocity Press, 2021.

Frieß, Peter, and Peter M. Steiner, eds. "Oskar Sala im Gespräch." In *Deutsches Museum Bonn: Forschung und Technik in Deutschland nach 1945*, 215–236. Berlin: Deutscher Kunstverlag, 1995.

Führer, Karl Christian. "A Medium of Modernity? Broadcasting in Weimar Germany, 1923–1932." *Journal of Modern History* 69 (1997): 722–53.

———. *Wirtschaftsgeschichte des Rundfunks in der Weimarer Republik*. Potsdam: Verlag für Berlin-Brandenburg, 1997.

Funkhusen, W. "Das Trautonium des Bastlers: Bauanleitung für ein einfaches elektrisches Musikinstrument." *Funk: Die Wochenschrift des Funkwesens* 22 (May 25, 1934): 385–86.

Funktechnischer Verein, ed. *Bestimmungen über den Unterhaltungs-Rundfunk. Funk-Taschenbuch*. 3 parts. Berlin: Weidmannsche Buchhandlung, 1924.

Gabriel, John Arthur. "Opera after Optimism: The Fate of Zeitoper at the End of the Weimar Republic." PhD diss., Harvard University, 2016.

Gagarin Records. *Historische Aufnahmen / Historical Recordings*, GR 2013, no. 1, 2009. http://www.gagarinrecords.de/.

Garcia, Megan. "Racist in the Machine: The Disturbing Implications of Algorithmic Bias." *World Policy Journal* 33 (2016–17): 111–17.

Garten, Siegfried, and Friedrich Kleinknecht. "Beiträge zur Vokallehre III: Die automatische harmonische Analyse der gesungenen Vokale." *Abhandlungen der mathematisch-physikalischen Klasse der Sächsischen Akademie der Wissenschaften* 38 (1921): 1–43.

Garvie, Clare, and Jonathan Frankle. "Facial-Recognition Software Might Have a Racial Bias Problem." *The Atlantic*, April 7, 2016. https://www.theatlantic.com/technology/archive/2016/04/the-underlying-bias-of-facial-recognition-systems/476991/.

Gerlach, Hans. "Wie Königs Wusterhausen zum ersten deutschen Rundfunksender wurde." In *Rundfunk Jahrbuch 1930*, edited by the Reichs-Rundfunk-Gesellschaft, 27–41. Berlin: Union Deutsche Verlagsgesellschaft, 1930.

Gerlach, Walther. "Über einen registrierenden Schallmesser und seine Anwendungen." *Zeitschrift für technische Physik* 8 (1927): 515–19.

Germann, Walter. "Das Trautonium." *Sonderdruck aus Telefunkenzeitung* 64 (1933): 1–12.

Gernsback, Hugo. "Electronic Music." *Radio-Craft* 4 (1933): 521.

Geyer, Martin H. "The Period of Inflation, 1919–1923." In *The Oxford Handbook of the Weimar Republic*, edited by Nadine Rossol and Benjamin Ziemann, 48–71. NY: Oxford University Press, 2020.

Giesecke, Heinrich. "Die Organisation des Deutschen Rundfunks." In *Rundfunk Jahrbuch 1929*, edited by the Reichs-Rundfunk-Gesellschaft, 29–41. Berlin: Union Deutsche Verlagsgesellschaft, 1929.

———. "Wie die Organisation des Deutschen Rundfunks vorbereitet wurde." In *Rundfunk Jahrbuch 1930*, edited by Reichs-Rundfunk-Gesellschaft, 59–67. Berlin: Union Deutsche Verlagsgesellschaft, 1930.

Gilfillan, Daniel. *Pieces of Sound: German Experimental Radio*. Minneapolis: University of Minnesota Press, 2009.

Gilling, Ted. "The Colour of Music: An Interview with Bernard Hermann." *Sight and Sound* 41 (1971–72): 37.

Gispen, Kees. *New Profession, Old Order: Engineers and German Society, 1815–1914*. NY: Cambridge University Press, 1989.

Goebbels, Joseph. "Über die Bedeutung der Technik." *Deutsche Technik: Technopolitische Zeitschrift der Architekten, Chemiker, Ingenieure, Techniker* (March 1939): 105–6.

Goldhammer, Klaus, Peter Laufer, and Sebastian Lehr. "'Achtung! Achtung! Hier ist die Sendestelle Berlin': German Radio between Regulation and Competition." In *The Palgrave Handbook of Global Radio*, edited by John Allen Hendricks, 74–91. London: Palgrave/Macmillian, 2012.

Goucher, F. S. "The Carbon Microphone: An Account of Some Researches Bearing on Its Action." *Bell System Technical Journal* 13 (1934): 163–94.

Gradenwitz, Alfred. "The Trautonium." *Wireless Constructor* (February 1931): 254

Graener, Paul. "Aufklang." *Deutsche Kultur-Wacht* 2 (1933): 1.

Gratis, H. G. "'Die Werke klingen im Rundfunk nicht!' Hermann Scherchen über Rundfunkprobleme." *Funk: Die Wochenschrift des Funkwesens* 8 (February 21, 1930): 32.

Gregersen, Friedrich. "Zur Psychologie des Rundfunkhörens." *Funk: Die Wochenschrift des Funkwesens* 10 (March 6, 1931): 78.

Grey, John M. "An Exploration of Musical Timbre, Center for Computer Research in Music and Acoustics." Report No. STAN-M-2. Palo Alto, CA: Department of Music, Stanford University, 1975.

Grosch, Nils. *Die Musik der neuen Sachlickeit*. Stuttgart: J. B. Metzler, 1999.

Grützmacher, Martin. "Eine neue Methode der Klanganalyse." *Zeischrift für technische Physik* 8 (1927): 506–9, expanded in *Elektrische Nachrichtentechnik* 4 (1927): 533–45 and *Sonderabdruck aus den Mittheilungen aus dem Telegraphentechnischen Reichsamt* 13 (1927): 216–28.

———. "Zur Analyse von Geräuschen." *Zeitschrift für technische Physik* 10 (1929): 570–72.

———. "Klanganalyse mit einem Einfadenelektrometer." *Zeitschrift für technische Physik* 10 (1929): 572–73.

Grützmacher Martin, and Erwin Meyer. "Eine Schallregistriervorrichtung zur Aufnahme der Frequenzkurven von Telephonen und Lautsprechern." *Sonderabdruck aus den Mittheilungen aus der Elektrischen Nachrichtentechnik* 4 (1927): 203–11.

Grunicke, W. "Die Sprachkunst im Rundfunk: Ein neuer Weg zum Sendespiel?" *Funk: Die Wochenschrift des Funkwesens* 13 (1925): 149–50.

Guicking, Dieter. *Erwin Meyer—ein bedeutender deutscher Akustiker: Biographische Notizen*. Göttingen: Universitätsdrücke Göttingen, 2012. http://www.guicking.de/dieter/Erwin-Meyer.pdf.

Gutzmann, Hermann. *Physiologie der Stimme und Sprache*. Braunschweig: Friedrich Vieweg and Son, 1909.

Guzatis, Heinz-Gert. "Berliner Programmstatistik vom Januar bis März 1930." *Funk: Die Wochenschrift des Funkwesens* 25 (June 20, 1930): 130.

Haas, Michael. *Forbidden Music: The Jewish Composers Banned by the Nazis*. New Haven, CT: Yale University Press, 2013.

Hadamovsky, Eugen. *Der Rundfunk im Dienste der Volksführung*. Leipzig: R. Noske, 1934.

Hagemann, Carl. "Wie kommt ein Rundfunk-Programm zustande?" In *Rundfunk Jahrbuch 1929*, edited by Reichs-Rundfunk-Gesellschaft, 301–6. Berlin: Union Deutsche Verlagsgesellschaft, 1929.

———. "Die künstlerisch-kulturelle Zielsetzung des Deutschen Rundfunks." In *Rundfunk Jahrbuch 1929*, edited by Reichs-Rundfunk-Gesellschaft, 121–37. Berlin: Union Deutsche Verlagsgesellschaft, 1929, reprinted in *Aus meinem Archiv: Probleme des Rundfunks*, edited by Hans Bredow, 227–37. Heidelberg: Vowinckel, 1950.

Hagen, Wolfgang. *Das Radio: Zur Geschichte und Theorie des Hörfunks—Deutschland/USA*. Munich: Wilhelm Fink, 2005.

Halefelt, Horst O. "Sendegesellschaft und Rundfunkordnungen." In *Programmgeschichte des Hörfunks in der Weimar Republik*, edited by Joachim-Felix Leonhard, 2 vols. Munich: Deutscher Taschenbuch Verlag, 1997.

Halliwell, Stephen. *The Aesthetics of Mimesis: Ancient Texts and Modern Problems*. Princeton, NJ: Princeton University Press, 2002.

Hårt, Mikael. "German Regulation: The Integration of Modern Technology into National Culture." In *The Intellectual Appropriation of Technology: Discourses on Modernity, 1900–1939*, edited by Mikael Hård and Andrew Jamison, 33–67. Cambridge, MA: MIT Press, 1998.

Heiber, Helmut. *Joseph Goebbels*. London: Hawthorn Books, Inc., 1972.

Heinitz, Wilhelm. *Klangprobleme im Rundfunk*. Berlin: Rothgiesser & Diesing AG, 1926.

Hempel, Manfred. "German Television Pioneers and the Conflict between Public Programming and Wonder Weapons." *Historical Journal of Film, Radio, and Television* 10 (1990): 123–62.

Henson, Karen. "Sound Recording and the Operatic Canon: 'Three Drops of the Needle.'" In *The Oxford Handbook of the Operatic Canon*, edited by Cormac Newark and William Weber, 495–508. NY: Oxford University Press, 2020.

Henson, Karen, ed. *Technology and the Diva. Sopranos, Opera, and Media from Romanticism to the Digital Age*. NY: Cambridge University Press, 2016.

Herf, Jeffrey. *Reactionary Modernism: Technology, Culture and Politics in Weimar and the Third Reich*. NY: Cambridge University Press, 1984.

Hermann, Ludimar, ed. *Handbuch der Physiologie*. 6 vols. Leipzig: F. C. W. Vogel, 1879–83.

Hernried, Robert. "Geistige Probleme der Rundfunkoper." In *Funk: Die Wochenschrift des Funkwesens* 10 (March 7, 1930): 39.

———. "Das Orchesterhören: Ein Vorschlag zur musikalischen Erziehung." In *Funk: Die wochenschrift des Funkwesens* 23 (June 6, 1930): 117–18.

Herrmann-Goldap, Erich. "Über die Klangfarbe einiger Orchesterinstrumente." In *Annalen der Physik* 23 of the new series, 328 of the entire series (1907): 979–85.

Herter, Hans, Joseph Schmidt-Görg, Leo Weisgerber, and Hanswilly Sendhoff. *In Memoriam Werner Meyer-Eppler*. Bonn: Peter Hanstein Verlag, 1962.

Hilmes, Michele. *Radio Voices: American Broadcasting 1922–1952*. Minneapolis: University of Minnesota Press, 1997.

Hobsbawm, Eric, and Terence Ranger, eds. *The Invention of Tradition*. New York: Cambridge University Press, 1983.

Höpfner, Karl. "Entwicklung des Kabelwesens in seiner Bedeutung für den Rundfunk." In *Rundfunk-Jahrbuch 1930*, edited by Reichs-Rundfunk-Gesellschaft, 261–67. Berlin: Union Deutsche Verlagsgesellschaft, 1930.

Holl, Karl. "Elektro-akustische Musik." *Das Illustriete Blatt* 30 (1930): 826.
Hollows, Joanne. *Domestic Cultures.* Maidenhead, UK: Open University Press and McGraw-Hill House, 2008.
Holmes, Thom. *Electronic and Experimental Music: Technology, Music, and Culture,* 3rd edition. NY: Routledge, 2008.
Horosko, Marian. "A Matter of Values." *Dance Magazine* 33 (1959): 24–25.
Hortleder, Gerd. *Das Gesellschaftsbild des Ingenieurs: Zum politischen Verhalten der technischen Intelligenz in Deutschland.* Frankfurt: Suhrkamp, 1970.
Hui, Alexandra. *The Psychophysical Ear: Musical Instruments, Experimental Sounds, 1840–1910.* Cambridge, MA: MIT Press, 2012.
Hui, Alexandra, Julia Kursell, and Myles W. Jackson, eds. *Music, Sound, and the Laboratory from 1750–1980.* Chicago: University of Chicago Press, 2013.
Hunt, Frederick V. *Electroacoustics: The Analysis of Transduction, and Its Historical Background.* Cambridge, MA: Harvard University Press, 1954.
Huth, Arno. "Elektrische Tonerzeugung: Zu den Erfindungen von Jörg Mager und Leo Theremin." *Die Musik* 21 (1927): 42–45.
Ikl. "Fernmeldetechnik." *Elektrotechnische Zeitschrift* 25 (1928): 954–55.
Iverson, Jennifer. *Electronic Inspirations: Technologies of the Cold War Musical Avant-Garde.* NY: Oxford University Press, 2018.
Jackson, Myles W. *Harmonious Triads: Physicists, Musicians, and Instrument Makers in Nineteenth-Century Germany.* Cambridge, MA: MIT Press, 2006.
Janovsky, J. "Elektrische Musikinstrumente, ihre Wirkungsweise und Aufgaben." *Elektrotechnische Zeitschrift* 54 (1933): 675–78, 727–30.
Jarausch, Konrad H. *The Unfree Professions: German Lawyers, Teachers, and Engineers, 1900–1950.* NY: Oxford University Press, 1990.
Jeans, James. *Science & Music.* NY: Macmillan, 1937.
Joerges, Bernward. "Do Artifacts Have Politics?" *Social Studies of Science* 29 (1999): 411–31.
Jones, Isaac H., and Vern O. Knudsen. "Facts of Audition." *Annals of Otology, Rhinology, and Laryngology* 34 (1925): 1013–25.
KAB. "Das Stuttgarter Rundfunkprogramm." *Der deutsche Rundfunk: Rundschau und Programm für alle Funkteilnehmer* 3 (1925): 2281.
Kämpfer, Frank. "Ein Leben für das Trautonium: Ein Besuch bei Oskar Sala zum 85." *Neue Zeitschrift für Musik* 156 (1995): 44–47.
Kalähne, Alfred. "Schwallwiedergabe durch Lautsprecher." In *Müller-Pouillets Lehrbuch der Physik,* edited by by Erich Waetzmann, 333–34. Braunschweig: Friedrich Vieweg & Sohn, 1929.
Kapeller, Ludwig. "Der enträtselte Rundfunk: Die erste Tagung für Rundfunkmusik in Göttingen." *Funk: Die Wochenschrift des Funkwesens* 19 (May 4, 1928): 145–46.
———. "Die 'Rede' im Rundfunk." *Funk: Die Wochenschrift des Funkwesens* 31 (1924): 469–470.
———. "Die Schwierigkeiten der Rundfunk-Aufnahme." *Funk: Die Wochenschrift des Funkwesens* 6 (February 3, 1928): 41–42.
———. "Die verwaiste Kultur in Rundfunk." *Funk: Die Wochenschrift des Funkwesens* 23 (4 June 1926): 177–79.
Karimi, Faith. "'Mom, These Bad Men Have Me': She Believes Scammers Cloned Her Daughter's Voice in a Fake Kidnapping." CNN, April 29, 2023. https://www.cnn.com/2023/04/29/us/ai-scam-calls-kidnapping-cec/index.html.
Katz, Mark. *Capturing Sound: How Technology Has Changed Music,* revised edition. Berkeley: University of California Press, 2010.
———. "Hindemith, Toch, und *Grammophonmusik.*" *Journal for Musicological Research* 20 (2001): 161–80.

———. *Music and Technology: A Very Short Introduction.* NY: Oxford University Press, 2022.
Kautny, Oliver. "Pionierzeit der elektronischen Musik: Werner-Meyer Epplers Einfluß auf Herbert Eimert." In *Musik im Spektrum von Kultur und Gesellschaft: Festschrift für Brunhilde Sonntag,* edited by Berhard Müßgens, Oliver Kautny, and Martin Gieseking, 315–37. Osnabrück: Electronic Publishing Osnabrück, 2001.
Khruschchev, N. S. "In hohem Ideengehalt und künstlerischer Meisterschaft liegen die Kraft der sowjetischen Literatur und Kunst." *Neues Deutschland* 18 (1963): 3–6.
Klotz, Sebastian. "Timbre, *Komplexeindruck,* and Modernity: Klangfarbe as a Catalyst of Psychological Research in Carl Stumpf, 1890–1926." In *The Oxford Handbook of Timbre,* edited by Emily I. Dolan and Alexander Rehding, 609–40. NY: Oxford University Press.
Kochka, Jürgen. "German History before Hitler: The Debate of the German Sonderweg." *Journal of Contemporary History* 23 (1988): 3–16
———. "Looking Back at the Sonderweg." *Central European History* 51 (2018): 137–42.
Koenecke, Alison, Andrew Nam, Emily Lake et al. "Racial Disparities in Automated Speech Recognition." *Proceedings of the National Academy of Sciences* 117 (2020): 7684–89.
König, Wolfgang. "Die Ingenieure und der VDI als Großverein in der wilhelminischen Gesellschaft 1900 bis 1918." In *Technik, Ingenieure und Gesellschaft: Geschichte des Vereins Deutscher Ingenieure 1856–1981,* edited by Karl-Heinz Ludwig with assistance from Wolfgang König, 235–87. Düsseldorf: VDI-Verlag, 1981
Koeßler, Paul. "Das Gespräch über die Technik. Entstehung und Arbeit der VDI-Hauptgruppe Mensch und Technik des Vereins Deutscher Ingenieure." *Humanismus und Technik* 9 (1964): 70.
Kolb, Richard, and Heinrich Siekmeier, eds. *Rundfunk und Film im Dienste nationaler Kultur.* Düsseldorf: Friedrich Floeder Verlag, 1933.
Kotowski, P., and W. Germann. "Das Trautonium." In *Elektrische Nachrichtentechnik* 11 (1934): 389–99.
Krigar-Menzel, O., and A. Raps. "Über Saitenschwingungen." *Annalen der Physik und Chemie* 280 (1891): 623–41.
Kromhout, Melle Jan. *The Logic of Filtering: How Noise Shapes the Sound of Recorded Music.* NY: Oxford University Press, 2021.
Kromhout, Melle Jan. "The Unmusical Ear: Georg Simon Ohm and the Mathematical Analysis of Sound." *Isis* 111 (2020): 471–92.
Kursell, Julia. "Carl Stumpf and the Beginnings of Research in Musicality." In *The Origins of Musicality,* edited by Henkjan Honing, 323–46. Cambridge, MA: MIT Press, 2018.
———. "Coming to Terms with Sound: Carl Stumpf's Discourse on Hearing Music and Language." *History of Humanities* 6 (2021): 35–59.
———. "Experiments on Tone Color in Music and Acoustics: Helmholtz, Schoenberg, and *Klangfarbenmelodie.*" In *Music, Sound, and the Laboratory from 1750–1980,* edited by Alexandra Hui, Julia Kursell, and Myles W. Jackson, 191–211. Chicago: University of Chicago Press, 2013.
———. "'False Relations': Hermann von Helmholtz's Study of Music and the Delineation of Nineteenth-Century Physiology." *Nineteenth-Century Music Review* 19 (2022): 85–96.
———. "A Gray Box: The Phonograph in Laboratory Experiments and Fieldwork, 1900–1920." In *The Oxford Handbook of Sound Studies,* edited by Trevor Pinch and Karin Bijsterveld, 176–97. NY: Oxford University Press, 2011.
———. "Listening to More Than Sounds: Carl Stumpf and the Experimental Recordings of the Berliner Phonogramm-Archiv." *Technology and Culture* 60 (2019): S39–S63.
———. "Musikwissenschaft am Berliner Institut für Psychologie: Carl Stumpf und der Interferenzapparat." In *Musikwissenschaft 1900–1930: Zur Institutionalisierung und Legitimierung einer jungen akademischen Disziplin,* edited by Wolfgang Auhagen, Wolfgang Hirschmann, and Tomi Mäkelä, 73–90. New York: Georg Olms Verlag, 2017.

———. "From Tone to Tune: Carl Stumpf and the Violin." *19th-Century Music* 4 (2019): 121–39.

Lacey, Kate. *Feminine Frequencies: Gender, German Radio, and the Public Sphere, 1923–1945*. Ann Arbor: University of Michigan Press, 1996.

Lade, Ludwig. "Zweite Tagung für Rundfunkmusik." *Melos* 10 (1931): 284–86.

Laemmli, Whitney E. "A Case in Pointe: Romance and Regimentation at the New York City Ballet." *Technology & Culture* 56 (2015): 1–27.

Lalli, Robert. "The Interplay of Theoretical Assumptions and Experimental Practice in the History of Ether-Drift Experiments." In *XXXIII Congresso Nazionale della Società Italiana degli Storici della Fisica e dell'Astronomia*, edited by Lucio Fregonese, 343–60. Pavia: Pavia University Press, 2016.

Landeen, A. G. "Analyzer for Complex Electric Waves." *Bell Systems Technical Journal* 6 (1927): 230–47.

Lane, C. E. "Minimal Sound Energy for Audition." *Physical Review* 19 (1922): 492–97.

Lawrence, Harold. "Electronics by Leaps and Bounds." *Audio* 45 (1961): 70–72.

Lebede, Hans. "Mundart und deutsche Hochsprache vor dem Mikrophon." *Funk: Die Wochenschrift des Funkwesens* 25 (1933): 97–98.

Leberke, Ministerialrat. "Wirtschaftsführung im Rundfunk" (originally published beginning of 1933). In *Aus meinem Archiv: Probleme des Rundfunks*, edited by Hans Bredow, 73–74. Heidelberg: Vowinckel, 1950.

Lee, Xiaochang, and Mara Mills. "Vocal Features: From Voice Identification to Speech Recognition by Machine." *Technology and Culture* 60 (2019): 129–60.

Leers, Peter W. "Elektrische Musik in der Kritik der Öffentlichkeit." *Funk: Die Wochenschrift des Funkwesens* (1932): 178.

Lehmann, Nina Ermlich, Sophie Fetthauer, Mathias Lehmann et al., eds. *Fokus* Deutsches Miserere *von Paul Dessau und Bertolt Brecht. Festschrift Peter Petersen zum 65. Geburtstag.* Hamburg: Von Bockel Verlag, 2005.

Leithäuser, Gustav. "Theorie und Praxis der neuen Wellenverteilung." *Funk: Die Wochenschrift des Funkwesens* 6 (February 8, 1929): 21–22.

———. "Wissenwertes über neuzeitliche Empfänger." In *Rundfunk Jahrbuch 1930*, edited by the Reichs-Rundfunk-Gesellschaft, 273–88. Berlin: Union Deutscher Verlagsgesellschaft, 1930.

Leppert, Richard. "Commentary." *Essays on Music by Theodor Adorno*, edited by Richard Leppert, 213–50. Berkeley and Los Angeles: University of California Press, 2002.

Lertes, Peter. *Elektrische Musik: Eine gemeinverständliche Darstellung ihrer Grundlagen, des heutigen Standes der Technik und ihrer Zukunftsmöglichkeiten.* Dresden & Leipzig: Theodor Steinkopff, 1933.

———. "Zur historischen Entwicklung der elektrischen Musikinstrumente." *Funk: Die Wochenschrift des Funkwesens* 16 (April 13, 1934): 297–98.

———. "Unser Programm!" *Radio-Umschau: Nachrichtenblatt der Südwestdeutschen Gesellschaft A.-G., Frankfurt-M.* 1 (1924): 1–2.

Levi, Erik. *Music in the Third Reich*. NY: St. Martin's Press, 1994.

Levin, Thomas Y., and Michael von der Linn. "Elements of a Radio Theory: Adorno and the Princeton Radio Research Project." *The Musical Quarterly* 78 (1994): 316–24.

Lichtenthal, Herbert. "Der Chorgesang im Rundfunk." *Funk: Die Wochenschrift für Funkwesen* 27 (July 4, 1930): 138.

Lima, Lanna, Elizabeth Furtado, and Vasco Furtado. "Empirical Analysis of Bias in Voice-based Personal Assistants." *WWW '19: Companion Proceedings of the World Wide Web Conference*. International World Wide Web Conference Committee, 2019: 533–38.

Lion, A. "Die technischen Grundlagen von Theremins Ätherwellenmusik." *Die Musik* 21 (1929): 357–58.

Lockheart, Paula. "A History of Early Microphone Singing, 1925–1939: American Mainstream Popular Singing at the Advent of Electronic Microphone Amplification." *Popular Music and Society* 26 (2003): 367–85.

Lombardi, Michael A., and Glenn K. Nelson. "WWVB: A Half Century of Delivering Accurate Frequency and Time by Radio." *Journal of Research of the National Institute of Standards and Technology* 118 (2014): 25–54.

Lommers, Suzanne. *Europe—On Air: Interwar Projects for Radio Broadcasting*. Amsterdam: University of Amsterdam Press, 2012.

Lubsznynski, Günther. "Rundfunk Aufnahme." In *Rundfunk Jahrbuch 1930*, edited by the Reichs-Rundfunk-Gesellschaft, 245–54. Berlin: Union Deutsche Verlagsgesellschaft, 1930.

Ludwig, Karl-Heinz. *Technik und Ingenieure im Dritten Reich*. Düsseldorf: Droste, 1974.

———. "Der VDI als Gegenstand der Parteipolitik 1933 bis 1945." In *Technik, Ingenieure und Gesellschaft—Geschichte des Vereins Deutscher Ingenieure 1856–1981*, edited by Karl-Heinz Ludwig with assistance from Wolfgang König, 407–26. Düsseldorf: VDI-Verlag, 1981.

———. "Vereinsarbeit im Dritten Reich 1933 bis 1945." In *Technik, Ingenieure und Gesellschaft—Geschichte des Vereins Deutscher Ingenieure 1856–1981*, edited by Karl-Heinz Ludwig with assistance from Wolfgang König, 429–54. Düsseldorf: VDI-Verlag, 1981.

Ludwig, Karl-Heinz with assistance from Wolfgang König, eds. *Technik, Ingenieure und Gesellschaft—Geschichte des Vereins Deutscher Ingenieure 1856–1981*. Düsseldorf: VDI-Verlag, 1981.

Luening, Otto. "An Unfinished History of Electronic Music." *Musical Educators Journal* 55 (1968): 42–49, 135–42, 145.

McClelland, Charles E. *The German Experience of Professionalization: Modern Learned Professions and Their Organizations from the Early Nineteenth Century to the Hitler Era*. NY: Cambridge University Press, 1991.

Mager, Jörg. *Eine neue Epoche der Musik durch Radio*. Berlin-Neukölln: Selbstverlag des Verfassers, 1924.

———. "Eine Rundfunkprophezeiung." *Der Deutsche Rundfunk* 2 (1924): 2952.

Magnus, Kurt. "Gegen Zentralisierung." In *Aus meinem Archiv: Probleme des Rundfunks*, edited by Hans Bredow, 58–61. Heidelberg: Vowinckel, 1950.

———. "Die Gründung der Berliner 'Radio-Stunde.'" In *Rundfunk Jahrbuch 1930*, edited by Reichs-Rundfunk-Gesellschaft, 53–58. Berlin: Union Deutsche Verlagsgesellschaft, 1930.

———. "Reichs-Rundfunk-Gesellschaft: Organisation und Aufgabenkreis." In *Aus meinem Archiv: Probleme des Rundfunks*, edited by Hans Bredow, 31–34. Heidelberg: Vowinckel, 1950.

Maier, Helmut. "Nationalsozialistische Technikideologie und die Politisierung des 'Technikerstandes': Fritz Todt und die Zeitschrift 'Deutsche Technik.'" In *Technische Intelligenz und "Kulturfaktor Technik." Kulturvorstellungen von Technikern und Ingenieuren zwischen Kaiserreich und früher Bundesrepublik Deutschland*, edited by Burkhard Dietz, Michael Fessner, and Helmut Maier, 253–68. Münster: Waxmann, 1996.

Manegold, Karl-Heinz. *Universität, Technische Hochschule und Industrie: Ein Beitrag zur Emanzipation der Technik im 19. Jahrhundert unter besonderer Berücksichtigung der Bestrebungen Felix Kleins*. Berlin: Duncker & Humblot, 1970.

Manning, Peter. *Electronic and Computer Music*, 4th edition. NY: Oxford University Press, 2013.

———. "The Influence of Recording Technologies on the Early Development of Electroacoustic Music." *Leonardo Music Journal* 13 (2003): 5–10

Marten, Reinhold. "Was ist Radio für die Kunst?" *Die Besprechung (Beilage der Radio-Umschau)* 1 (1924): 33–35.

Martin, John. "The Dance: Synthetics—Balanchine Utilizes Electronic Score." *New York Times*, April 9, 1961.

Matthews, M. V. "The Evolution of Real-time Music Synthesizers." *Journal of the Acoustical Society of America* 77 (1985): S74.

Meissner, A. “Die Entwicklung des Röhrensenders.” *Radio-Umschau. Wochenschrift über die Fortschritte im Rundfunkwesen* 1 (1924): 629–32.

Mendelssohn, M. Felix. “Das Kompromiß zwischen Kunst und Technik.” *Funk: Die Wochenschrift des Funkwesens* 36 (September 2, 1927): 294.

———. “Die Versuchsanstalt für den Rundfunk.” *Funk: Die Wochenschrift des Funkwesens* 47 (November 18, 1927): 385–86.

Menzel, Werner. “Vor Mikrophon und Lautsprecher.” *Funk: Die Wochenschrift des Funkwesens* 41 (October 9, 1931): 321–22.

Mersmann, Hans. “Dr. Trautweins elektrische Musik.” *Sonderabdruck aus Melos: Zeitschrift für Musik* 5/6 (1930): 228–29.

Merten, Reinhold. “Was ist Radio für die Kunst?” *Die Besprechung (Beilage der Radio-Umschau)* 1 (1924): 33–35.

Meyer, Erwin. “Analysis of Noises and Musical Sounds.” In *Report of a Discussion on Audition Held on June 19, 1931 at the Imperial College of London*, edited by the Physical Society (London), 53–61. Cambridge, UK: Cambridge University Press, 1932.

———. “Beiträge zur Untersuchung des Nachhalles.” *Elektrische Nachrichtentechnik* 4 (1927): 135–39.

———. *Electro-Acoustics*, with a foreword by C. L. Fortescue. London: G. Bells and Sons Ltd., 1939.

———. “Das Gehör.” In *Handbuch der Physik*, vol. 8, *Akustik*, edited by H. Geiger and Karl Scheel, 477–543. Berlin: Julius Springer, 1927.

———. “Die Klangspektren der Musikinstrumente.” *Zeitschrift für technische Physik* 12 (1931): 606–11.

———. “Die Prüfung von Lautsprechern.” *Elektrische Nachrichtentechnik* 3 (1926): 290–96.

———. “Technische Grundlagen und Bedingungen in der mechanischen Musik.” In *Kunst und Technik*, edited by Leo Kestenberg 97–114. Berlin: Volksverbund der Bücherfreunde / Wegweiser-Verlag, GmbH, 1930.

———. “Über eine einfache Methode der automatischen Klanganalyse und der Messung der Nichtlinearität von Kohlemikrophonen.” *Elektrische Nachrichtentechnik* 5 (1928): 398–403.

———. “Über die nichtlineare Verzerrung von Lautsprechern und Fernhörern.” *Elektrische Nachrichten-Technik* 4 (1927): 509–15.

Meyer, Erwin, and Gerhard Buchmann. “Die Klangspektren der Musikinstrumente.” *Sitzungsberichte der Preußischen Akademie der Wissenschaften 1931, XXXII. Sitzung der Physikalisch-Mathematischen Klasse* (1931): 735–78.

Meyer Erwin, and Paul Just. “Messung von Nachhalldauer und Schallabsorption.” *Elektrische Nachrichtentechnik* 5 (1928): 293–300.

Meyer, Erwin, and Ernst-Georg Neumann. *Physical and Applied Acoustics: An Introduction*, translated by John M. Taylor, Jr. NY: Academic Press, 1972.

Meyer-Eppler, Werner. “Über die Anwendung elektronischer Klangmittel im Rundfunk.” *Technische Hausmitteilungen des Nordwestdeutschen Rundfunks, Sonderdruck* 7/8 (1952): 130–35.

———. *Elektrische Klangerzeugung: Elektronische Musik und synthetische Sprache*. Bonn: Ferdinand Dümmlers Verlag, 1949.

———. “Elektronische Kompositionstechnik.” *Melos* 1 (1953): 5–9.

———. “Fortschritte in der Akustik.” In *Antrittsvorlesunguen der Rheinischen Friedrich-Wilhelms-Universität Bonn am Rhein*, edited by Karl F. Chudoba, vol. 19. Bonn: Scheur Brothers, 1943.

———. “Informationstheorie.” In *Die Naturwissenschaften* 39 (1952): 341–47.

Miller, Dayton Clarence. “The Henrici Harmonic Analyzer and Devices for Extending and Facilitating its Use.” *Journal of the Franklin Institute* 182 (1916): 285–322.

———. *The Science of Musical Sounds*. NY: The Macmillan Company, 1916.

Mills, Mara. “Do Signals Have Politics? Inscribing Abilities in Cochlear Implants.” In *The Oxford Handbook of Sound Studies*, edited by Trevor Pinch and Karin Bijsterveld, 320–46. NY: Oxford University Press, 2011.

Mills, Mara. "Hearing Aids and the History of Electronics Miniaturization." *IEEE Annals of the History of Computing* 33 (2011): 24–45.
Mills, Mara, and John Tresch, eds. *Audio/Visual, Grey Room* 43 (2011).
Mitcham, Carl. *Thinking through Technology: The Path between Engineering and Philosophy*. Chicago: University of Chicago Press, 1994.
Mönch, C. O. Werner. *Mikrophon und Telephon einschliesslich der Lauthörer (Lautsprecher). Ihre Geschichte, Ihr Wesen und Ihre Bedeutung in Nachrichtenwesen besonders im Rundfunk*. Berlin: Hermann Meusser, 1925.
Mol, H. *Fundamentals of Phonetics. II: Acoustical Models Generating the Formants of the Vowel Phonemes*. The Hague: Mouton, 1970.
Mommsen, Wolfgang J. *Bürgerliche Kultur und künstlerische Avantgarde Kultur und Politik im deutschen Kaiserreich, 1870 bis 1918*. Berlin: Propyläen-Studienausgabe, 1994.
———. *Bürgerliche Kultur und politische Ordnung: Künstler, Schriftsteller und Intellektuelle in der deutschen Geschichte 1830–1933*. Frankfurt: Fischer, 2000.
Moog-Koussa, Michelle. "Uncovering the Moogtonium." Bob Moog Foundation, https://moogfoundation.org/from-the-archives-moogtonium-discovered/.
Moore, C. R., and A.S. Curtis. "An Analyzer for Voice Frequency Range." *Bell System Technical Journal* 6 (1927): 217–29.
Moral, Tony Lee. *The Making of Hitchcock's "The Birds."* Harpenden, UK: Oldcastle Books Limited, 2013.
Morawski-Büngeler, Marietta. *Schwingende Elektronen. Eine Dokumentation über das Studio für Elektronische Musik des Westdeutschen Rundfunks in Köln 1951–1982*. Cologne-Rodenkirchen: P. J. Tonger, 1988.
Mn. "Hörspielkursus in der Rundfunkversuchsstelle." *Funk: Die Wochenschrift des Rundfunkwesens* 24 (June 13, 1930): 126.
Mueller, Friedrich. "Berlin baut sein neues Funkhaus." In *Rundfunk Jahrbuch 1930*, edited by the Reichs-Rundfunk-Gesellschaft, 219–24. Berlin: Union Deutsche Verlagsgesellschaft, 1930.
Müller, Philipp. "Liberalism." In *The Oxford Handbook of the Weimar Republic*, edited by Nadine Rossol and Benjamin Ziemann, 317–39. NY: Oxford University Press, 2020.
Müller-Hartmann, Robert. "Klang- und Stilprobleme der Rundfunkmusik." In *Aus meinem Archiv: Probleme des Rundfunks*, edited by Hans Bredow, 205–7. Heidelberg: Vowinckel, 1950.
Müller-Uri, Irmengart. "Oskar Sala und das Trautonium: ein Gedenken zu seinem 100. Geburtstag." *Der Heimatbote* 56 (2010): 27–31.
Mulvin, Dylan. *Proxies: The Cultural Work of Standing In*. Cambridge, MA: MIT Press, 2022.
Mumford, Lewis. "Authoritarian and Democratic Technics." *Technology and Culture* 5 (1964): 1–8.
MZL. "Ein enttäuschender Vortragsabend: Vom Sinn und Unsinn der Funkversuchsstelle." *Funk: Die Wochenschrift des Funkwesens* 27 (July 5, 1929): 120.
Nelson, Edward L. "Radio Broadcasting Transmitters and Related Transmission Phenomena." *Bell System Technical Journal* 9 (1930): 121–140, reprinted in *Proceedings of the Institute of Radio Engineers* 17 (1931): 1949–68.
Nesper, Eugen. *Kompendium für Funktechnik: Ein Funklexikon*. Berlin: Union Deutsche Verlagsgesellschaft, 1931.
Nestel, W. "Der gegenwärtige Stand der Rundfunktechnik." *Funk: Die Wochenschrift des Funkwesens* 16 (August 15, 1935): 489–91.
Noack, Fritz. "More Information on the Trautonium." *Radio-Craft* 4 (April 1933): 590–91, 625.
Obermayer, Hans Peter. "'Yes, to Nothingness!' 'The Condemnation of Lucullus': An Opera by Bertolt Brecht and Paul Dessau." *International Journal of the Classical Tradition* 8 (2001): 217–33.

Odendahl, F. W. "Rundfunk: Von Morgens bis Mitternacht." In *Rundfunk Jahrbuch 1930*, edited by the Reichs-Rundfunk-Gesellschaft, 345–54. Berlin: Union Deutsche Verlagsgesellschaft, 1930.

Ohm, Georg Simon. "Ueber die Definition des Tones, nebst daran geknüpfter Theorie der Sirene und ähnlicher tonbildender Vorrichtungen." Poggendorff's *Annalen der Physik* 59 (1843): 513–65.

Olson, Harry F. "A History of High-Quality Studio Microphones." *Journal of the Audio Engineering Society* 24 (1976): 798–807.

———. "Mass Controlled Electrodynamic Microphones: The Ribbon Microphone." *Journal of the Acoustical Society of America* 3 (1931): 56–68.

Olson, Harry F., and Irving Wolff. "Sound Concentration for Microphones." *Journal of the Acoustical Society of America* 1 (1930): 410–17.

Ortega y Gasset, José. *Betrachtungen über die Technik*. Stuttgart: Deutsche Verlags-Anstalt, 1949.

Ottmann, Solveig. *Im Anfang war das Experiment. Das Weimarer Radio bei Hans Flesch und Ernst Schoen*. Berlin: Kunstverlag Kadmos, 2013.

Pantalony, David. "Seeing a Voice: Rudolph Koenig's Instruments for Studying Vowel Sounds." *American Journal of Psychology* 117 (2004): 425–42.

Patteson, Thomas. *Instruments for New Music: Sound, Technology, and Modernism*. Los Angeles and Berkeley: University of California Press, 2015.

Paulding, James E. "Mathis der Maler—The Politics of Music." *Hindemith Jahrbuch* 5 (1976): 104–9.

Pemberton, Cecilia, Paul McCormack, and Alison Russell. "Have Women's Voices Lowered across Time? A Cross Sectional Study of Australian Women." *Journal of Voice* 12 (1998): 208–13.

Pesic, Peter. *Music and the Making of Modern Science*. Cambridge, MA: MIT Press, 2014.

———. *Sounding Bodies: Music and the Making of Biomedical Science*. Cambridge, MA: MIT Press, 2022.

Petersen, Peter. "Nachwort," Paul Dessau, *Deutsches Miserere für gemischten Chor, Sopran-, Alt-, Tenor- und Baß-Solo großes Orchester, Orgel und Trautonium. Text von Bertolt Brecht*. Munich: C. F. Peters Verlag and Suhrkamp Verlag, 2006.

Peukert, Detlev. *Die Weimarer Republik: Krisenjahre der Moderne*. Frankfurt: Suhrkamp, 1987.

Pinch, Trevor, and Karin Bijsterveld, eds. *The Oxford Handbook of Sound Studies*. NY: Oxford University Press, 2011.

Pinch, Trevor, and Frank Trocco. *Analog Days: The Invention and Impact of the Moog Synthesizer*. Cambridge, MA: Harvard University Press, 2004.

Pohle, Heinz. *Der Rundfunk als Instrument der Politik: Zur Geschichte des deutschen Rundfunks von 1923/38*. Hamburg: Verlag Hans Bredow-Institut, 1955.

Poirier, Alain. "Die Avantgarde in Deutschland zwischen den Weltkriegen." In *Das "Dritte Reich" und die Musik*, edited by der Stiftung Schloss Neuhardenberg in conjunction with Cité de la Musique, Paris, 61–71. Berlin: Nicolai, 2006.

Pompino-Marschall, Bernd. "Carl Stumpf und die Phonetik." In *Musik und Sprache: Zur Phänomenologie von Carl Stumpf*, edited by Margaret Kaiser-El-Safi and Matthias Ballod, 131–50. Würzburg: Königshausen & Neumann, 2003.

Potter, Pamela M. *The Art of Suppression: Confronting the Nazi Past in Histories of the Visual and Performing Arts*. Los Angeles and Berkeley: University of California Press, 2016.

Preuß, Thorsten. *Brechts "Lukullus" und seine Vertonungen durch Paul Dessau und Roger Sessions*. Würzburg: Ergon Verlag, 2007.

Quincke, Georg Hermann. "Ueber Interferenzapparate für Schallwellen." *Annalen der Physik und Chemie* 204 (128 of the new series) (1866): 177–92.

Raven-Hart, R. "Electronic Musical Instruments." *Electronics* (1931): 18–19, 42.

———. "Neon Musical Oscillator: A New Electro-musical Instrument and a New Theory." *Wireless World* 27 (1930): 648–50.

———. "Radio, and a New Theory of Tone-Quality." *Musical Quarterly* 17 (1931): 380–88.

Raz, Carmel. "The Lost Movements of Ernst Toch's *Gesprochene Musik*." *Contemporary Music* 97 (2014): 37–59.

Rehding, Alexander. "Instruments of Music Theory." *Music Theory Online: A Journal of the Society for Music Theory* 22 (2016). http://mtosmt.org/issues/mto.16.22.4/mto.16.22.4.rehding.html.

———. *Music and Monumentality: Commemoration and Wonderment in Nineteenth-Century Germany*. NY: Oxford University Press, 2009.

Reich, Herbert J. "A New Method of Testing for Distortion in Audio-Frequency Amplifiers." In *Proceedings of the Institute of Radio Engineers* 19 (1931): 401–15.

Reichs-Rundfunk-Gesellschaft, ed. *Rundfunk Jahrbuch 1929*. Berlin: Union Deutsche Verlagsgesellschaft, 1929.

———. *Rundfunk Jahrbuch 1930*. Berlin: Union Deutsche Verlagsgesellschaft, 1930.

———. *Rundfunk Jahrbuch 1931*. Berlin: Union Deutsche Verlagsgesellschaft, 1931.

Reuter, Christoph. "Erich Schumann's Laws of Timbre as an Alternative." *Systematische Musikwissenschaft / Systematic Musicology* 4 (1996): 185–200.

Révész, Geza. *Einführung in der Musikpsychologie*. Bern: A. Francke, 1946, translated as *Introduction to the Psychology of Music*, translated by G. I. C. de Courcy. Mineola, NY: Dover, 2001.

Rice, Chester W., and Edward W. Kellogg. "Notes on the Development of a New Type of Hornless Loudspeaker." *Transactions of the American Institute of Electrical Engineers* 44 (1925): 461–75.

Rider, John F. "Why Is a Radio Soprano Unpopular? Present-Day Radio Laws and Radio Equipment Make Proper Reproduction of Soprano's High Notes Impossible." *Scientific American* (October 1928): 334–37.

Riegger, Hans. "Zur Theorie des Lautsprechers." *Wissenschaftliche Veröffentlichungen aus dem Siemens-Konzern* 3, part 2 (1924): 67–100.

Rienäcker, Gerd. "Fibel-Musik? Anmerkungen zu Hanns Eislers und Paul Dessaus Vertonung der Kriegsfibel." In *Fokus* Deutsches Miserere *von Paul Dessau und Bertolt Brecht: Festschrift für Peter Petersen zum 65.* Geburtstag, edited by Nima Ehrlich Lehmann, Sophie Fetthauer, Mattias Lehmann et al., 87–110. Neumünster: Von Bockel Verlag, 2005.

Riley, Joseph. "The 'Thyratron': An Addition to the Vacuum Tube Family." *Radio-Craft* 2 (1930): 150.

Rindfleisch, Hans. *Technik um Rundfunk: Ein Stück deutscher Rundfunkgeschichte von den Anfängen bis zum Beginn der achtizger Jahre*. Norderstedt: Mensing GmbH & Co., 1985.

Ringer, Fritz. *The Decline of the German Mandarins: The German Academic Community*. Cambridge, MA: Harvard University Press, 1960.

Roeseler, Hans. "Die Frau als Rundfunkhörerin." In *Rundfunk Jahrbuch 1929*, edited by Reichs-Rundfunk-Gesellschaft, 342–46. Berlin: Union Deutsche Verlagsgesellschaft, 1929.

Rösser, Thomas. *Bilder zum Hören: Die Zusammenarbeit von Alfred Hitchcock mit dem Komponisten Bernard Hermann*. Hamburg: Verlag Dr. Kovac, 2013.

Rosenthal, Lecia. "Introduction." In *Radio Benjamin*, edited by Lecia Rosenthal, translated by Jonathan Lutes with Lisa Harries Schumann and Diana K. Reese. NY: Verso, 2014.

Ross, Corey. *Media and the Making of Modern Germany: Mass Communications, Society, and Politics from the Empire to the Third Reich*. NY: Oxford University Press, 2008.

Rossol, Nadine, and Benjamin Ziemann. "Introduction." In *The Oxford Handbook of the Weimar Republic*, edited by Nadine Rossol and Benjamin Ziemann, 1–24. NY: Oxford University Press, 2020.

Rühle-Gerstel, Alice. "Back to the Good Old Days?" In *The Weimar Republic Sourcebook*, edited by Anton Kaes, Martin Jay, and Edward Dimendberg. Los Angeles and Berkeley: University of California Press, 1994.

Sacia, C. F. "Speech Power and Energy." *Bell System Technical Journal* 4 (1925): 627–41.
Sala, Oskar. "Anfänge." In Oskar-Sala-Fonds, Deutsches Museum, München. http://www.oskar-sala.de/oskar-sala-fonds/oskar-sala/interview/anfaenge/.
———. "Auf den Wegen II: Elektrische Musik und Trautonium." In *Der Musikstudent: Mitteilungsblätter der Deutschen Musikstudentengesellschaft* 3. Berlin: Max Hess, 1933.
———. "Biografie." In Oskar-Sala-Fonds, Deutsches Museum, München. http://www.oskar-sala.de/oskar-sala-fonds/oskar-sala/biografie/1933-1935/index.html.
———. "Elektronische Klanggestaltung mit dem Mixtur-Trautonium." In *Gravesano: Musik, Raumgestaltung, Elektroakustik*, edited by Werner Meyer-Eppler, 78–87. Mainz: Araviva Verlag, 1955.
———. "Experimentelle und theoretische Grundlagen des Trautoniums." *Frequenz: Zeitschrift für Schwingungs- und Schwachstromtechnik* 2 (1948): 315–22.
———. "Experimentelle und theoretische Grundlagen des Trautoniums, Zweiter Teil." *Frequenz: Zeitschrift für Schwingungs-und Schwachstromtechnik* 3 (1949): 13–19.
———. "50 Jahre Trautonium." In *Für Augen und Ohren: Berliner Musiktage 2. Januar—14. Februar 1980*. Berlin: Berliner Festspieler, 1980, 78–82.
———. "Die Geschichte des Trautoniums." *Erdenklang, The Catalogue, 1982–1992*. Hamburg: The Eclectic-Musik Company, 1992, 11.
———. "Historische Übersicht über die Verwendung elektronischer Musikinstrumente im Konzertsaal und Theater." In *Gravesano: Musik, Raumgestaltung, Elektroakustik*, edited by Werner Meyer-Eppler, 94–96. Mainz: Arsviva Verlag (Hermann Scherchen GmbH), 1955.
———. "Klanggestaltung mit dem Mixtur-Trautonium." In *Musik. Raumgestaltung. Elektroakustik. Internationaler Kongreß "Musik und Elektroakustik," Gravesano, August 1954*, edited by Werner Meyer-Eppler, 78–87. Mainz: Arsviva Verlag (Hermann Scherchen GmbH), 1954.
———. "Klanggestaltung mit dem Mixtur-Trautonium." *Das Ton-Magazin für Freunde des Tonbandes und der Schallplatte* (March/April 1959): 28–30.
———. "Die Kollegen von damals." Interview on September 1, 1989. http://www.klangspiegel.de/trautonium/kollegen-damals.
———. "Das Mixtur-Trautonium." In *Klangstruktur der Musik. Neue Erkenntnisse musikelektronischer Forschung*, 91–108. Berlin-Borsigwalde: Verlag für Radio-Foto-Kinotechnik GmbH, 1955.
———. "Mixtur-Trautonium." *Melos: Zeitschrift für Neue Musik* 17 (1950): 247–51.
———. "Das Mixtur-Trautonium." *Physikalische Blätter* 6 (1950): 390–98.
———. "Das Mixturtrautonium." *Die Musik-Woche* 30–31 (1951): 245–46.
———. "Mixture-Trautonium and Studio Technique." *Gravesaner Blätter* 6 (1962): 53–61.
———. "My Fascinating Instrument." In *Neue Musiktechnologie: Vorträge und Berichte vom KlangArt-Kongress 1991*, edited by Bernd Enders with the assistance of Stefan Hanheide, 75–93. Osnabrück: KlangArt-Kongress Universität Osnabrück, 1991.
———. "Das neue Mixtur-Trautonium." *Das Musikleben Monatschrift* 10 (1953): 346–48.
———. "Neue Möglichkeiten der Musikausübung durch elektrische Instrumente." *Deutsche Tonkünstler-Zeitung* 29 (1931): 136–37.
———. "Ein neues elektrisches Soloinstrument." *Neues Musikblatt* 17 (1938): 5–6.
———. "Objektive und subjektive Resonanzeffekte bei kurzdauernden Impulsfolgen." *Frequenz: Zeitschrift für Schwingungs- und Schwachstromtechnik* 5 (1951): 250–58.
———. "Psycho-physische Konsequenzen elektro-akustischer Klangsynthesen." *Frequenz: Zeitschrift für Schwingungs- und Schwachstromtechnik* 5 (1951): 13–20.
———. "Das Rundfunktrautonium." In Oskar-Sala-Fonds, Deutsches Museum, München. http://www.oskar-sala.de/oskar-sala-fonds/trautonium/rundfunktrautonium/index.html.
———. "Das Trautonium." *Der Rundfunk* (August, 8–14 1948): 21.
———. "Das Trautonium: Begriff und Aufgabe." *Theater der Zeit: Blätter für Bühne, Film und Musik* 4 (1949): 25–28.

———. "Das Trautonium, ein Instrument der Zukunft." *Das Notenpult* 2 (1949): 162–63.
Salinger, H. "Physikalische Grundlagen der Empfangstechnik." In *Die wissenschaftlichen Grundlagen des Rundfunksempfangs*, edited by Karl Willy Wagner, 142–80. Berlin: Julius Springer, 1927.
Salinger, H. "Zur Theorie der Frequenzanalyse mittels Suchton." *Elektrische Nachrichtentechnik* 6 (1929): 293–302.
Sander, Tobias. *Die doppelte Defensive: Lage, Mentalitäten und Politik der Ingenieure in Deutschland 1890–1933*. Wiesbaden: VS Verlag für Sozialwissenschaften, 2009.
Saraga, W. "Das Trautonium: Anleitung zum Aufbau und zu Versuchen." *Funk Bastler: Fachblatt des Deutschen Funktechnischen Verbandes e.V.* 16 (1933): 241–45.
Sattelberg, O. *Dictionary of Technological Terms Used in Electrical Communications / Wörterbuch der Elektrischen Nachtrichtentechnik*, 2 parts. Berlin: Springer Verlag, 1926.
Savage, Mark. "Sir Paul McCartney Says Artificial Intelligence has Enabled a 'Final' Beatles Song," BBC News, June 13, 2023. https://www.bbc.com/news/entertainment-arts-65881813.
Schäfer, Walter Erich. *Wieland Wagner: Persönlichkeit und Leistung*. Tübingen: R. Wunderlich, 1970.
Schenk, Dietmar. "Die Berliner Rundfunkversuchsstelle (1928–1935): Zur Geschichte und Rezeption einer Institution aus der Frühzeit von Rundfunk und Tonfilm." *Rundfunk und Geschichte: Mitteilungen des Studienkreises Rundfunk und Geschichte, Information aus dem Deutschen Rundfunkarchiv* 23 (1997): 124–27.
———. *Die Hochschule für Musik zu Berlin. Preußens Konservatorium zwischen romantischem Klassizismus und Neuer Musik, 1869–1932/33*. Stuttgart: Franz Steiner, 2004.
Schmidgen, Henning. "The Donders Machine." *Configurations* 13 (2005): 211–56.
Schneider, Albrecht. "Change and Continuity in Sound Analysis: A Review of Concepts in Regard to Musical Acoustics, Music Perception, and Transcription." In *Sound-Perception-Performance*, edited by Rolf Bader, 71–111. NY: Springer, 2013.
Schneider, John. "The Woman Who Overcame Radio's Earliest Glass Ceilings." *Radioworld*, August 1, 2020. https://www.radioworld.com/news-and-business/headlines/the-women-who-overcame-radios-earliest-glass-ceilings.
Schoen, Ernst. "Aufgaben und Grenzen des Rundfunk-Programs." *Süddeutsche Rundfunk-Zeitung* 6 (1930): 14–18.
Schuder, Kurt. "Der Techniker Goethe." *Deutsche Technik* (August–September 1941): 417–18.
Schünemann, Georg. "Die Arbeit der Rundfunksversuchsstelle." *Funk: Die Wochenschrift des Funkwesens* 11 (March 14, 1930): 41–42.
———. "Von der Arbeit der Rundfunkversuchsstelle." *Funk: Die Wochenschrift des Funkwesens* 30 (July 26, 1929): 130–131.
———. "Die Aufgaben der Funkversuchsstelle." In *Staatliche akademische Hochschule für Musik in Berlin zu Charlottenburg: Jahresbericht vom 1 Oktober 1927 bis 30. September 1928*, edited by Georg Schünemann, 7–16. Berlin: Staatliche akademische Hochschule für Musik in Berlin, 1929.
———. "Die Funkversuchsstelle bei der Staatlich Akademischen Hochschule für Musik in Berlin." In *Rundfunk Jahrbuch 1929*, edited by the Reichs-Rundfunk-Gesellschaft, 277–93. Berlin: Union Deutsche Verlagsgesellschaft, 1929.
———. "Die neue Funkversuchsstelle." *Funk: Die Wochenschrift des Funkwesens* 20 (May 11, 1928): 153–54.
———. "Neue Musik Berlin 1930." *Funk: Die Wochenschrift des Funkwesens* 4 (June 13, 1930): 121–22.
———. "Produktive Kräfte der mechanischen Musik." *Die Musik* 24 (1932): 246–49.
———. "Tonfilm und Rundfunk im Musikunterricht." *Rheinische Musik- und Theaterzeitung* 30 (November 1, 1929): 421–23, reprinted in *Jahresbericht der Hochschule für Musik* 50 (1929): 7–12.

Schütte, Uwe. *Kraftwerk: Future Music from Germany.* London: Penguin Books, 2020.
Schütte, Wolfgang. *Regionalität und Föderalismus im Rundfunk: Die geschichtliche Entwicklung in Deutschland 1923–1945.* Frankfurt: Josef Knecht, 1971.
Schumann, Karl Erich. *Akustik.* Breslau: Ferdinand Hirt, 1925.
———. *Die Physik der Klangfarben,* 2 vols. Leipzig: Breitkopf & Härtel, 1940.
Schwandt, Erich. "Kraftverstärker—Lautsprecher—Elektroakustische Geräte: Neuerungen an Einzelteilen und Zubehör. Schlußbericht über die Große Deutsche Funkaustellung 1934." *Funk: Die Wochenschrift des Funkwesens* 35 (September 15, 1934): 649–56.
Scriven, E. O. "Amplifiers." *Bell System Technical Journal* 13 (1934): 278–84.
Seidler-Winkler, Bruno, and Wilhelm Buschkötter. "Instrumente und Singstimmen im Rundfunk." In *Sitzung des Programmausschusses der deutschen Rundfunkgesellschaften* (1928), reprinted in *Aus meinem Archiv: Probleme des Rundfunks,* edited by Hans Bredow, 247–53. Heidelberg: Vowinckel, 1950.
Siebs, Theodor. *Deutsche Bühnenaussprache. Hochschule.* Cologne: Albert Ahn, 1930.
Siemens, Daniel. "National Socialism." In *The Oxford Handbook of the Weimar Republic,* edited by Nadine Rossol and Benjamin Ziemann, 382–403. NY: Oxford University Press, 2020.
Siemens, Frederick. "New 'Velocity' Microphone." *Radio News* XIV (1933): 406 and 431.
Silvan, L. J., H. K. Dunn, and S. D. White. "Absolute Amplitudes and Spectra of Certain Musical Instruments and Orchestras." *Bell Telephone System Technical Publications,* Monograph B-551 (April 1931): 1–42, reprinted in *Journal of the Acoustical Society of America* 2 (1931): 330–71.
Simonite, Tom. "The Best Algorithms Struggle to Recognize Black Faces Equally." *Wired,* July 22, 2019. https://www.wired.com/story/best-algorithms-struggle-recognize-black-faces-equally/.
Singer, Natasha. "Amazon Is Pushing Facial Technology that a Study Says Could be Biased." *New York Times,* January 24, 2019. https://www.nytimes.com/2019/01/24/technology/amazon-facial-technology-study.html.
Skelton, Geoffroy. *Paul Hindemith: The Man Behind the Music.* London: Gollancz, 1975.
Skinner, D. Dixon, "Music Goes into Mass Production." *Harper's Magazine,* April 1939: 487.
Slonimsky, Nicolas, ed. *Music since 1900.* New York: W.W. Norton & Company, 1971.
Slotten, Hugh Richard. "Radio Engineers: The Federal Radio Commission, and the Social Shaping of Broadcast Technology: Creating 'Radio Paradise.'" *Technology and Culture* 36 (1995): 950–86.
Smith, Helmut Walser. "When the Sonderweg Left Us." *German Studies Review* 31 (2008): 225–40.
Spix, Jörg. "The Digital Trautonium" (1994). https://archive.ph/plDuL.
Stapelfeldt, Kurt. "Der Rundfunk als Träger und Erhalter der Heimatkultur." In *Rundfunk Jahrbuch 1929,* edited by Reichs-Rundfunk-Gesellschaft, 233–43. Berlin: Union Deutsche Verlagsgesellschaft, 1929.
Statistisches Reichsamt, ed. *Die Lebenshaltung von 2000 Arbeiter- Angestellten-, und Beamtenhaushaltungen: Erhebungen von Wirtschaftsrechnungen im Deutschen Reich vom Jahre 1927/28,* 2 vols. Berlin: Verlag von Reimar Hobbing, 1932.
Stauder, Wilhelm. "Der Tonmeister: Mittler zwischen Musik und Technik." *Zeitschrift für Musik* 112 (1952): 79–81.
Steege, Benjamin. *An Unnatural Attitude: Phenomenology in Weimar Musical Thought.* Chicago: University of Chicago Press, 2021.
Stege, Fritz. "Neue Musik Berlin 1930." *Neue Zeitschrift für Musik* 97 (1930): 645–55.
Steidle, Hans Carl. "Vortragskunst und große Musik in elektrischer Übertragung durch Fernsprecher." *Elektrische Nachrichtentechnik* 2 (1925): 309–31.
Steinberg, John C. "The Relationship between the Loudness of a Sound and Its Physical Stimulus." *Physical Review* 26 (1925): 507–23.

———. "Understanding Women." *Bell Laboratories Record* 3 (1927): 153–54.
Stephani, Otfried. "Ein Trautonium neuerer Ausführung." *Funk: Die Wochenschrift des Funkwesens* 11 (June 1, 1940): 167–72.
Sterne, Jonathan. *The Audible Past: Cultural Origins of Reproduction.* Durham, NC: Duke University Press, 2003.
Stewart, John Q. "An Electrical Analogue of the Vocal Organs." *Nature* 110 (1922): 311–12.
Stockhausen, Karlheinz. "Arbeitsbericht 1953: Die Entstehung der elektronischen Musik." *Texte* 1 (1953): 39–44.
———. "Electronic and Instrumental Music." *Die Reihe* 5 (1961): 60.
———. "Komposition 1953, No. 2." *Technische Hausmitteilungen des Nordwestdeutschen Rundfunks* 6 (1954): 46–51.
———. "The Origins of Electronic Music." *The Musical Times* 112 (1971): 649–50.
Stoffels, Ludwig. "Rundfunk als Erneuerer und Förderer." In *Programmgeschichte des Hörfunks in der Weimarer Republik,* edited by Joachim-Felix Leonhard, 2 vols. Munich: Deutscher Taschenbuch Verlag, 1997.
Strobel, Heinrich. "Neue Sachlichkeit in der Musik." *Musikblätter des Anbruch* 8 (1926): 3–4.
Stuckenschmidt, Hans Heinz. "Die Mechanisierung der Musik." *Pult und Taktstock* 2 (1925): 1–8, reprinted in Stuckenschmidt, *Die Musik eines halben Jahrhunderts,* 9–15.
———. *Die Musik eines halben Jahrhunderts.* Munich: R. Piper & Co. Verlag, 1976, 187–201.
———. "Neue Sachlichkeit in der Musik." *Der Auftakt* 8 (1928): 3–6, reprinted in *Die Musik eines halben Jahrhunderts,* 36–41.
———. "Die Ordnung der Freiheit," (1961), republished in Stuckenschmidt, *Die Musik eines halben Jahrhunderts,* 187–201.
———. "The Third Stage: Some Observations on the Aesthetics of Electronic Music." *Die Reihe* 1 (1955): 11–13.
Stumpf, Carl. "Zur Analyse geflüsterter Vokale." *Beiträge zur Anatomie, Physiologie, Pathologie und Therapie des Ohres, der Nase und des Halses* 12 (1919): 234–54.
———. "Zur Analyse der Konsonanten." *Beiträge zur Anatomie, Physiologie, Pathologie und Therapie des Ohres, der Nase und des Halses* 17 (1921): 151–81.
———. *Die Sprachlaute: Experimentell-Phonetische Untersuchungen.* Berlin: Julius Springer, 1926.
———. "Sprachlaute und Instrumentalklänge." *Zeitschrift für Physik* 38 (1926): 745–58.
———. "Die Struktur der Vokale." *Sitzungsberichte der Königlich-Preußischen Akademie der Wissenschaft* (1918): 333–58.
———. "Über die Tonlage der Konsonanten und die für das Sprachverständnis entscheidende Gegend des Tonreiches." *Sitzungsberichte der Preußischen Akademie der Wissenschaften* 1921: 636–40.
———. "Veränderungen des Sprachverständnisses bei Abwärts fortschreitender Vernichtung der Gehörsempfindungen." *Beiträge zur Anatomie, Physiologie, Pathologie und Therapie des Ohres, der Nase und des Halses* 17 (1921): 182–90.
Szendrei, Alfred. "Instrumentalmusik in Rundfunk: Nach einem Vortrag, gehalten auf der 'Ersten Tagung für Rundfunkmusik.'" *Funk: Die Wochenschrift des Funkwesens* 28 (July 6, 1928): 209–10.
Tallon, Tina. "A Century of 'Shrill': How Bias in Technology Has Hurt Women's Voices." *New Yorker,* September 2, 2019. https://www.newyorker.com/culture/cultural-comment/a-century-of-shrill-how-bias-in-technology-has-hurt-womens-voices.
Temmer, Stephen F. "In Memoriam of Georg Neumann." *Journal of the Audio Engineering Society* 24 (1976): 708.
Thadeusz, Frank. "Nazi-Labor in Oberfranken: Geheimwaffen aus Burgverlies." *Spiegel-Geschichte,* April 21, 2011. https://www.spiegel.de/geschichte/nazi-labor-in-oberfranken-a-947186.html.

Thérberge, Paul. *Any Sound You Can Image: Making Music/Consuming Technology*. Middletown, CT: Wesleyan University Press, 1997.

Thierfelder, Franz. *Sprachpolitik und Rundfunk*. Berlin: R. von Decker's Verlag G. Schenk, 1941.

Thomas, H. A. "The Performance of Amplifiers." *Institution of Electrical Engineers—Proceedings of the Wireless Section of the Institution* 1 (1926): 253–78.

Thompson, Emily. *The Soundscape of Modernity: Architectural Acoustics and the Culture of Listening in America*. Cambridge, MA: MIT Press, 2004.

Thomson, Virgil. "Beethoven in the House." In *Musical Science*. NY: Greenwood, 1945/1968.

Thun, H. "Die Vorgeschichte des deutschen Rundfunks." In *Rundfunk Jahrbuch 1929*, edited by Reichs-Rundfunk-Gesellschaft, 11–17. Berlin: Union Deutsche Verlagsgesellschaft, 1929.

Thurmann, Kira. *Singing Like Germans: Black Musicians in the Land of Bach, Beethoven, and Brahms*. Ithaca, NY: Cornell University Press, 2021.

Tkaczyk, Viktoria. "Radio Voices and the Formation of Applied Research in the Humanities." *History of Humanities* 6 (2021): 85–109.

———. *Thinking with Sound: A New Program in the Sciences and Humanities Around 1900*. Chicago: University of Chicago Press, 2023.

Toop, Richard. "Stockhausen and the Sine-Wave: The Story of an Ambiguous Relationship." *Musical Quarterly* 65 (1979): 379–91.

Trautwein, Friedrich. "Besinnung auf das Gehör." *108. Niederrheinisches Musikfest in Duisburg 2. bis 6. Juli 1954: Jahresbuch*. Darmstadt: Mykenae Verlag, 1954, 92–100.

———. *Drahtlose Telephonie und Telegraphie in gemeinverständlicher Darstellung*. Leipzig: Akademische Verlagsgesellschaft, 1925.

———."Dynamische Probleme der Musik bei Feiern unter freiem Himmel." *Deutsche Musikkultur: Zweimonatshefte für Musikleben und Musikforschung* 2 (1937–38): 33–44.

———. "Über elektrische Analogien der Sprachwerkzeug und der Musikinstrumente." In *Die deutschen Vorträge auf dem Allgemeinen Stimmkongreß in Paris vom 19. bis 30 September 1937*, edited by Hans-Joachim vom Braunmühl, 39–47. Berlin: Metten & Co., 1938.

———. "Elektrische Klangbildung und elektrische Musikinstrumente." *Funk: Die Wochenschrift des Funkwesens* 24 (June 3, 1930): 123–24.

———. *Elektrische Musik*. Berlin: Wiedmann, 1930.

———. "Über elektrische Synthese von Sprachlauten und musikalischen Tönen." *Archives Néerlandaises de Phonétique Expérimentale* 7 (1932): 291–92.

———. "Elektroakustische Mittel in der aktiven Tonkunst." *Acta Acoustica united with Acustica* 4 (1954): 256–59.

———. "Elektronische Klangerzeugung und Musikästhetik: Grundlagen der elektronischen Klangerzeugung und ihre Anwendung auf den Musikinstrumentenbau." *Aus dem Jahrbuch 1954 der Technischen Hochschule Aachen* 12 (1954): 176–81.

———."Die Elektronenröhre in der elektrischen Meßtechnik." *Telegraphen- u. Fernsprech-Technik* 9 (1929): 101–4, 119–23.

———. "Das Elektronische Monochord." *NWDR-technische Hausmittelungen* 7 (1954): 345–52.

———. "Modulation und Übertragungsgüte in der Hochfrequenztechnik." *Zeitschrift für technische Physik* 7 (1926): 343–52.

———. "Perspektiven der musikalischen Elektronik." In *Gravesano: Musik, Raumgestaltung, Elektroakustik*, edited by Werner Meyer-Eppler, 103–10. Mainz: Arsviva Verlag/Hermann Scherchen, 1955.

———. "Die technische Entwicklung der elektrischen Musik." *Deutsche Tonkünstler-Zeitung* 29 (1931): 133–34.

———. "Toneinsatz und elektrische Musik." *Zeitschrift für technische Physik* 13 (1932): 244–46.

———. *Trautonium-Schule*. Mainz: B. Schott's Söhne, 1933.
———. "Über Verlustmessung bei hohen Frequenzen." *Mitteilung aus dem Telegraphentechnischen Reichsamt* 9 (1921): 235–64.
———. "Wesen und Ziele der Elektromusik." *Zeitschrift für Musik* 103 (1936): 694–99.
Trendelenburg, Ferdinand. *Klänge und Geräusche: Methoden und Ergebnisse der Klangforschung—Schallwahrnehmung—Grundlegende Fragen der Klangübertragung*. Berlin: Julius Springer, 1935.
———. "Klangspektren von Musikinstrumenten." *Die Naturwissenschaften* 21 (1933): 11–14.
———. "Objektive Klangaufzeichnung mittels des Kondensatormikrophons." *Wissenschaftliche Veröffentlichungen aus dem Siemens-Konzern* 3 (1924): 43–66, and 4 (1926–29): 1–13.
———. "Die physikalischen Eigenschaften der Sprachklänge." In *Handbuch der Physik*, edited by H. Geiger and Karl Scheel, 24 volumes, vol. 8, 471–72. Berlin: Julius Springer, 1926–29.
Trendelenburg, Wilhelm "Physiologische Untersuchungen über die Stimmklangbildung." *Sitzungsberichte der Preußischen Akademie der Wissenschaften* 31 (1935): 525–73.
Trendelenburg, W., and H. Wullstein, "Untersuchungen über die Stimmbandschwingungen." *Sitzungsberichte der Preußischen Akademie der Wissenschaften* 31 (1935): 399–426.
Tresch, John. *The Romantic Machine: Utopian Science and Technology after Napoleon*. Chicago: University of Chicago Press, 2012.
Tresch, John, and Emily L. Dolan. "Toward a New Organology: Instruments of Music and Science." In *Music, Sound, and the Laboratory from 1750–1980*, edited by Alexandra Hui, Julia Kursell, and Myles W. Jackson, 278–98. Chicago: University of Chicago Press, 2013.
Turner, R. Steven. "The Ohm-Seebeck Dispute, Hermann von Helmholtz, and the Origins of Physiological Acoustics." *British Journal for the History of Science* 10 (1977): 1–24.
Ulrich, Volker. *Deutschland 1923: Das Jahr am Abgrund*. Munich: C. H. Beck, 2022.
Ungeheuer, Elena. *Wie die elektronische Musik "erfunden" wurde: Quellenstudie zu Werner Meyer-Epplers Entwurf von 1949 und 1953*. Mainz: Schott, 1992.
Vail, Mark. *The Synthesizer: A Comprehensive Guide to Understanding, Programming, Playing, and Recording the Ultimate Electronic Music*. NY: Oxford University Press, 2014.
Van B. Roberts, Walter. "The Uses of the Three Electrode Tube: What Makes the Wheels Go 'Round' Better Radio: IV." *Radio Broadcast* (June 1924): 168–71.
Varése, Edgard. "Musik auf neuen Wegen." *Stimmen* 15 (1949): 401–4.
Verein deutscher Ingenieure, ed. *Der deutsche Ingenieur in Beruf und Gesellschaft: Ergebnis einer Erhebung*, VDI-Information, Nr. 5. Düsseldorf: Deutscher Ingenieur-Verlag, 1959.
Viefhaus, Erwin. "Ingenieure in der Weimarer Republik: Bildungs-, Berufs- und Gesellschaftspolitik 1918 bis 1933." In *Technik, Ingenieure und Gesellschaft—Geschichte des Vereins Deutscher Ingenieure 1856–1981*, edited by Karl-Heinz Ludwig with assistance from Wolfgang König, 289–346. Düsseldorf: VDI-Verlag, 1981.
Vierling, Oskar. "Elektrische Musik." *Elektrotechnische Zeitschrift* 53 (1932): 155–59.
———. *Das elektroakustische Klavier*. Berlin: VDI Verlag, 1936.
———. "Der Formantbegriff." *Annalen der Physik* 418 (1936): 219–32.
Von Békésy, Georg. "Über die nichtlinearen Verzerrungen des Ohres." *Annalen der Physik* 412 (1934): 809–27.
Von Boeckmann, Kurt. January 1929. "Grundlegende Fragen der allgemeinen Programmgestaltung," as quoted in Konrad Dussel and Edgar Lersch, eds., *Quellen zur Programmgeschichte des deutschen Hörfunks und Fernsehens*. Göttingen: Muster-Schmidt 1999, 39–51.
———. "Organisation des deutschen Rundfunks." In *Kunst und Technik*, edited by Leo Kestenberg, 219–42. Berlin: Volksverband der Bücherfreunde/Wegweiser-Verlag, GmbH, 1930.
Von Essen, Otto. "Werner Meyer-Eppler." *Zeitschrift für Phonetik* 13 (1960): 189–93.
Von Helmholtz, Hermann. "Ueber die Klangfarbe der Vocale." *Annalen der Physik* 184 (1859): 280–90.

———. *Die Lehre von den Tonempfindungen als physiologische Grundlage der Theorie der Musik.* Braunschweig: Friedrich Vieweg & Sohn, 1863, translated as *On the Sensations of Tone as a Physiological Basis for the Theory of Music,* translation of 4th edition (1877) by Alexander J. Ellis. NY: Dover Books, 1954.

Von Raman, C. "Musikinstrumente und ihre Klänge." In *Handbuch der Physik,* edited by H. Geiger and Karl Scheel, 24 volumes. Berlin: Julius Springer, 1926–29.

Voskuhl, Adelheid. "Engineering Philosophy: Theories of Technology, German Idealism, and Social Order in High-Industrial Germany." *Technology and Culture* 57 (2016): 721–52.

Wachsmuth, R., ed. "Klangaufnahme an Blasinstrumenten, eine Grundlage für das Verhältnis der menschlichen Stimme.Nachgelassenes Manuskript von Georg Meissner." *Pflügers Archiv für die gesammte Physiologie des Menschen und der Thiere* 116 (1907): 543–99.

Waetzmann, Erich. "Die Entstehungsweise von Kombinationstönen im Mikrophon-Telephonkreis." *Annalen der Physik* 347 (1913): 729–44.

Wagner, Karl Willy. "Der Frequenzbereich von Sprache und Musik." *Elektrotechnische Zeitschrift* 19 (1924): 451–56.

———. "Das Heinrich-Hertz-Institut für Schwingungsforschung an der Technischen Hochschule zu Berlin." In *Rundfunk Jahrbruch 1929,* edited by Reichs-Rundfunk-Gesellschaft Berlin, 296–99. Berlin: Union Deutsche Verlagsgesellschaft, 1929.

———. "Die Kunstaufgabe des Rundfunks: seine Organisation und Technik. Inhalt und Ziele der Vortragsreihe." In *Die wissenschaftlichen Grundlagen des Rundfunkempfangs,* edited by Karl Willy Wagner, 1–17. Berlin: Julius Springer, 1927.

———. "Der Lichtbogen als Wechselstromerzeuger. Inaugural-Dissertation zur Erlangung der Doktorwürde." PhD diss., Department of Physics, Georg-August-Universität in Göttingen, 1910.

———. "Sprache und Musik im Rundfunk." In *Mittheilung aus dem Heinrich-Hertz-Institut für Schwingungsforschung, Sonderdruck des "Funk,"* 1–6. Berlin: Weidmannsche Buchhandlung, 1928.

Warschauer, Frank. "Die Zukunft der Technisierung." In *Kunst und Technik,* edited by Leo Kestenberg, 409–46. Berlin: Volksverband der Bücherfreunde/Wegweiser-Verlag, GmbH, 1930.

Warwick, Andrew C. "The Laboratory of Theory or What's Exact about the Exact Sciences?" In *The Values of Precision,* edited by M. Norton Wise, 311–52. Princeton, NJ: Princeton University Press, 1995.

Weber, Max. *Gesammelte Aufsätze zur Religionssoziologie,* 5. Aufl. Tübingen: J. C. B. Mohr, 1963.

Weck, Ursula. *Nur Einer kann es spielen: Oskar Sala, Meister des Trautoniums. Bericht über einen Pionier der elektronischen Klänge und sein Instrument,* originally broadcast on Schweizer Radio und Fernsehen on August 15, 2010. https://www.srf.ch/audio/passage/nur-einer-kann-es-spielen-oskar-sala-meister-des-trautoniums?id=10144241.

Wegel, R. L., and C. E. Lane. "The Auditory Masking of One Pure Tone by Another and its Probable Relation to the Dynamics of the Inner Ear." *Physical Review* 23 (1924): 266–85.

Wegel, R. L., and C. R. Moore. "An Electrical Frequency Analyzer." *Bell System Technical Journal* 3 (1924): 299–323.

Weichart, F. "In 14 Tagen einen Sender für Berlin." In *Rundfunk Jahrbuch 1930,* edited by Reichs-Rundfunk-Gesellschaft, 43–52. Berlin: Union Deutsche Verlagsgesellschaft, 1930.

———. "Das Mikrophon auf der Bühne: die unmittelbare Übertragung von Opern, Konzerten und Theaterstücken. Mikrophon oder Kathodophon?—Kabel oder Freileitung?—Die ersten Versuche." *Funk: Die Wochenschrift des Funkwesens* 1 (1924): 1–4.

Weickert, W. "Neue Darbietungen auf dem Trautonium." *Funkschau: Zeitschrift für Funktechniker* 14 (1941): 22.

Weihe, Carl. "Geistige Sozialisierung, Technik und Volksbildung." *Zeitschrift des Vereines deutscher Ingenieure* 63 (1919): 86–87.

Weihe, Carl. "Spengler und die Maschine." *Technik und Kultur* 18 (1927): 37–38.

———. "Technik und Kultur." *Technik und Kultur* 13 (1922): 2–3.

Weill, Kurt. "Über die Möglichkeiten einer Rundfunkversuchsstelle (1927)." Reprinted in *Kurt Weill: Musik und Theater. Gesammelte Schriften. Mit einer Auswahl von Gesprächen und Interviews*, edited by Stephen Hinton and Jürgen Schebera, 243–45. Berlin: Henschelverlag Kunst und Gesellschaft, 1990.

———. "Übertragungsbedingungen für Orchesterklang" (1928) in *Kurt Weill: Musik und Theater. Gesammelte Schriften. Mit einer Auswahl von Gesprächen und Interviews*, edited by Stephen Hinton and Jürgen Schebera, 278–88. Berlin: Henschelverlag Kunst und Gesellschaft, 1990.

Weinberger, Julius, Harry F. Olson, and Frank Massa. "A Uni-Directional Ribbon Microphone." *Journal of the Acoustical Society of America* 5 (1933): 139–47.

Weiskopf, H. "Das Sphärophone." *Musik und Maschine, Sonderheft zu den Musikblättern des Anbruch* 8 (1926): 388.

Weissmann, Adolf. "Mensch und Maschine." *Die Musik* 20 (1927): 103.

Wente, Edward Christopher. "A Condenser Transmitter as a Uniformly Sensitive Instrument for the Absolute Measurement of Sound Intensity." *Physical Review* 10 (1917): 39–63.

———. "The Sensitivity and Precision of the Electrostatic Transmitter for Measuring Sound Intensities." *Physical Review* 19 (1922): 498–503.

Wente, E. C. and A. L. Thuras. "A High Efficiency Receiver for a Horn-Type Loudspeaker of Large Power Capacity." *Bell System Technical Journal* 7 (1928): 140–53.

———. "An Improved Form of Moving Coil Microphones." *Journal of the Acoustical Society of America* 3 (1931): 8.

———. "Loud Speakers and Microphones." *Bell System Technical Journal* 13 (1934): 259–77.

———. "Moving-Coil Telephone Receivers and Microphones." *Journal of the Acoustical Society of America* 3 (1931): 44–55.

Werbeprospekt SH 3981. "Der bewährte Lautsprecher Siemens 072 (Protos), 1926," as reproduced at https://www.medienstimmen.de/chronik/1926-1930/1926-siemens-halske-protos-lautsprecher-siemens-072/.

Weyer, Rolf-Dieter. "Probleme der Analyse und Synthese von musikalischen Klängen am Beispiel der Trautweinischen Hallformanten-Theorie." *Studien zur Musikgeschichte des Rheinlandes* 4 (1975): 123–30.

Wiener, Karl. "Die musikalische Stimme im Rundfunk." *Der deutsche Rundfunk* 2 (1924): 1180–81.

———. "Welche Instrumente eignen sich am besten für das Radio?" *Der deutsche Rundfunk* 2 (1924): 758–59.

Wierzbicki, James. "Shrieks, Flutters, and Vocal Curtains: Electronic Sound/Electronic Music in Hitchcock's 'The Birds.'" *Music and the Moving Image* 1 (2008): 10–36.

Williams, Alfred L. "Piezo-electric Loudspeakers and Microphones." *Electronics* 4 (1932): 166–67.

Williams, Raymond. "Base and Superstructure in Marxist Cultural Theory." In *Problems in Materialism and Culture*, by Raymond Williams. London: Verso, 1980.

Winkelmann, Joachim. *Das Trautonium: Ein neues Radio-Musikinstrument*. Berlin: Deutsch Literarisches Institut J. Schneider, 1931.

Winner, Langon. "Do Artifacts Have Politics?" *Daedalus* 109 (1980): 121–36.

Winzheimer, Bernhard. *Das musikalische Kunstwerk in elektrischer Fernübertragung*. Augsburg: Dr. Benno Filser Verlag, G.m.b.H, 1930.

Winzheimer, Rudolf. *Übertragungstechnik*. Munich and Berlin: Oldenbourg, 1929.

Wise, M. Norton. *Aesthetics, Industry and Science: Hermann von Helmholtz and the Berlin Physical Society*. Chicago: University of Chicago Press, 2018.

Wittje, Roland. *The Age of Electroacoustics: Transforming Science and Sound*. Cambridge, MA: MIT Press, 2016.
———. "Karl Willy Wagner." In *Neue Deutsche Biographie*, edited by Bayerische Akademie der Wissenschaften, Historische Kommission, 27 (2019): 253–54.
Wolf, Rebecca. "Haltbarkeit. Zeit erleben und Klang erforschen mit Instrumenten." *MusikTheorie. Zeitschrift für Musikwissenschaft* 34 (2019): 63–82.
Wolf, Rebecca. "Musical Instruments as Material Culture." In *Towards the Rebuilding of an Italian Renaissance-Style Wooden Organ*, edited by Walter Chinaglia, 7–11. Deutsches Museum Studies 5. Munich: Deutsches Museum Studies, 2020.
Wormbs, Nina. "Technology-dependent Commons: The Example of Frequency Spectrum for Broadcasting in Europe in the 1920s." *International Journal of the Commons* 5 (2011): 92–109.
Wulf, Joseph. *Musik im Dritten Reich. Eine Dokumentation*. Frankfurt: Ullstein, 1983.
Würtburger, K. "Rundfunkvortrag-Dialog." In *Aus meinem Archiv: Probleme des Rundfunks*, edited by Hans Bredow, 59–74. Heidelberg: Vowinckel, 1950.
Yeang, Chen-Pang. *Transforming Noise: A History of Its Science and Technology from Disturbing Sounds to Informational Errors, 1900–1955*. NY: Oxford University Press, 2023.

INDEX

Note: Page numbers followed by an *f* refer to figures.

A NOTE ON THE TYPE

This book has been composed in Arno, an Old-style serif typeface in the classic Venetian tradition, designed by Robert Slimbach at Adobe.